Performance En   and
Control of Photovoltaic Systems

Performance Enhancement and Control of Photovoltaic Systems

Edited by

Saad Motahhir
ENSA, SMBA University, Fez, Morocco

Mamdouh El Haj Assad
Department of Sustainable and Renewable
Energy Engineering, University of Sharjah,
Sharjah, United Arab Emirates

ELSEVIER

Elsevier
Radarweg 29, PO Box 211, 1000 AE Amsterdam, Netherlands
125 London Wall, London EC2Y 5AS, United Kingdom
50 Hampshire Street, 5th Floor, Cambridge, MA 02139, United States

Notices

ISBN: 978-0-443-13392-3

For Information on all Elsevier publications
visit our website at https://www.elsevier.com/books-and-journals

Publisher: Megan Ball
Acquisitions Editor: Edward Payne
Editorial Project Manager: Helena Beauchamp
Production Project Manager: Sharmila Kirouchenadassou
Cover Designer: Vicky Pearson

Typeset by MPS Limited, Chennai, India

Dedication

I dedicate this book to my uncle Abderrahmane Ouali, whose
memories are still with us.
Dear readers, I kindly ask you to a make Dua and prayer for his soul.

—Saad Motahhir

To our families and friends, sources of love, inspiration, and joy.

—Mamdouh El Haj Assad

Contents

List of contributors

A. Abbassi Laboratory of Research in Physics and Engineering Sciences, Sultan Moulay Slimane University, Polydisciplinary Faculty, Beni Mellal, Morocco

Ibtihal Ait Abdelmoula Green Energy Park research platform (GEP, IRESEN/UM6P), Ben Guerir, Morocco

Mounir Abraim Laboratory of Signals, Systems, and Components, Sidi Mohamed Ben Abdellah University, Fez, Morocco; Green Energy Park research platform (IRESEN/UM6P), Ben Guerir, Morocco

Mohamed Said Adouairi Laboratoire LIMAS, Faculté des Sciences Dhar El Mahraz, Université Sidi Mohammed Ben Abdellah, Fès, Morocco

Shabbir Ahmad Institute of Geophysics and Geomatics, China University of Geosciences, Wuhan, P.R. China; Department of Basic Sciences and Humanities, Muhammad Nawaz Sharif University of Engineering and Technology, Multan, Pakistan

Fawzi Mohammed Munir Al-Naima Department of Computer Engineering, Al-Nahrain University, Baghdad, Iraq; Department of Electrical Power Engineering, Al-Kut University College, Wasit, Iraq

Ghassane Aniba Mohammadia School of Engineers, Mohammed V University in Rabat, Rabat, Morocco

Yashar Aryanfar Department of Electric Engineering and Computation, Autonomous University of Ciudad Juárez, Ciudad Juárez, Chihuahua, México

Mamdouh El Haj Assad Department of Sustainable and Renewable Energy Engineering, University of Sharjah, Sharjah, United Arab Emirates

Mohamed M. Awad Mechanical Power Engineering Department, Faculty of Engineering, Mansoura University, Mansoura, Egypt

Alae Azouzoute National Renewable Energy Laboratory, PV Materials Reliability and Durability Group, Denver, CO, United States

T. Sudhakar Babu Department of Electrical and Electronics Engineering, Chaitanya Bharati Institute of Technology, Hyderabad, Telangana, India

Amina Bendaoudi Laboratory E.P.O, University of S.B.A., Sidi Bel Abbes City, Algeria

Kaddour Benkhallouk Laboratory E.P.O, University of S.B.A., Sidi Bel Abbes City, Algeria

El Ghali Bennouna Green Energy Park research platform (IRESEN/UM6P), Ben Guerir, Morocco

Mohammed Berka Department of Electrotechnic, University of Mustapha Stambouli Mascara, Mascara City, Algeria; Laboratory E.P.O, University of S.B.A., Sidi Bel Abbes City, Algeria

Badre Bossoufi Laboratoire LIMAS, Faculté des Sciences Dhar El Mahraz, Université Sidi Mohammed Ben Abdellah, Fès, Morocco

Abdallah Bouabidi Mechanical Modeling, Energy and Material (MEM), National School of Engineering of Gabes (ENIG), University of Gabes, Gabes, Tunisia; Higher Institute of Industrial Systems of Gabes (ISSIG), University of Gabes, Gabes, Tunisia

Ibtissam Bouarfa Green Energy Park research platform (IRESEN/UM6P), Ben Guerir, Morocco; Laboratory of Innovative Technologies, Sidi Mohamed Ben Abdellah University, Fez, Morocco

Mohamed Boujoudar Laboratory of Signals, Systems, and Components, Sidi Mohamed Ben Abdellah University, Fez, Morocco; Green Energy Park research platform (IRESEN/UM6P), Ben Guerir, Morocco

Mohammad Waqas Chandio Department of Mechanical Engineering, Mehran University of Engineering and Technology, Jamshoro, Pakistan

José Roberto Díaz-Reza Department of Electric Engineering and Computation, Autonomous University of Ciudad Juárez, Ciudad Juárez, Chihuahua, México

Minh Quan Duong The University of Danang-University of Science and Technology, Danang, Vietnam

Omaima El Alani Laboratory of Signals, Systems, and Components, Sidi Mohamed Ben Abdellah University, Fez, Morocco; Green Energy Park research platform (IRESEN/UM6P), Ben Guerir, Morocco

Massaab El Ydrissi Laboratory of Signals, Systems, and Components, Sidi Mohamed Ben Abdellah University, Fez, Morocco; Green Energy Park research platform (IRESEN/UM6P), Ben Guerir, Morocco

Julio Blanco Fernandez Department of Mechanical Engineering, University of La Rioja, Logroño, La Rioja, Spain

Jorge Luis García Alcaraz Department of Industrial Engineering and Manufacturing, Autonomous University of Ciudad Juárez, Ciudad Juárez, Chihuahua, México

Abdellatif Ghennioui Laboratory of Signals, Systems, and Components, Sidi Mohamed Ben Abdellah University, Fez, Morocco; Green Energy Park research platform (IRESEN/UM6P), Ben Guerir, Morocco

Hicham Ghennioui Laboratory of Signals, Systems, and Components, Sidi Mohamed Ben Abdellah University, Fez, Morocco

Felicia Iacomi Faculty of Physics, Alexandru Ioan Cuza University of Iasi, Iasi, Romania

Abdalrhman A. Kandil Mechanical Power Engineering Department, Faculty of Engineering, Mansoura University, Mansoura, Egypt

Hicham Karmouni National School of Applied Cadi Ayyad University, Marrakech, Fez, Morocco

Ali Keçebaş Department of Energy Systems Engineering, Technology Faculty, Muğla Sıtkı Koçman University, Muğla, Turkey

Muhammad Khalid Electrical Engineering Department, King Fahd University of Petroleum and Minerals, Dhahran, Saudi Arabia; Interdisciplinary Research Center for Sustainable Energy Systems, King Fahd University of Petroleum and Minerals, Dhahran, Saudi Arabia

Laveet Kumar Department of Mechanical Engineering, Mehran University of Engineering and Technology, Jamshoro, Pakistan

Najwa Lamdihine Mohammadia School of Engineers, Mohammed V University in Rabat, Rabat, Morocco

Nassim Lamrini Green Energy Park research platform (GEP, IRESEN/UM6P), Ben Guerir, Morocco

Thi Minh Chau Le Hanoi University of Science and Technology-School of Electrical and Electronic Engineering, Hanoi, Vietnam

Tuan Le Roberval Laboratory, University of Technology of Compiegne, Compiegne, France

Muhammad Maaruf Control and Instrumentation Engineering Department & Center for Smart Mobility and Logistics, King Fahd University for Petroleum and Minerals, Dhahran, Saudi Arabia

Zoubir Mahdjoub Laboratory E.P.O, University of S.B.A., Sidi Bel Abbes City, Algeria

B. Manaut Laboratory of Research in Physics and Engineering Sciences, Sultan Moulay Slimane University, Polydisciplinary Faculty, Beni Mellal, Morocco

Babay Mohamed-Amine Department of Physics, Laboratory of Industrial Engineering, Sultan Moulay Slimane University, Beni Mellal, Morocco

Saad Motahhir ENSA, SMBA University, Fez, Morocco

Ali Faisal Murtaza Director Research, University of Central Punjab, Lahore, Pakistan

Adar Mustapha Department of Physics, Laboratory of Industrial Engineering, Sultan Moulay Slimane University, Beni Mellal, Morocco

Mabrouki Mustapha Department of Physics, Laboratory of Industrial Engineering, Sultan Moulay Slimane University, Beni Mellal, Morocco

Hicham Oufettoul Mohammadia School of Engineers, Mohammed V University in Rabat, Rabat, Morocco; Green Energy Park research platform (GEP, IRESEN/UM6P), Ben Guerir, Morocco

Raha Ranaei Department of Architecture Engineering, Arak Branch, Islamic Azad University, Arak, Iran

Challa Krishna Rao Department of Electrical and Electronics Engineering, Aditya Institute of Technology and Management, Tekkali, Andhra Pradesh, India; Department of Electrical Engineering, Parala Maharaja Engineering College, Berhampur, Affiliated to Biju Patnaik University of Technology, Rourkela, Odisha, India

A. Razouk LGEM, FST, Sultan Moulay Slimane University, Beni-Mellal, Morocco

Ahmed Yacine Rouabhi Laboratory E.P.O, University of S.B.A., Sidi Bel Abbes City, Algeria

Hussam Khalil Ibrahim Rushdi Department of Computer Engineering, Al-Nahrain University, Baghdad, Iraq

Sarat Kumar Sahoo Department of Electrical Engineering, Parala Maharaja Engineering College, Berhampur, Affiliated to Biju Patnaik University of Technology, Rourkela, Odisha, India

Sudhansu S. Sahoo School of Mechanical Sciences, Odisha University of Technology and Research, Bhubaneswar, Odisha, India

Mohamed S. Salem Mechanical Power Engineering Department, Faculty of Engineering, Mansoura University, Mansoura, Egypt

Hadeed Ahmed Sher Faculty of Electrical Engineering, Ghulam Ishaq Khan Institute of Engineering Sciences and Technology, Topi, Pakistan

C.H. Siow Department of Electrical & Electronic Engineering, UCSI University, Kuala Lumpur, Malaysia

Gamal I. Sultan Mechanical Power Engineering Department, Faculty of Engineering, Mansoura University, Mansoura, Egypt

Rodney H.G. Tan Department of Electrical & Electronic Engineering, UCSI University, Kuala Lumpur, Malaysia

Sanju John Thomas School of Mechanical Sciences, Odisha University of Technology and Research, Bhubaneswar, Odisha, India

Sheffy Thomas Department of Electronics and Instrumentation, Federal Institute of Science and Technology (FISAT), Ernakulam, Kerala, India

Ngoc Thien Nam Tran Delta Electronic Inc., Tainan, Taiwan

The Hoang Tran The University of Auckland, Auckland, New Zealand

Franco Fernando Yanine School of Engineering of Universidad Finis Terrae, Providencia, Santiago, Chile

Preface

The use of solar energy, especially photovoltaic (PV), is expanding across industries and other economic sectors, and it is growing year over year. Even so, there is a dearth of information on this industry, in part because so few colleges and universities offer degrees in this field. Furthermore, there is little knowledge on the subject, and much of solar energy technology is still in its infancy.

This requirement is addressed in the *Performance Enhancement and Control of Photovoltaic Systems* book by offering a deeper comprehension of the possibilities and key difficulties in this developing area. Finding appropriate references that condense or summarize the PV technology as well as the most pertinent PV materials and tools for their characterization can be difficult for students, researchers, and engineers who are interested in PV technology.

In addition to providing information on PV energy systems, this book also discusses ways to assess PV performance using thermodynamics and models of PV modules. Moreover, this book discusses in detail the parameters affecting the performance of PV systems such as soiling, dust removal, and climate conditions.

The content in the book is arranged to give students, engineers, and researchers interested in PV systems an easily accessible and comprehensive source of information. Thus the book offers readers a thorough overview of the fundamental ideas and current state of the art of the many aspects of PV systems, as well as what is required to advance the study of PV technology and improve the efficiency of PV systems. For these systems to be optimized, a thorough analysis, including all the parameters affecting the PV performance, is discussed, and modeling of PV modules under shading and different climate conditions is also considered.

The book's primary goal is to propose the design concepts for PV systems as well as their control techniques. This book contains information on the use of PV systems for freshwater production, PV soiling loss, electrical models of PV modules, digital metasurfaces as a new technology for enhancing PV performance, optimization algorithms to estimate PV cell circuit parameters, maximum power point tracking technique, Internet of things for PV applications, control of electric ships powered by PV, and challenges of modern PV systems under large-scale forms.

Saad Motahhir
Mamdouh El Haj Assad

Current challenges in nanomaterials for photovoltaics: 2D graphene layers and double perovskites

1

A. Abbassi[1], A. Razouk[2], B. Manaut[1] and Felicia Iacomi[3]
[1]Laboratory of Research in Physics and Engineering Sciences, Sultan Moulay Slimane University, Polydisciplinary Faculty, Beni Mellal, Morocco, [2]LGEM, FST, Sultan Moulay Slimane University, Beni-Mellal, Morocco, [3]Faculty of Physics, Alexandru Ioan Cuza University of Iasi, Iasi, Romania

1.1 Introduction

Double perovskites with a general formula $A_2B'B''O_6/A'A''B'B''O_6$ represent a large family of compounds, which until now have been extensively studied due to their interesting properties and applications in microelectronic circuits, as active photocatalytic material, thin film substrate, solid oxide fuel cell electrolytes, contactless potentiometer, and recently in spintronic devices (Abbassi et al., 2023; Mitchel, 2000; Tejuca & Fierro, 1993; Westerburg et al., 2002; Kobayashi et al., 1999; Kobayashi et al., 1998). In the chemical formula $A_2B'B''O_6/A'A''B'B''O_6$, A/A'/A'' are 12-fold coordinated alkaline earth metal ions like barium, strontium, calcium, or lanthanide while B'/B'' are sixfold coordinated transition metal ions or light alkaline earth ions like magnesium. It has been shown that the structure of double perovskite can be not only cubic but also tetragonal, orthorhombic, and monoclinic (Barnes et al., 2006). Among a large number of double perovskites, the Ba_2MgWO_6 crystal was also examined, mainly experimentally. Ba_2MgWO_6 has shown great potential as a ceramic component for temperature sensors, for example, in petroleum wells (Salgado & Filho, 2008). This ceramic was produced by a solid-state reaction process and sintered at different sintering conditions (Khalyavin et al., 2003; Rangel de Aguiar et al., 2005). Raman spectroscopy of Ba_2MgWO_6 and some other perovskites have been performed by (Hardcastle & Wachs, 1995).

Despite the recent progress in the physics of the double perovskites, there is always an open debate on adequate theoretical modeling. In particular, the details of the interplay between structural, electronic, and magnetic degrees of freedom in the double perovskites are not yet clearly understood. As far as we are concerned, electronic and optical properties of mixed alkaline earth tungstate double perovskites $BaSrMgWO_6$ have not been reported. In continuation with these types of ceramics (Ezzahi et al., 2011), an attempt is made to synthesize a double perovskite with the general formula $BaSrMgWO_6$.

Performance Enhancement and Control of Photovoltaic Systems. DOI: https://doi.org/10.1016/B978-0-443-13392-3.00001-3

In this chapter, the BaSrMgWO$_6$ compound was investigated in order to understand its behaviors in terms of electrical and optical properties by using first-principle calculation based on the density function theory.

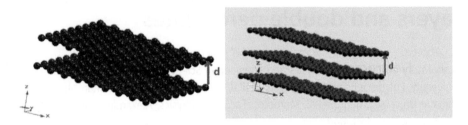

The revolutionary character of graphene lies mainly in its two dimensionality. Physically, it is a layer of single carbon atoms arranged in a hexagonal pattern that visually resembles a honeycomb. Graphene is therefore an allotrope of carbon. The C—C bond length is equal to 1.42 Å similar to graphite. The unit cell is given by $a = 2.4612$ Å and $b = 2.4612$ Å (Latham et al., 2015). In this study, the calculation was made with the vacuum layer $c = 3.5$ Å that was taken along the normal to graphene layer in order to ignore the effect of interaction between layers. The electronic, optical, thermal, and mechanical properties of graphene have opened the opportunity to many applications, which will develop dynamically in the coming decades. Already today, graphene is considered the best material instead of silicon in the electronic field. This transparent and flexible conductor can be used to manufacture photovoltaic cells, as well as LED lights. It also greatly increases the frequency of electromagnetic signals, allowing the production of faster transistors. Graphene sensors are also attracting considerable interest. We will use the ab initio method to study the optoelectronic properties of bilayers and trilayers of these systems in order to explain the interaction of radiation with these materials and to check the ability to use these layers as window layers in solar cells.

1.2   Method and calculation

In order to investigate deeply the optoelectronic properties of these materials, we proceed to calculate the band structure, total/partial DOS, and optical properties (absorption and transparency parameters) of BaSrMgWO$_6$ and graphene Bi and trilayers within a self-consistent scheme by solving the Kohn-Sham equation based on the first principles using DFT with the local density approximation (LDA), generalized gradient approximation (GGA) method, and the exchange—correlation function realized by Perdew-Burke-Ernzerhof (PBE) implemented in the WIEN2K package (Blaha & Schwarz, 2006); the spin polarization has been taken into account. We take the energy cutoff of -8.0 Ryd to describe the wave functions in the interstitial region for the LAPW calculation. The integrals over the Brillouin zone are performed up to 400 k-points in the irreducible Brillouin zone (IBZ). The self-consistent calculations are considered to converge when the total energy of the system is stable within 10^{-5} Ryd.

1.3 Result and discussion

1.3.1 Electronic properties of double perovskite system

In order to have more information about the electronic structure and to determine the origin of the charge transfer that is responsible for the band gap value that is observed for this compound, we have calculated the total and partial DOS. Fig. 1.1 presents the results of the total (DOS) calculations in the BaSrMgWO$_6$ compound as well as the partial DOS related to Ba, Sr, Mg, W, and O.

From this figure, the valence band (VB) of such a compound is composed of two regions. The first region, located between −5 and 0 eV, corresponds to the hybridization of s, p states of Mg, d of W, and s, p of oxygen. The second region, ranging from −20 to −15 eV, corresponds to the hybridization between p-states of Sr and s-states of oxygen. The figure also demonstrates that the p-states of Ba contribute to the TDOS by the occurrence of a peak that appears at −10 eV. The conduction band (CB) is composed of d-states of Ba and Sr, s, p-states of Mg, d of W, and finally p of oxygen. Moreover, the corresponding maximum and minimum energy levels of the VB and the

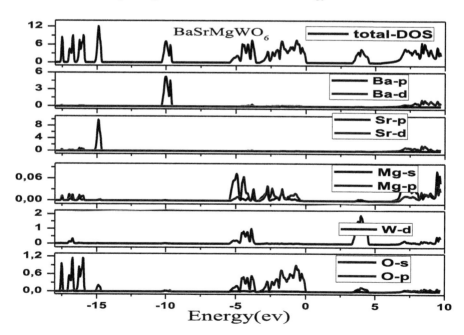

Figure 1.1 Total and partial density of state of BaSrMgWO$_6$.

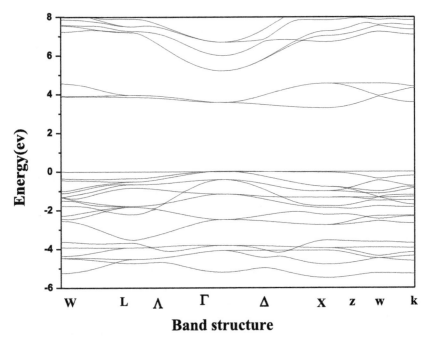

Band structure

Figure 1.2 Band structure of $BaSrMgWO_6$.

Table 1.1 Gap values of $BaSrMgWO_6$.

$BaSrMgWO_6$ Gap	LDA 3.22 eV	GGA 3.35 eV

CB are 0 eV and 3.35 eV, respectively. At 0 eV, there is contribution of only p-state of oxygen, and at 3.35 eV, contribution of only d-state of W. This result confirms the explanation of the band gap origin suggested by (Blasse & Corsmit, 1973). We have determined the band structure of the $BaSrMgWO_6$ compound using the DFT method. The obtained results, as presented in Fig. 1.2, show that the bottom of the conduction band and the top of the valence band are at the same point "X." This indicates that the gap of our compound $BaSrMgWO_6$ is direct. The obtained values of the band gap depend on the type of approximation used. Indeed, we found that the gap value is equal to 3.22 eV and 3.35 eV for LDA and GGA, respectively (see Table 1.1).

These results demonstrate that the value obtained by GGA approximation is closer to the experimental one than those obtained by LDA. Indeed, it is known that the latter approximation underestimates the band gap value. This fact can be explained by the incorrect interpretation of Kohn-Sham eigenvalues, associated with the exchange and correlation in LDA functional. It is known that the optical band gap gives the accurate value of band energy. For this reason, we will calculate the absorption coefficient in order to estimate with more accuracy the gap value, using the GGA approximation.

1.3.2 Electronic properties of graphene layers

The band structure of the graphene bilayer (Fig. 1.3) shows a direct aspect of the gap obtained; the valence and conduction bands constitute intense internal bands from -2 eV to -6 eV and from 1.5 eV up to 6 eV, respectively. These bands are mainly due to the s and p orbitals of carbon; this is also observed in the calculation made for the total density.

By changing the distance d (Table 1.2) between layers that is chosen so as to avoid the effects of interactions between layers, gap openings are found. With the increase in distance d, the gap also increases; this is mainly due to the weak mutual interaction between the two layers. The effect of surface 1 is assumed to be weak on surface 2 and vice versa. The graphene trilayer was carried out with a fixed

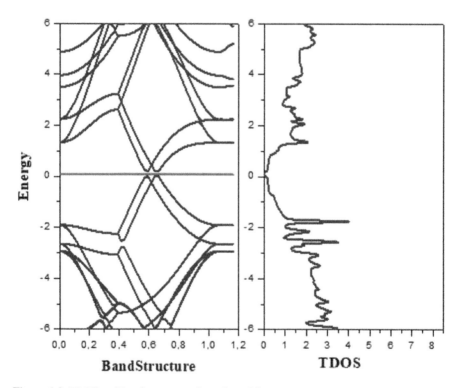

Figure 1.3 TDOS and band structure of graphene bilayer.

Table 1.2 Gap values of graphene bilayer.

Distance between layers (d)	gap
3.34 Å	0.20 eV
3.36 Å	0.22 eV
3.37 Å	0.23 eV

distance d, which corresponds to the stability of the system; the figure below gives the variation of the energy according to the distance d taken. The stability that gives a minimum energy is obtained for $d = 3.5$ Å (Fig. 1.4).

The graphene trilayer after calculation shows the presence of several energy levels in both bands, with intense intrabands consisting essentially of s and p of carbon. A direct gap is observed with the value 0.6 eV. These intense internal bands are due to the grouping of three layers. Even with a weak interaction, the energies are added to make these internal bands appear (Fig. 1.5).

1.3.3   Optical properties of double perovskites

In order to calculate an accurate value of the optical band gap of our material, we proceed to study the optical absorption. The optical band gap can be deduced using the following equation (Abbassi et al., 2015):

$$(\alpha h\upsilon)^m = A\left(h\upsilon - E_g\right)$$

It is noted that m is a constant that equals 2 for a direct gap and ½ for an indirect gap; A is a constant depending on the transition probability. The plot $(\alpha h\upsilon)^m$ versus the incident radiation leads us to extract the band gap energies by the linear extrapolation of the coefficient of absorption α to zero. Based on the band structure of BaSrMgWO$_6$, we take m equal to 2 in order to present $(\alpha h\upsilon)^m$ in terms of energy $(h\upsilon)$ (see Fig. 1.6).

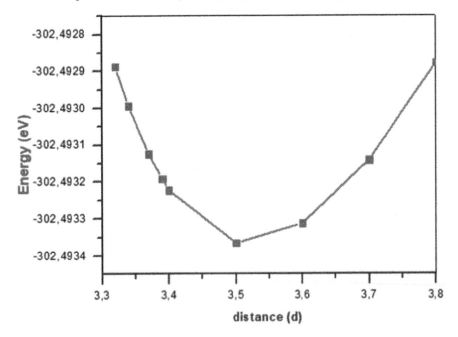

Figure 1.4 Optimization of distance d between layers.

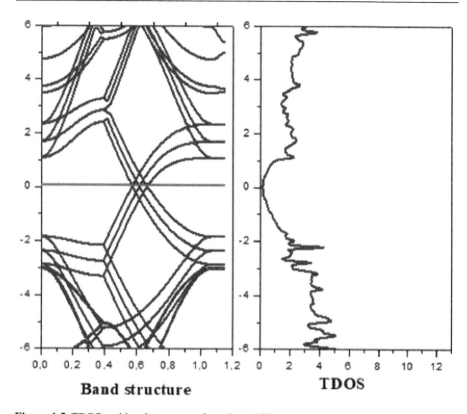

Figure 1.5 TDOS and band structure of graphene trilayer.

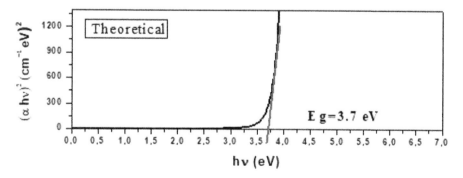

Figure 1.6 Optical band gap of BaSrMgWO$_6$.

The absorption coefficient was obtained using the ratio of $\alpha d = \ln(1/T)$, where d is the optical path traveled by electromagnetic radiation and T is the stand for transmittance. The value obtained is precise and in agreement with the experimental results (Fig. 1.6).

From Fig. 1.7, we can observe a stable behavior of absorbance and can notice that there is high absorption in the ultraviolet region and low absorption in the visible and infrared regions. Such low absorption is due to the large band gap of such material.

Several computational and experimental investigations performed on bulk and 2D-BaSrMgWO$_6$ prove a direct band-to-band transition with an E_g of 3.7 eV. Due to the transitions between bands and the excitonic aspect, an important absorption can appear when light penetrates the matter. In a recently published work (Punga et al., 2022), we have elaborated the thin films of BaSrMgWO$_6$ with different temperatures; this study allowed us to visualize the transmittance of this perovskite in 2D form. For two temperatures, we have estimated after measurement that the value of the transparency is around 90% for a crystallization of 850°C (Fig. 1.8).

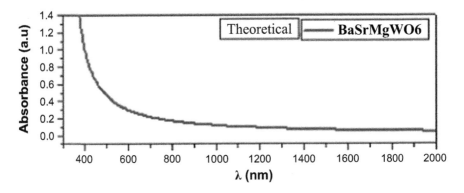

Figure 1.7 Absorption spectra versus wavelength for BaSrMgWO$_6$.

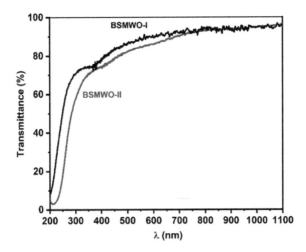

Figure 1.8 Transmittance spectra versus wavelength of 2D-BaSrMgWO$_6$ thin films for different annealing processes.

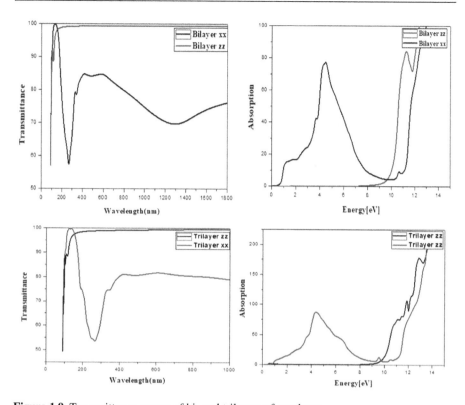

Figure 1.9 Transmittance curves of bi- and trilayers of graphene.

1.3.4 Optical properties of 2D graphene

The graphene bilayer presents an anisotropy with respect to the dispersion of light in the material; the 2D structure of the graphene bilayer is a structure that has no privileged directions for the propagation of light radiation. The zz direction in these variations presents a good behavior; the absorption in zz is almost null in visible light and starts after the excitation of 8 eV. This is proven by the variation of the transmittance whose average is estimated at more than 97%. The same behavior is observed for the trilayer; a negligible decrease in transmittance is presented by this structure compared to the bilayer. The absorption is also very low in visible light. These results obtained for the bulk/2D of double perovskite $BaSrMgWO_6$ and the 2D of graphene bi- and trilayers constitute an important step toward the establishment of a new generation of solar electrodes based on transparent nanomaterials, which also have significant conductivity (Fig. 1.9).

1.4 Conclusion

In this study, we treated two categories of nanomaterials that can be considered a real challenge in the manufacturing technologies of solar cells. We have studied the optoelectronic properties of the following forms: bulk/2D of double perovskite BaSrMgWO$_6$ and graphene bilayer and trilayer. The results obtained present an important transmittance for these nanomaterials, which is around 97%. These materials can be used as transparent coatings and surfaces in solar electrode applications.

References

Abbassi, A., Agouri, M., Iacomi, F., et al. (2023). Magneto-thermal, mechanical, and opto-electronic properties of Sr$_2$MW06(M = V,Rh,Ru): ab initio study. *Journal of Superconductivity and Novel Magnetism, 36*, 995−1001. Available from https://doi.org/10.1007/s10948-023-06537-0.

Abbassi, A., Ez-Zahraouy, H., & Benyoussef, A. (2015). First principles study on the electronic and optical properties of Al- and Si-doped ZnO with GGA and mBJ approximations. *Optical and Quantum Electronics, 47*, 1869−1880. Available from https://doi.org/10.1007/s11082-014-0052-7.

Barnes, P.W., Lufaso, M.W.,& Woodward, P.M. (2006). Structure determination of A$_2$M^{3+}TaO$_6$ and A$_2$M^{3+}NbO$_6$ ordered perovskites: octahedral tilting and pseudosymmetry, *Acta Crystallographica B62*, 384-396. https://doi.org/10.1107/S0108768106002448.

Blaha, P., & Schwarz, K. (2006). *WIEN2k*. Austria: Vienna University of Technology.

Blasse, G., & Corsmit, A. F. (1973). Electronic and vibrational spectra of ordered perovskites. *Journal of Solid State Chemistry, 6*(4), 513−518. Available from https://doi.org/10.1016/S0022-4596(73)80008-8.

Ezzahi, A., Bouchaib Manoun., Ider, A., Bih, L., Benmokhtar, S., Azrour, M., Azdouz, M., Igartua, J. M., & Lazor, P. (2011). X-ray diffraction and Raman spectroscopy studies of BaSrMWO6 (MNi, Co, Mg) double perovskite oxides. *Journal of Molecular Structure, 985*(2−3), 339−345. Available from https://doi.org/10.1016/j.molstruc.2010.11.017.

Hardcastle, F. D., & Wachs, I. E. (1995). Determination of the molecular structures of tungstates by Raman spectroscopy. *Journal of Raman Spectroscopy, 26*(6), 397−405. Available from https://doi.org/10.1002/jrs.1250260603.

Khalyavin, D. D., Han, J., Senos, A., et al. (2003). Synthesis and dielectric properties of tungsten-based complex perovskites. *Journal of Materials Research, 18*, 2600−2607. Available from https://doi.org/10.1557/JMR.2003.0364.

Kobayashi, K.-I., Kimura, T., Tomioka, Y., Sawada, H., Terakura, K., & Tokura, Y. (1999). Intergrain tunneling magnetoresistance in polycrystals of the ordered double perovskite Sr2FeReO6. *Physical Review, B 59*, 11159. Available from https://link.aps.org/doi/10.1103/PhysRevB.59.11159.

Kobayashi, K. I., Kimura, T., Sawada, H., et al. (1998). Room-temperature magnetoresistance in an oxide material with an ordered double-perovskite structure. *Nature, 395*, 677−680. Available from https://doi.org/10.1038/27167.

Latham, C. D., McKenna, A. J., Trevethan, T. P., Heggie, M. I., Rayson, M. J., & Briddon, P. R. (2015). On the validity of empirical potentials for simulating radiation damage in graphite: A benchmark. *Journal of Physics. Condensed Matter: An Institute of Physics Journal, 27*(31), 316301−316312.

Mitchel, R. H. (2000). *Perovskites modern and ancient*. Ontario: Almaz Press.

Punga, L., Abbassi, A., Toma, M., Alupului, T., Doroftei, C., Dobromir, M., Timpu, D., Doroftei, F., Hrostea, L., Rusu, G. G., et al. (2022). Studies of the structure and optical properties of BaSrMgWO6 thin films deposited by a spin-coating method. *Nanomaterials*, *12*, 2756. Available from https://doi.org/10.3390/nano12162756.

Rangel de Aguiar, L. A., Lapa, C. M., Sanguinetti, R. A. F., Aguiar, J. A., da Silva, C. L., Souza, D. P. F., & Yadava, Y. P. (2005). Production, Sintering and Microstructural Characteristics of Ba2MgWO6 Ceramics. *Materials Science Forum*, *498–499*, 523–528. Available from https://doi.org/10.4028/www.scientific.net/msf.498-499.523.

Salgado, L., & Filho, F. A. (2008). Advanced Powder Technology VI. *Materials Science Forum*, *448*, 591. Available from https://doi.org/10.4028/b-2g12BT.

Tejuca, L. G., & Fierro, J. L. G. (1993). *Properties and applications of perovskite type oxides*. New York: Marcel Decker.

Westerburg, W., Lang, O., Ritter, C., Felser, C., Tremel, W., & Jakob, G. (2002). Magnetic and structural properties of the double-perovskite Ca2FeReO6. *Solid State Communications*, *122*(3–4), 201–206. Available from https://doi.org/10.1016/S0038-1098(02)00079-0.

A thorough review of PV performance, influencing factors, and mitigation strategies; advancements in solar PV systems

2

Yashar Aryanfar[1], Mamdouh El Haj Assad[2], Jorge Luis García Alcaraz[3], Julio Blanco Fernandez[4], José Roberto Díaz-Reza[1], Shabbir Ahmad[5,6], Raha Ranaei[7] and Ali Keçebaş[8]

[1]Department of Electric Engineering and Computation, Autonomous University of Ciudad Juárez, Ciudad Juárez, Chihuahua, México, [2]Department of Sustainable and Renewable Energy Engineering, University of Sharjah, Sharjah, United Arab Emirates, [3]Department of Industrial Engineering and Manufacturing, Autonomous University of Ciudad Juárez, Ciudad Juárez, Chihuahua, México, [4]Department of Mechanical Engineering, University of La Rioja, Logroño, La Rioja, Spain, [5]Institute of Geophysics and Geomatics, China University of Geosciences, Wuhan, P.R. China, [6]Department of Basic Sciences and Humanities, Muhammad Nawaz Sharif University of Engineering and Technology, Multan, Pakistan, [7]Department of Architecture Engineering, Arak Branch, Islamic Azad University, Arak, Iran, [8]Department of Energy Systems Engineering, Technology Faculty, Muğla Sıtkı Koçman University, Muğla, Turkey

2.1 Introduction

Solar photovoltaic (PV) systems have made significant advancements in recent years, revolutionizing how we harness solar energy. These advancements have increased efficiency, affordability, and scalability, making solar PV systems a viable and sustainable alternative to traditional energy sources. One of the significant advances in solar PV systems is improving solar cell efficiency. Traditional solar cells have an average efficiency of around 15−20, meaning they can convert only a fraction of the sunlight they receive into electricity. However, recent breakthroughs in solar cell technologies have pushed efficiencies well beyond 20%. For example, silicon-based heterojunction solar cells have achieved efficiencies exceeding 26, while perovskite-based solar cells have demonstrated efficiencies above 25%. These higher efficiency solar cells enable more electricity generation from the same amount of sunlight, thereby increasing the overall energy output of solar PV systems. Solar panels, an inverter, AC and DC cables, a backup power source, a supply grid, and a monitoring system are the critical elements of a PV system. Solar

Performance Enhancement and Control of Photovoltaic Systems. DOI: https://doi.org/10.1016/B978-0-443-13392-3.00002-5

radiation is captured by the solar panels, transforming it into DC electrical power (Agrawal et al., 2022). DC power is transformed into AC electricity and supplied to the load by the inverter. When solar energy is insufficient or there is no backup power source, excess solar energy is injected into the grid and used to generate electricity. The monitoring system displays the PV system's status in real time (Høiaas et al., 2022). Fig. 2.1 depicts the basic schematic layout of a PV system.

Another significant advancement in solar PV systems is the development of thin-film solar cells. Thin-film solar cells are made using fragile layers of semiconductor materials that can be deposited on flexible substrates such as plastic or metal. This flexibility allows for integrating solar cells into various applications, including building-integrated photovoltaics (BIPV solar roof tiles) and solar-powered clothing. Thin-film solar cells are also lightweight and less expensive to produce compared to traditional silicon-based solar cells.

Furthermore, the cost of solar PV systems has been steadily declining. The development of new manufacturing processes and economies of scale have significantly reduced the production costs of solar panels over the years. As a result, the cost of solar PV systems has become increasingly competitive with fossil fuel-based electricity generation. Additionally, advancements in solar cell efficiency

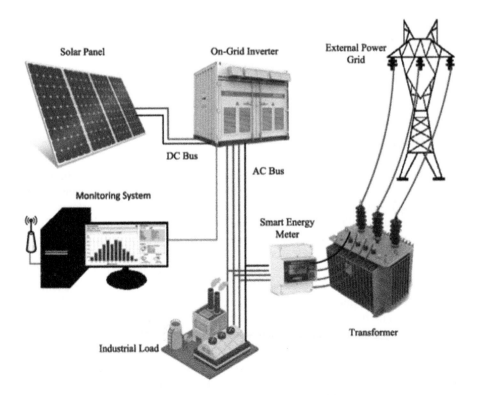

Figure 2.1 Schematic of a photovoltaic facility (Ahmed et al., 2021).

have contributed to maximizing energy output, making solar PV systems more cost-effective and attractive for residential, commercial, and utility-scale applications. Integrating storage solutions such as batteries with solar PV systems has also significantly improved. Energy storage allows for capturing and storing excess solar energy during peak production periods, which can be utilized when sunlight is limited, such as nighttime or cloudy days. Incorporating energy storage into solar PV systems enables greater self-consumption of solar energy, reducing reliance on the grid and providing reliable power supply during grid outages. Fig. 2.2 displays the installed PV plant capacity globally. In 2022, PV systems of 240 GWdc were installed worldwide. Analysts predict yearly global PV installations will rise: 372 GWdc in 2024, 418 GWdc in 2025, and 458 GWdc in 2026. Analysts who made worldwide estimations last year boosted those predictions this year (34%−38% for projections for 2025, for example). Over the indicated period, nearly 70% of all PV installations worldwide were made in China, Europe, the United States, and India (Feldman et al., 2023).

Lastly, monitoring and control technology advances have improved the performance and operational efficiency of solar PV systems. Intelligent inverters, for example, allow for real-time monitoring and optimization of the solar PV system's performance. This enables better fault detection, remote troubleshooting, and improved system management. Additionally, advanced control algorithms can optimize the power output of solar PV systems by tracking the maximum power point (MPP) and mitigating the effects of shading or partial module failure.

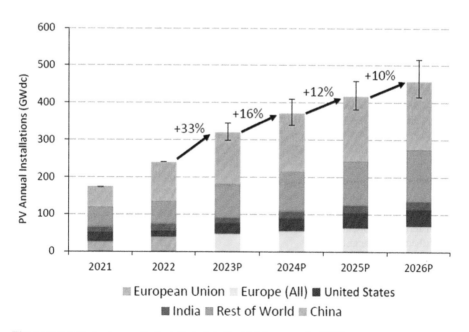

Figure 2.2 PV plant capacity installed globally (Feldman et al., 2023).

In summary, advances in solar PV systems have led to increased efficiency, affordability, scalability, and integration with storage solutions. These advancements have propelled solar energy toward becoming a mainstream and sustainable energy source, contributing to a greener and more sustainable future.

2.2   PV power plant characteristic parameters

Photovoltaic power plant characteristic parameters refer to the critical details that define the performance and specifications of a solar power plant. These parameters are crucial for evaluating the PV system's efficiency capacity and overall effectiveness. Here are some essential characteristic parameters:

1. Capacity: The capacity of a PV power plant refers to the maximum power output it can generate. It is typically measured in kilowatts (kW) or megawatts (MW). The capacity determines the amount of electricity the plant can produce and supply to the grid.
2. Efficiency: PV plant efficiency measures the conversion efficiency of sunlight into electricity. It is the ratio of the actual power output of the plant to the incident solar irradiation. Higher efficiency indicates better utilization of solar energy.
3. Performance ratio (PR): The performance ratio is the ratio of the actual energy output of the system to the theoretical energy output under ideal conditions. It takes into account various losses, such as temperature shading and soiling. A higher performance ratio indicates better plant performance.
4. Sunlight irradiance: The total solar energy incident on the PV panels is measured in watts per square meter (W/m^2). This parameter depends on various factors like location, weather conditions, and tilt angle of the panels.
5. Orientation and tilt angle: The PV panels' orientation (azimuth) and tilt angle affect the amount of sunlight they receive. Optimal angles are typically based on the latitude of the installation location.
6. Temperature coefficients: Temperature coefficients define how the performance of the PV panels is affected by temperature variations. They indicate the percentage change in output power per degree Celsius change in panel temperature.
7. Levelized cost of energy (LCOE): LCOE is an economic parameter that assesses the cost of generating electricity from the PV power plant over its operational lifetime. It considers factors such as installation costs, operation and maintenance costs, financing, and expected energy yield.
8. Degradation rate: PV panels may experience degradation over time, leading to decreased efficiency. The degradation rate indicates the average annual reduction in power output over the plant's lifetime.
9. Mounting system: The mounting system used to install the PV panels can vary in design, material, and orientation. It should be chosen based on wind loads, site conditions, and esthetic considerations.
10. Grid connection: PV power plants are typically connected to the grid to supply electricity. Parameters related to grid interconnection, such as voltage-level connection codes and power quality requirements, are essential for seamless integration and operation.

These characteristic parameters play a crucial role in designing, evaluating, and maximizing the performance of a PV power plant, ensuring efficient and sustainable solar energy generation.

2.3 Comparative performance studies of different PV plants

Comparative performance studies of different PV plants are crucial to evaluate and optimize the efficiency and productivity of solar power generation systems. We can identify the most efficient PV technologies and plant configurations by comparing the performance metrics, such as energy production capacity factors and system losses. There are several factors to consider when conducting comparative performance studies of PV plants:

1. Location: The geographic location of the PV plant plays a vital role in its performance. Factors such as solar irradiance, temperature, and climate patterns vary across regions. Therefore, comparing plants in similar environmental conditions is essential to obtain accurate performance evaluations.
2. PV technology: Various PV technologies exist in the market, including crystalline silicon (monocrystalline and polycrystalline thin film) and concentrator photovoltaics. Each technology has advantages and disadvantages regarding efficiency, cost, and performance under different environmental conditions. Comparing the performance of different PV technologies can help determine the most suitable one for a particular application.
3. Plant configuration: The configuration of a PV plant, such as the layout orientation and tracking systems, can significantly impact its performance. Comparisons can be made between fixed-tilt installations and tracking systems (single axis or dual axis) and various module orientations (horizontal or tilted). Additionally, the effects of shading soiling and interrow spacing should be considered.
4. Performance metrics: Key performance metrics such as energy yield capacity factor performance ratio and specific energy yield are essential for comparing the performance of different PV plants. Energy yield represents the total amount of electricity the PV system generates, while the capacity factor indicates the plant's average output relative to its maximum potential. The performance ratio compares the actual energy yield to the predicted energy yield, while specific energy yield considers the energy produced per unit of installed capacity. These metrics provide insights into the overall efficiency and productivity of the PV plant.
5. Long-term performance: When conducting comparative studies, it is essential to consider the long-term performance of PV plants. Factors such as degradation rates, module aging, and maintenance requirements can significantly affect the system's overall performance and economic viability. Thus, comparing performance over several years can provide a more accurate assessment.

Researchers typically collect data from multiple PV plants and analyze the performance metrics mentioned earlier to conduct comparative studies. This data can be gathered using on-site monitoring systems, weather stations, and various remote sensing technologies. Statistical analyses such as regression models can be employed to identify the key factors impacting performance and assess the significance of any differences observed. The graph in Fig. 2.3 displays the performance ratio year over year. The vertical axis shows the average performance ratio of the plants observed during the same year, while the horizontal axis shows the plant observation years. According to observations, the plant's performance ratio is improving yearly (Ameur et al., 2022). This is because more accurate and efficient systems are created as technology develops, leading to a more excellent performance ratio.

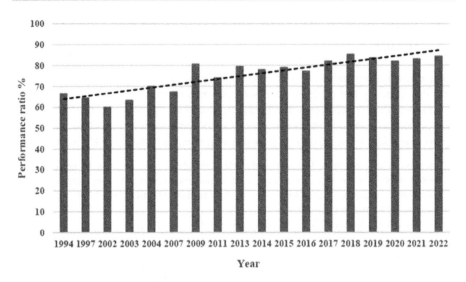

Figure 2.3 Trend in performance ratio over time (year) (Aslam et al., 2022).

Comparative performance studies are essential for advancing the development of PV technology and optimizing the design and operation of PV systems. They provide valuable insights into the performance characteristics of different PV plants under varying conditions, facilitating informed decision-making for investors, policymakers, and project developers in the renewable energy sector.

2.4 Different factors affecting PV power system performance

A variety of elements affect a solar system's output power and life span. One of the most important elements is the type of PV technology being utilized, along with the amount of solar radiation received, ambient temperature, cell temperature, shading effect, dust accumulation, module orientation, weather conditions, and geographic location (Farahmand et al., 2021). The various elements that affect PV efficiency are depicted in Fig. 2.4. This essay looks at these crucial elements that influence the efficiency of PV systems.

2.4.1 PV technology

Several types of PV technologies are used to convert sunlight into electricity. Here are some of the most common types:

1. Poly passivated emitter rear contact (PERC): Poly PERC technology is a PV technology that utilizes polycrystalline silicon cells with passivated emitter and rear contact architecture. This design

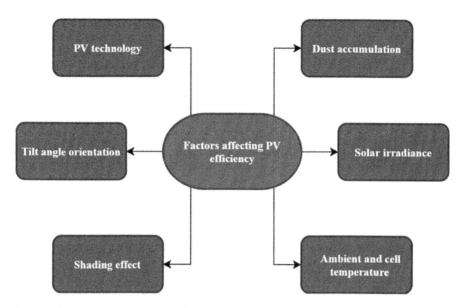

Figure 2.4 PV system efficiency influencing factors (Aslam et al., 2022).

reduces electronic losses and increases cell efficiency by improving light capture and minimizing the recombination of charge carriers.

2. Mono PERC: Mono PERC technology is similar to poly PERC but uses monocrystalline silicon cells instead of polycrystalline silicon cells. Mono PERC cells offer higher efficiency than poly PERC cells due to the higher purity and uniformity of the monocrystalline silicon material.

3. Shingled mono cells: Shingled mono cells are a type of PV technology where the solar cells are arranged in an overlapping shingle-like fashion. This allows for a higher packing density and eliminates the need for busbars, resulting in lower resistive losses. Shingled mono cells are known for their improved performance and better shade tolerance.

4. Half-cut mono PERC: Half-cut mono PERC technology involves dividing a monocrystalline solar cell into two halves, later connected in series. This design reduces resistive losses and improves overall module performance. It also enhances the module's shade tolerance and reduces the impact of potential-induced degradation (PID).

5. Half-cut mono PERC multibusbar: Half-cut mono PERC multibusbar (MBB) technology combines the benefits of half-cut cells with multiple smaller busbars for electrical current conduction. The multibusbar design reduces resistive losses and enhances power output, thereby increasing the module-level efficiency.

6. Shingled mono PERC: Shingled mono PERC combines the shingling technology mentioned earlier with the advantages of monocrystalline PERC cells. The overlapping arrangement of cells enables higher power output and improved temperature coefficient, resulting in higher energy yields.

7. Half-cut MBB heterojunction: Half-cut MBB heterojunction technology combines the benefits of half-cut cells with multiple smaller busbars similar to the half-cut mono PERC MBB design. The heterojunction technology incorporates layers of different semiconducting materials to reduce recombination losses and increase the overall efficiency of solar cells.

8. N-type interdigitated back contact: N-type interdigitated back contact (IBC) technology uses n-type silicon cells with an interdigitated back contact design. In this design, the contact points are located on the cell's rear side, which reduces shading and improves light absorption. N-type IBC cells typically offer higher efficiency, lower temperature coefficients, and better performance in low-light conditions than other cell technologies.

The most recent technologies with high-performance efficiencies are shown in Fig. 2.5, along with details about them (Atsu et al., 2021). Monocrystalline silicon technologies (m-si) have the best performance in terms of energy production and performance ratio (77%), according to authors who compared three different types of PV technology in Morocco (i.e., m-si, p-si, and a-si); however, they also note that polycrystalline technology is the most cost-effective technology when compared to others (Ameur et al., 2019).

Six distinct technologies were examined in Brazil under various climatic conditions (m-si, p-si, CdTe, CIGS, A-si, and c-si). The crystalline technologies (m-si and p-si) have the highest temperature coefficient. Still, thin-film technologies (i.e., a-si) have the best PR, reaching 90% despite having the lowest temperature

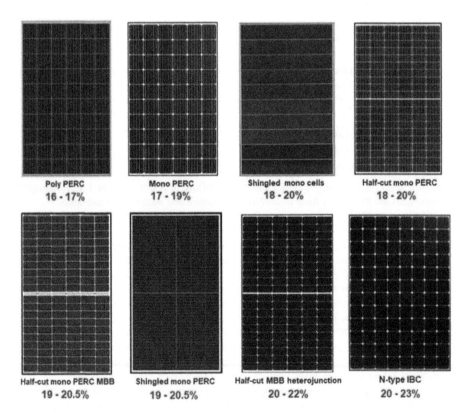

Figure 2.5 Various contemporary PV technologies are offered on the market (Aslam et al., 2022).

coefficient (do Nascimento et al., 2020). Four rooftop PV systems in Abu Dhabi that used monocrystalline and polycrystalline technologies were assessed. The results of the study's analysis demonstrated that monocrystalline technologies outperform polycrystalline ones (Emziane & Al Ali, 2015). The graphs in Figs. 2.6–2.8 compare the performance of various PV technologies. Performance comparison graphs for several PV technologies are shown in Figs. 2.7 and 2.8. Figs. 2.7 and 2.8 provide a graphical representation of the I-V and P-V parameters for monocrystalline silicon and polycrystalline silicon PV technologies, as well as the effects of climate on various PV technologies. These two figures thus provide a graphical representation of the efficiency traits of different PV technologies. It is concluded that thin-film technology is more effective and produces better outcomes in higher temperature zones than crystalline technology, which is less suited to these areas. Crystalline technologies perform better in areas that have moderate or low temperatures.

2.4.2 Solar irradiance

Solar irradiance plays a significant role in PV efficiency. PV panels are designed to convert sunlight into electricity, and the sunlight received directly affects their performance. Solar irradiance refers to the amount of solar power per unit area that reaches the surface of the Earth. When solar irradiance is high, such as on a clear and sunny day, PV panels receive more sunlight. This leads to higher power generation and increased efficiency. Conversely, when solar irradiance is low, such as on a cloudy day or at night, the amount of sunlight reaching the PV panels is reduced, resulting in lower power output and decreased efficiency. The efficiency of PV panels is typically measured as a percentage representing the ratio of electricity output to the incoming solar energy. Higher solar irradiance levels generally result in higher PV efficiency because the panels can convert more sunlight into electricity.

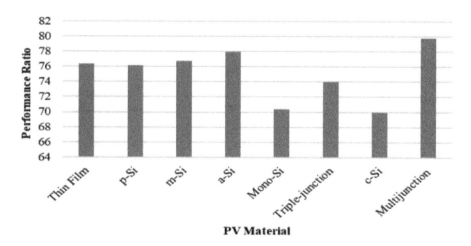

Figure 2.6 Performance ratio versus various PV technologies (Srivastava et al., 2020).

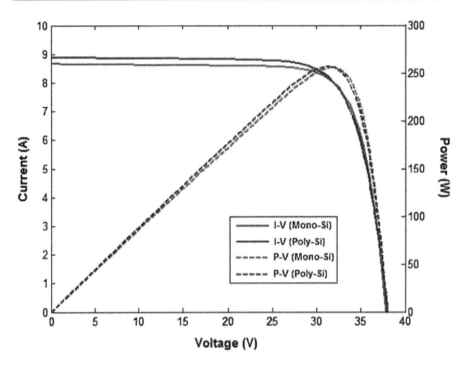

Figure 2.7 Comparison of the characteristic curves for Poly-Si and Mono-Si (Allouhi et al., 2016).

However, while higher irradiance generally leads to increased efficiency, there is a point of saturation where increasing irradiance levels may not significantly improve efficiency. Additionally, solar irradiance can vary depending on geographic location, time of day, season, and atmospheric conditions. For example, regions closer to the equator or at higher altitudes tend to receive higher solar irradiance levels throughout the year, leading to greater PV efficiency. Conversely, areas with frequent cloud cover or pollution may experience lower solar irradiance, reducing PV efficiency.

To optimize PV system performance, it is crucial to consider solar irradiance levels when selecting the location for PV installations. By analyzing historical solar radiation data and considering local climatic conditions, solar panel orientation tilt angles, and shading effects, the PV system can be designed to maximize solar irradiance and thus increase overall efficiency. In summary, solar irradiance directly affects the efficiency of PV panels. Higher solar irradiance levels generally lead to increased power generation and improved efficiency, while lower levels can reduce power output. Designing PV systems to account for solar irradiance variations is critical to maximizing their efficiency and overall energy production.

A study examines how various external factors, such as solar irradiation, affect electricity output. The regression analysis results indicated that sun irradiation is the most critical factor, with a coefficient of determination of 96.5%

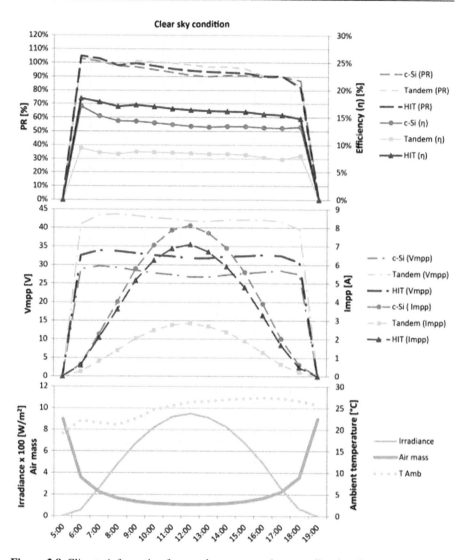

Figure 2.8 Climate information from a clear summer day as well as hourly parameter variation (Aste et al., 2014).

(Al–Bashir et al., 2019). Fig. 2.9 illustrates various irradiation levels according to location. The impact of radiation on the efficiency of the PV system is depicted in Fig. 2.10. The performance ratios for monocrystalline cells were determined in 1994, 1997, and 2010 and are displayed in Fig. 2.11 as a function of annual irradiation on the module plane. Monocrystalline cells have the lowest performance ratio in 1994 and the greatest in 2010 when measured against the annual irradiation. As a result, since the relationship between module current

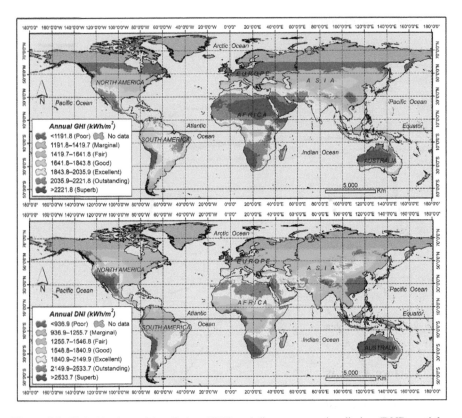

Figure 2.9 Global horizontal irradiation (GHI) and direct average irradiation (DNI) spatial representation (Prăvălie et al., 2019).

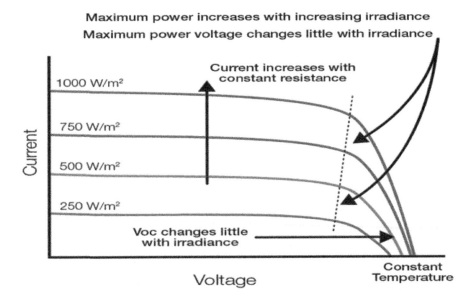

Figure 2.10 Performance of PV systems as a result of irradiance (Agrawal et al., 2022).

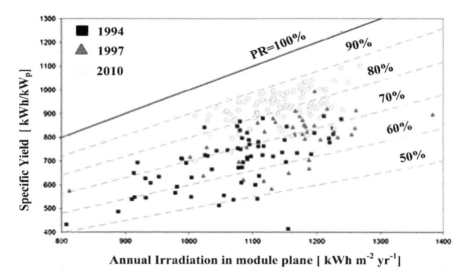

Annual Irradiation in module plane [kWh m⁻² yr⁻¹]

Figure 2.11 Irradiance yearly function and performance ratio (Fouad et al., 2017).

and irradiance value is roughly linear, the literature claims that the impact of solar irradiance on PV panel performance cannot be expressed in terms of a precise percentage increase.

2.4.3 Ambient and cell temperature

Various factors influence PV efficiency, and one important consideration is the temperature at which the PV modules operate. Both ambient and cell temperatures play a significant role in determining the overall efficiency of a PV system.

Ambient temperature refers to the temperature of the surrounding air, while cell temperature refers to the temperature of the PV cells. These temperatures differ because the cells can heat up due to several factors, including solar radiation, electrical resistance, and ambient temperature.

As temperature increases, the efficiency of PV modules tends to decrease. This is mainly due to the properties of the semiconductor materials used in PV cells. The most common material used is silicon, which experiences a decrease in electrical performance as its temperature rises.

The decrease in efficiency with increasing temperature can be attributed to several factors:

1. Increased resistance: The electrical resistance of the semiconductor material in the PV cells increases with temperature. This results in higher energy losses within the system, reducing overall efficiency.
2. Increased dark current: Dark current, also known as leakage current, is the current that flows through a semiconductor even when no light is present. As temperature increases, the dark current in PV cells increases, leading to additional energy losses.

3. Changes in bandgap: The bandgap determines the energy level at which electrons can move within the semiconductor material, which decreases as temperature increases. This can result in a decrease in the voltage output of the PV system.

To mitigate the adverse effects of temperature on PV efficiency, several techniques can be employed:

1. Temperature compensation: Some PV systems incorporate temperature sensors to monitor the cell temperature and adjust the operating parameters accordingly. This compensation helps optimize the system's performance under varying temperature conditions.
2. Active cooling: Techniques such as using fans, heat sinks, or liquid cooling can be implemented to cool the PV cells actively. By dissipating excess heat, these methods help maintain lower cell temperatures and improve overall efficiency.
3. Module design: PV modules can be designed with features that enhance cooling, such as incorporating ventilation channels or using materials with better thermal conductivity. These design strategies help dissipate heat more effectively, reducing the cell temperature rise.

It is important to note that although temperature negatively affects PV efficiency, it also influences the module's output power. In some cases, the increase in solar irradiance with higher temperatures can partially offset the decline in efficiency, resulting in a relatively constant power output.

Overall, the impact of ambient and cell temperature on PV efficiency highlights the need to consider temperature effects during system design operation and performance optimization. By implementing temperature monitoring and appropriate cooling strategies, the adverse effects of temperature on PV efficiency can be minimized, ensuring maximum energy production from solar power systems.

2.4.4   Tilt angle orientation

Tilt angle orientation is an essential factor that can significantly affect the efficiency of PV systems. PV panels are typically installed on rooftops or ground-mounted structures, and the tilt angle refers to the angle at which these panels are positioned relative to the horizontal plane. The optimal tilt angle for PV panels depends on various factors such as location, seasonal variations, and the specific objectives of the PV system. In general, the tilt angle of PV panels should be set to maximize the amount of solar energy captured throughout the year.

One important consideration is the latitude of the installation site. For example, PV panels are often tilted at smaller angles in locations closer to the equator where the sun's path is more perpendicular to the surface. Conversely, in higher latitudes, steeper tilt angles are typically used to capture sunlight at lower angles better. The tilt angle also affects the performance of PV panels during different seasons. In regions with distinct seasons, adjusting the tilt angle of the panels can help optimize energy capture. For example, a steeper tilt angle in winter can help compensate for the lower sun angle. In comparison, a shallower tilt angle in summer can help maximize energy capture during the longer daylight hours.

Additionally, the tilt angle can impact the self-cleaning capability of PV panels. Rainfall can help remove dust and debris that may accumulate on the surface, improving the panels' efficiency. The optimal tilt angle can facilitate self-cleaning by allowing water to flow more efficiently over the panel surface. Setting the tilt angle is typically based on a trade-off between maximizing energy capture and optimizing system costs. Installation costs, structural considerations, and esthetics may influence the best tilt angle for a specific PV system. Sometimes, there may be more efficient solutions than fixed panel orientations. Some PV systems incorporate tracking mechanisms that allow the panels to follow the sun's path throughout the day. These tracking systems can increase energy capture by dynamically adjusting the panel orientation based on the sun's position.

The experimental setups built at various tilt degrees are shown in Fig. 2.12. One study used MATLAB® to maximize sun radiation to determine the ideal tilt angle (Kaddoura et al., 2016). The findings show that when the tilt angles are altered six times annually, 99.5% of solar light is caught.

In summary, the tilt angle orientation of PV panels plays a significant role in determining the overall efficiency of a PV system. By carefully selecting the optimal tilt angle considering factors such as latitude seasonal variations and system objectives, the energy capture potential of a PV system can be maximized, leading to improved efficiency and performance.

2.4.5 Dust accumulation

Dust accumulation is a significant factor that can affect the efficiency of PV systems. When dust settles on the surface of PV modules, it forms a layer that reduces the sunlight reaching the solar cells. This directly impacts the system's power output and overall performance. Dust accumulation on PV modules can lead to a decrease in energy production as it obstructs the absorption of sunlight. Dust particles act as a barrier, reducing the light that can penetrate the solar cells. This, in

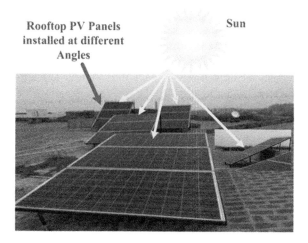

Figure 2.12 Varied tilt angles for solar panels (Sharma et al., 2020).

turn, reduces the amount of electricity the system generates. The impact of dust accumulation on PV efficiency depends on several factors, including the type and amount of dust, location, climate, and cleaning frequency. In arid and dusty environments such as deserts, the problem becomes more critical as the dust accumulates more rapidly and in larger quantities. However, regular cleaning is necessary in less dusty areas to maintain optimal system performance.

Regular cleaning and maintenance of PV modules are recommended to mitigate the adverse effects of dust accumulation. The cleaning process typically involves using water and mild detergents to remove the dirt and dust from the surface. Different cleaning methods, such as manual cleaning or automated cleaning systems, can be employed depending on the scale and accessibility of the PV installation. It is important to note that not all dust particles have the same impact on PV efficiency. Coarser particles like sand may cause shading and reduce the overall power output more significantly than finer dust particles. Additionally, the angle at which the dust settles on the PV surface can also affect the efficiency as it may create localized shadows or uneven coverage. In recent years, researchers and engineers have been exploring various strategies to reduce the impact of dust on PV efficiency. This includes the development of specialized coatings that repel dust, innovative cleaning techniques, and installing self-cleaning mechanisms on PV modules. These advancements aim to enhance the longevity and performance of PV systems, particularly in environments prone to dust accumulation.

In conclusion, dust accumulation is a crucial factor that can significantly impact PV efficiency. Regular cleaning and maintenance, along with technological advancements, are essential to mitigate the adverse effects of dust on solar energy production and ensure optimal performance of PV systems.

2.4.6 Shading

Shading is a significant factor that can impact the efficiency of PV systems. When a portion of a PV module is shaded, it reduces the overall power generation of the system. This is because shading interrupts the flow of sunlight, decreasing electricity output. Shading can occur due to various factors, such as nearby trees, building panels, or even dust accumulation on the PV modules. The impact of shading depends on the extent and location of the shaded area. How PV systems are designed can help somewhat mitigate the effects of shading. Shading multiple bypass diodes on one part of a module does not affect the entire array, enabling other areas to produce power. Additionally, optimizers or microinverters can help overcome the adverse effects of shading by maximizing the output of each panel.

However, it is essential to note that shading can still significantly impact PV system efficiency. A fully shaded module may produce little to no electricity, impacting the overall energy yield and return on investment. Therefore, it is crucial to carefully analyze the shading patterns at a site and consider these factors during the design and placement of the PV system. Regular maintenance and cleaning of the PV modules can also minimize the impact of shading. By removing dust, dirt, and debris from the surface, the modules can capture more sunlight and maintain higher efficiency.

Shadows on the panels may impact PV power generation (Fouad et al., 2017). Different constructions like poles and trees near the PV plant site may provide shade. Fig. 2.13 shows multiple PV systems with different PV module shadings. Additionally, the panels may get covered with bird nests, bird droppings, and leaves, creating a shade effect. The current flow ceases in the shaded cells as they are serially connected, stopping the current flow in the unshaded cells. Numerous connecting solutions have been put out in the literature to mitigate the power losses from shadings (Rathinadurai Louis et al., 2016). According to one study, the panel's performance was reduced by 70% even though only 2% of its surface was shaded (Kawamura et al., 2003). Another study revealed that if 5%−10% of the array is shaded, the array's performance can be lowered by up to 80% (Aslam et al., 2022). According to their features, different types of cells were shaded in a study, and the findings revealed that different power losses occurred, ranging from 59% to 73% (Alonso-Garcia et al., 2006).

In conclusion, shading is a critical factor affecting PV systems' efficiency. It is crucial to consider shading patterns during the design phase and implement strategies like bypass diode optimizers and regular maintenance to mitigate its impact. By doing so, the PV system's overall performance and energy production can be optimized.

Figure 2.13 PV panels cast a variety of shadows (Aslam et al., 2022).

2.5   Different techniques to mitigate performance degradation

Many areas with high radiation potential for installing solar panels are areas with dry weather, and dust in these areas dramatically affects the performance of solar panels. This problem is more severe in climates where dust storms and particulate emissions are expected. Dust impact is the most common factor, especially for large-scale power plants installed in plains and open areas. Accumulation of dust, fine dust, and pollution on solar panels leads to a decrease in the efficiency of solar power plants. It will directly affect the efficiency of the solar system. The impact of losses due to pollution, such as dust, depends mainly on the system environment and precipitation conditions. For example, in humid climates such as Central Europe and residential areas, pollution losses are estimated to be less than 1%. This means that from the amount of energy that reaches the surface of the panels from sunlight, 1% of it is absorbed and wasted due to pollution on its surface. Also, in areas where solar panels are installed next to trees, the energy that reaches the surface of the solar panels is wasted by bird droppings. Therefore, solar panels must be cleaned according to a specific schedule. Otherwise, the solar panels will be damaged, and the solar system's performance will be significantly reduced. In areas close to industrial areas, factories, and busy roads, pollution losses occur, reducing the efficiency of solar systems.

Clogging can be evaluated as the accumulation of harmful substances on the surface of the module and as a decrease in production. We have to reduce the performance of the system due to pollution. If other factors are constant, comparing production values between a controlled and a contaminated module can determine contamination damage in situ. To simulate damage over time, we must determine the soil contamination or accumulation rate or extent. Although there are various methods to calculate the amount of pollution, the amount of pollution as a percentage of daily production reduction is the most valuable approach to model the PV power generation system. Once the extent of pollution is known, it can be used with rainfall data to estimate past, present, and future damage. The estimated contamination level shows the slope curve applied to the yield data between rainfall events.

2.5.1   Cleaning method

Different dust buildup mitigation approaches have been developed due to the increased awareness of PV panel cleaning (Smith et al., 2014). PV panel cleaning can be broadly split into two categories: artificial cleaning (which is further separated into manual and self-cleaning) and natural cleaning (via wind, rain, snow, etc.). To clean the panels manually, labor must be provided by humans. Active and passive cleaning are further subcategories of self-cleaning. Fig. 2.14 provides a flowchart illustrating the many PV cleaning techniques. This section reviews every cleaning method utilized to lessen the effects of dust buildup.

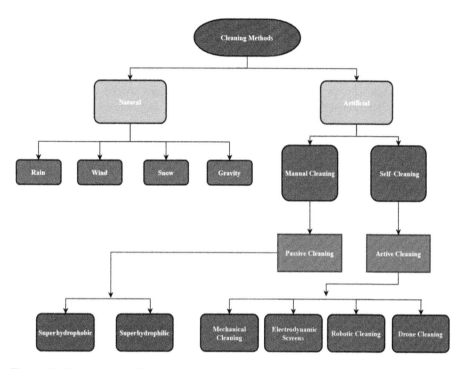

Figure 2.14 A flowchart illustrating various cleaning methods (Aslam et al., 2022).

2.5.1.1   Natural cleaning methods

Natural cleaning of PV panels results from various environmental factors such as rain, wind, snow, and gravity. These elements work together to keep the panels relatively clean and functioning optimally. Let's take a closer look at how these factors contribute to the cleaning process.

1. Rain: Rainwater acts as a natural cleaner for PV panels. When it rains, the water effectively rinses away dirt, dust, and debris that may have accumulated on the panels' surface. Rainwater also helps to remove bird droppings, pollen, and other organic matter that can obstruct sunlight from reaching the PV cells. This natural cleaning process ensures the panels can capture maximum sunlight and efficiently generate electricity.
2. Wind: Wind plays a significant role in keeping PV panels clean. It helps to blow away loose dirt, dust leaves, and other debris that might have settled on the panel's surface. Constant air movement prevents the accumulation of particles, which could decrease the panels' efficiency over time.
3. Snow: While snow accumulation can temporarily block the panels from absorbing sunlight, it eventually melts and trickles off, cleaning the panels. Additionally, as the snow slides off the slippery surface of the panels due to gravity, it often takes most of the accumulated dirt and debris, further aiding in the cleaning process.
4. Gravity: Gravity contributes to the natural cleaning of PV panels by facilitating the downward movement of rainwater, melted snow, and other cleaning agents. As these substances

flow downward, they carry away the accumulated dirt, dust, and debris, leaving the panels cleaner and more able to capture sunlight efficiently.

However, it is essential to note that in certain regions with minimal rainfall or high pollution levels, more than natural cleaning may be required to keep the PV panels completely clean. In such situations, periodic maintenance and cleaning by professionals may be required to ensure optimal performance and longevity of the panels. Regular inspection of the PV panels can also help identify potential issues such as bird droppings, leaves, or other debris that might need manual cleaning or removal.

In conclusion, the natural cleaning process of PV panels via rain, wind, snow, and gravity helps to minimize the accumulation of dirt and debris, ensuring optimal performance of the panels. However, additional maintenance and cleaning measures may be necessary in certain circumstances to maintain their efficiency.

2.5.1.2 Manual cleaning method

This method needs professional labor, water, and high-quality brushes or soft fabric. Low-quality brushes can lower the PV modules' efficiency, which might cause scratches. A study revealed that the performance and life span of PV modules will be harmed if poor-quality brushes are not utilized to clean their surfaces (Green et al., 2006). Only low-capacity PV plants can use this technique (Mohamed & Hasan, 2012). Large-scale PV facilities should use pressured jets and a brushing approach to recover efficiency (Moharram et al., 2013). On PV modules, a study compared water cleaning with brushing versus water cleaning without brushing. When the modules are cleaned with water and a brush, the power output rises by 6.9%, but only by 1.1% when the modules are cleaned with water but not a brush (Pavan et al., 2011). Due to the sensitivity of the modules and the height at which PV modules are put, this cleaning method would require more work for workers.

2.5.1.3 Mechanical cleaning method

A mechanical automated cleaning system uses brushes, a blowing system, and other controllers and sensors. PV system cleaning mechanical setups are shown schematically in Fig. 2.15. This technique works best where it is impossible to clean with water. Although several research studies showed prototypes and designs that perform better when employing this system, this technique is inefficient and not cost-effective due to the sophisticated mechanical design and the involvement of controllers. A mechanical device automatically cleaned PV panels using injected water and a brushing system in one investigation. The system demonstrated a 15% increase in power output while employing this cleaning process (Anderson et al., 2010). A study on a single-axis mechanism with solar tracking and self-cleaning, in which the PV module was cleaned twice daily, revealed that the tracking system contributed to an even more significant increase in power generated. This self-cleaning device also had a stepper motor, gearbox, and microcontroller. The highly intricate mechanical design and the high price of this technology are its limitations (Tejwani & Solanki, 2010). Two structures were created in an experimental

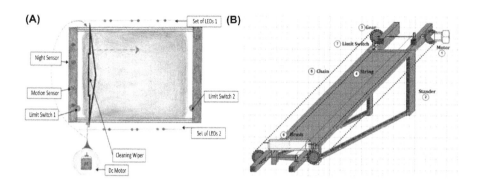

Figure 2.15 An electromechanical cleaning system's schematic diagram (Lamont & El Chaar, 2011).

investigation that significantly decreased the dust deposition factor. A PLC-controlled movement sensor, a dark-activated sensor, and an alert for signaling functions are included in one construction. A PIC controller, a roller brush, and several alarms were all part of the second structure (Lamont & El Chaar, 2011).

2.5.1.4 Electrodynamic display cleaning method

The electrodynamic display cleaning method cleans electrodynamic displays such as electrostatic or electrophoretic display panels commonly found in e-readers, electronic shelf labels, and some digital signage. These display technologies utilize electrically charged particles or pigments to create images on the screen. The cleaning process typically involves applying an electric field to the display surface, which helps remove dust, debris, and other contaminants that may have accumulated on the screen. The electric field helps attract and repel the charged particles or pigments, effectively redistributing them on the display and away from dirt or smudges. Different approaches to electrodynamic display cleaning depend on the specific display technology. For example, in electrophoretic displays that use charged particles to create images, a cleaning waveform can be applied to move the particles away from the contaminated areas. This waveform is designed to generate a force that pushes the particles across the screen.

A similar cleaning method can be employed in electrostatic displays that rely on electrically charged pigments. An electric field can be applied to repel the charged pigments, causing them to move away from the dirt or smudges and toward designated collection areas. The electrodynamic cleaning method offers several advantages. It is noncontact, meaning it does not require physical contact with the screen, which helps prevent scratches or damage to the display surface. Additionally, this method can be automated and integrated within the display device, allowing self-cleaning capabilities. The schematic diagram and prototype in Figs. 2.16 and 2.17 are typical examples. With the help of electricity produced by a high-voltage supply on a screen during this cleaning procedure, dust particles can charge and migrate

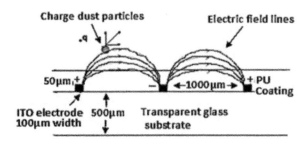

Figure 2.16 Schematic for the EDS (Mazumder et al., 2006).

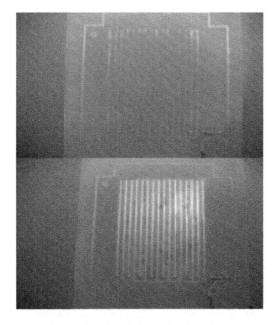

Figure 2.17 EDS before (up) and after (down) the application of voltage (Mazumder et al., 2011).

across the edge of a module surface. Within the first two minutes of operation, this device can eliminate 90% of the dry dust deposits (Mazumder et al., 2011).

Overall, electrodynamic display cleaning is an effective technique for maintaining the visual quality and longevity of electrodynamic displays by removing contaminants and preserving the screen's integrity.

2.5.1.5 Superhydrophobic cleaning method

Superhydrophobic cleaning is a specialized method for cleaning PV panels. PV panels accumulate dust, dirt, and other contaminants over time, reducing efficiency. Cleaning these panels is crucial to maintain optimal energy production. Superhydrophobic cleaning involves applying a superhydrophobic coating to the surface of the PV panels. This coating forms a protective layer that repels water and prevents dust and dirt particles from sticking to the surface. The water droplets that come into contact with the coated surface bead up and roll off, carrying away any accumulated debris. This self-cleaning effect reduces the need for frequent manual cleaning and increases the efficiency and life span of the panels. The superhydrophobic coating used in this method is typically a nanotechnology-based material that forms microscopic structures on the surface of the PV panels. These structures create a lotus leaf-like effect, making the surface extremely water repellent.

Superhydrophobic cleaning offers several advantages over conventional cleaning methods. Firstly, it reduces the water required for cleaning, making it a more sustainable and environmentally friendly approach. Additionally, the self-cleaning effect of the coating can help save time and labor costs since less manual cleaning is needed. Furthermore, the coating protects the PV panels against scratching and corrosion. However, it is essential to note that superhydrophobic cleaning is not a one-time solution. The coating may wear off over time due to weathering and exposure to UV radiation. Therefore, periodic reapplication or maintenance of the coating may be necessary to ensure continued performance. The superhydrophobic cleaning method is an innovative and practical approach to keep PV panels clean and maximize their energy output. It combines the benefits of reduced water usage, self-cleaning properties, and enhanced panel protection, contributing to solar installations' long-term performance and durability.

2.5.1.6 Superhydrophilic cleaning method

The superhydrophilic cleaning method is used to clean PV panels. PV panels are devices that convert sunlight into electricity, and they are typically installed in outdoor environments. Over time, dust, dirt, and other debris can accumulate on the surface of the panels, reducing their efficiency and energy production. The superhydrophilic cleaning method uses a hydrophilic or water-attracting surface coating on the panels. This coating helps to repel dust particles and allows water to spread evenly on the surface, forming a thin and uniform film. When it rains, or the panels are washed with water, the hydrophilic surface ensures that water sheets are off the panels uniformly, removing the accumulated dirt and debris.

One of the advantages of the superhydrophilic cleaning method is that it requires minimal water usage compared to traditional cleaning methods. The hydrophilic coating helps to reduce water consumption by allowing the water to spread evenly across the panel's surface, ensuring efficient cleaning without wastage. Additionally, this method is environmentally friendly as it minimizes the need for chemical cleaning agents and reduces water wastage. It is a cost-effective way of maintaining and cleaning PV panels as it reduces the frequency of manual cleaning required. Regularly cleaning PV panels is essential to ensure optimal performance and energy production. The superhydrophilic cleaning method helps maintain PV panels' efficiency by keeping them free from dust, dirt, and other contaminants without the need for excessive water or cleaning agents. It is worth noting that while the superhydrophilic cleaning method improves the self-cleaning properties

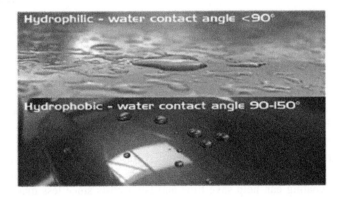

Figure 2.18 Comparing water droplets on hydrophilic and hydrophobic surfaces (Aslam et al., 2022).

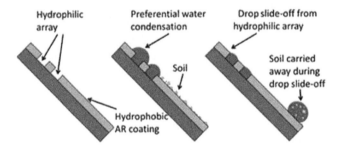

Figure 2.19 Schematic for a hybrid hydrophobic-hydrophilic substance (Aslam et al., 2022).

of PV panels, it may not eliminate the need for occasional manual cleaning, especially in areas with heavy pollution or if there are specific contaminants on the panels. However, it significantly reduces the frequency and effort required for manual cleaning, making it an attractive solution for PV panel maintenance. Fig. 2.18 depicts the difference between surfaces that are highly hydrophobic and highly hydrophilic: On the hydrophilic surface, water droplets are flattened and widely dispersed, whereas on the hydrophobic surface, water droplets are primarily circular and do not disperse, as shown in Fig. 2.19. According to a study, the superhydrophilic surface requires rain to clean itself, making it unsuitable for solar PV modules in desert conditions. As a result, this cleaning method may be appropriate in regions with moderate to high rainfall (He et al., 2011).

2.5.1.7 Drone-based cleaning method

The drone-based cleaning method of PV panels involves using drones equipped with specific cleaning systems to clean solar panels. PV panels convert sunlight into electricity, and their efficiency can be affected by dirt, dust, and other debris accumulating on their surface over time. In the traditional method, manual cleaning of PV panels requires personnel to climb onto rooftops or use elevated platforms, which can be time-consuming, labor intensive, and potentially dangerous. Drone-based cleaning offers a safer and more efficient alternative.

This method uses specially designed drones with brush nozzles and water sprayers to fly over the solar panels. The drones are equipped with sensors and cameras to detect the level of dirt on the surface and determine the optimal cleaning approach. They can be programmed to follow predefined paths to cover the entire panel array. The cleaning system on the drone sprays water or a cleaning solution onto the surface, while the brushes scrub away the dirt and debris. The water and loosened debris are then collected or allowed to dry and fall off the panels naturally. Some drones may also use high-pressure air to blow away the dirt. Advantages of the drone-based cleaning method include the following:

1. Efficiency: Drones can quickly cover a large area of PV panels in a relatively short time, reducing cleaning time and operational costs.
2. Safety: With drones, there is no need for manual labor involving climbing or working at heights, reducing the risk of accidents or injuries.
3. Accessibility: Drones can access areas that are difficult to reach or inaccessible for manual cleaning, such as rooftops or panels installed on steep slopes.
4. Precision: Drones can be programmed to follow precise paths, ensuring uniform and thorough cleaning of the entire panel surface.
5. Environmental impact: Using drones eliminates or reduces the need for chemicals and excessive water usage, making it an environmentally friendly cleaning option.

Overall, the drone-based cleaning method offers an efficient, cost-effective, and safe solution for maintaining the cleanliness and optimal performance of PV panels.

2.5.1.8 Ultrasonic self-cleaning method

The ultrasonic self-cleaning method of PV panels is used to effectively remove dust, dirt, and other contaminants from the surface of solar panels. This method utilizes ultrasonic waves, high-frequency sound waves beyond the range of human hearing, to generate vibrations that can dislodge and remove foreign particles from the panel surface. The process typically involves installing ultrasonic transducers on the backside of the PV panel. These transducers emit ultrasonic waves that create resonant vibrations on the front surface of the panel. These vibrations shake off the accumulated dust, sand bird droppings, pollen, and other debris, preventing them from obstructing sunlight and reducing panel performance. The ultrasonic waves produced by the transducers are guided toward the front surface of the PV panel through a coupling medium such as water or gel. The panel surface acts as a resonating membrane, amplifying the vibrations and effectively dislodging contaminants. The self-cleaning system can be automated or controlled remotely to activate the cleaning process based on predetermined schedules or environmental conditions. For example, the system can be set to clean the panels when a certain amount of dust has accumulated or when a decrease in power output is detected. The benefits of using an ultrasonic self-cleaning method for PV panels include improved efficiency, increased energy generation, and reduced maintenance costs. By keeping the panel surface clean, more

Figure 2.20 Water condensation over time on surfaces that are (A) hydrophobic and (B) hybrid hydrophobic-hydrophilic (Vasiljev et al., 2013).

sunlight can reach the PV cells, maximizing power output. Moreover, the method is gentle and does not require abrasive cleaning or chemicals that could damage the panels. PV panels are shown in Fig. 2.20 before and after being cleaned using ultrasonic technology. A study using ultrasonic cleaning as a cleaning technique found that surface immersion in a separate bath was the best way to achieve a successful result. Studies have shown that to create the cavities needed for the surface cleaning procedure to be successful on PV surfaces, a thin liquid layer (less than 1 mm) is needed (Vasiljev et al., 2013).

Overall, the ultrasonic self-cleaning method offers an efficient and environmentally friendly approach to maintaining the optimal performance of PV panels, ensuring that they operate at their highest efficiency and provide a consistent output of clean, renewable energy.

2.5.1.9 Robot cleaning method

The "Robot cleaning method of PV panels" refers to using robotic technology to clean PV panels. PV panels, also known as solar panels, convert sunlight into electricity through the PV effect. As PV panels are exposed to various elements such as dust, dirt, pollen, bird droppings, and other debris, their efficiency can decrease over time. Regular cleaning is essential to maintain the optimal performance of PV panels and maximize their energy production. The robot cleaning method involves deploying specialized robots designed to clean PV panels automatically. These robots are equipped with various cleaning mechanisms, sensors, and control systems to efficiently and effectively clean the panels. The cleaning robots typically use brushing, scrubbing, and rinsing techniques to remove the accumulated debris from the surface of the PV panels. They move along the rows of panels, ensuring thorough and consistent cleaning. Some key features of robot cleaning methods for PV panels include the following:

1. Autonomous operation: The robots are programmed to operate independently, reducing the need for human intervention.

Figure 2.21 Unclean PV array on the left and robotic cleaning on the right (Azouzoute et al., 2021).

2. Precision cleaning: The robots are designed to clean panels gently without causing any damage or scratching.
3. Water conservation: Many robot cleaning systems use minimal water, reducing water consumption.
4. Monitoring and data collection: Some robot cleaning systems are equipped with sensors to monitor the condition of the panels and collect data on energy production efficiency and maintenance needs.
5. Remote control and scheduling: Robot cleaning systems can be controlled remotely and scheduled to clean at specific times to optimize energy production.

As shown in Fig. 2.21, a robot was created, tested, and introduced in 2017 to clean a 1 MW solar power facility. Every day, the electricity produced was gathered, and it was compared to the power produced by panels that had yet to be cleaned within the same time frame. The results showed that the cleaning method effectively reduced the effect of dust on the solar panel's power output. Power generation increased as a result, on average, by 32.27% (Hassan et al., 2017). A mobile robotic cleaning system with a flexible platform that moves along a panel is depicted in Fig. 2.22. An Arduino microcontroller was used to create the robot's control system. The robot's initial testing phase produced successful results, demonstrating the viability of such a device. Future design improvements have been considered, especially about how the robot can be moved from one panel to the next.

The adoption of robot cleaning methods for PV panels offers several benefits. It improves the efficiency and performance of the panels, increases energy generation, reduces labor costs, saves water, and eliminates the risk of injuries associated with manual cleaning. Overall, the robot cleaning method is an efficient and effective way to maintain and preserve the performance of PV panels, ensuring optimal energy production from solar installations.

2.5.2 Cooling methods for PV systems

Cooling is an essential aspect of the performance and longevity of PV systems. Solar panels are designed to convert sunlight into electricity, but heat can adversely affect

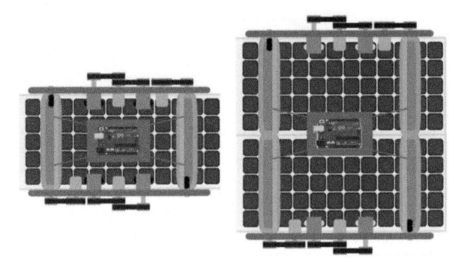

Figure 2.22 Robotic cooling device for solar panel prototype (Jaradat et al., 2015).

them. When panels get too hot, their efficiency decreases, and this can result in reduced power output. Here are a few key points to understand about cooling PV systems:

1. Temperature and efficiency: Solar panels operate more efficiently at lower temperatures. As the temperature increases, the efficiency of the panels decreases. This is known as the temperature coefficient, which represents the percentage drop in efficiency per degree increase in temperature.
2. Heat generation: Solar panels generate heat as a by-product when converting sunlight into electricity. The amount of heat generated depends on the sunlight's intensity and the panels' efficiency. Excessive heat buildup can lead to performance degradation and potential damage to the panels.
3. Cooling methods: There are different methods of cooling PV systems to control temperature and maximize performance:
   a. Passive cooling: This method utilizes natural heat dissipation through convection radiation and conduction. The panels are designed with proper spacing and ventilation to allow air to flow around them, dissipating heat. This can be achieved through tilted mounting systems, air gaps between panels, or raised mounts.
   b. Active cooling: Active cooling involves actively using external devices or systems to reduce the PV panels' temperature. Some standard techniques include the following:
      i. Water cooling: Water is circulated through pipes behind the solar panels, absorbing excess heat and dissipating it through a heat exchanger.
      ii. Air cooling: Fans or blowers direct airflow across the panels, removing heat through convection. This can be combined with passive cooling methods to improve overall performance.
      iii. Phase-change materials (PCMs): These materials absorb heat when the temperature rises and release it when it decreases. PCMs can be integrated into the backside of solar panels to help regulate temperature.
4. Monitoring and maintenance: Monitoring PV system performance and temperature ensures optimal efficiency. Temperature sensors can be installed to measure panel temperatures, and data loggers can be used to record performance parameters. Maintenance measures can be taken to address the issue promptly if any signs of overheating are detected.

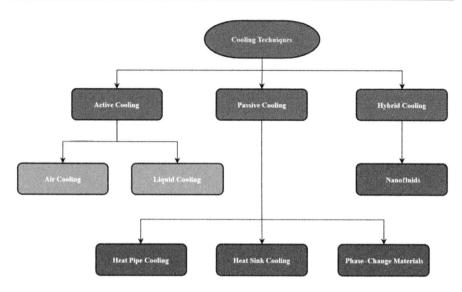

Figure 2.23 Diagram of a flowchart for cooling methods (Aslam et al., 2022).

Implementing cooling strategies for PV systems can help maintain efficiency and extend their life span. By controlling the temperature, the overall performance of the panels can be improved, leading to increased electricity generation. It is essential to consider the specific climate conditions, system setup, and budget when selecting the most suitable cooling method for a PV system. A flow chart of several PV system cooling methods is shown in Fig. 2.23.

2.5.2.1 Air cooling

Air cooling is a standard method to cool PV systems. It involves using air as a cooling medium to remove excess heat from the solar panels. There are a few different approaches to implementing air cooling for PV systems:

1. Natural convection: This method uses natural airflow around the PV panels to dissipate heat. By mounting the panels with a small gap between them and allowing for adequate airflow, heat is naturally transferred away from the panels. This method is relatively simple and cost-effective but may need to be improved in areas with high ambient temperatures or limited natural airflow.
2. Forced convection: In this method, fans or blowers actively move air over the PV panels' surface. This increases the rate of heat transfer and enhances cooling efficiency. Forced convection systems can be designed with air ducts or channels to direct the airflow more effectively. They are commonly used in larger PV installations and areas with limited natural airflow.
3. Liquid cooling: This approach involves circulating a liquid coolant, such as water or a glycol mixture, through a closed-loop system that is in contact with the back surface of the PV panels. The liquid absorbs the heat from the panels and carries it to a heat exchanger, where it is cooled before recirculating. Liquid cooling is highly efficient and

effective at removing heat, but it requires additional infrastructure, including pump pipes and a heat exchanger. The choice of air-cooling method depends on several factors, including the size of the PV system, available space ambient temperature, and cost considerations. In general, forced convection and liquid cooling methods offer better cooling performance and are suitable for larger PV installations or those operating in hot climates. On the other hand, natural convection is a more straightforward and cost-effective solution for smaller PV systems or areas with moderate ambient temperatures. It is important to note that proper maintenance and regular cleaning of the PV panels are essential for maximizing the effectiveness of any cooling method. Accumulated dirt, dust, or debris can reduce airflow and impede heat transfer, compromising the PV system's overall performance.

An illustration of their experimental setup is shown in Fig. 2.24. Their study examined varied flow rates and ambient air temperatures of 35°C, 40°C, and 45°C to see how they affected the module's effectiveness. It was found that a heat exchanger could be used to alter the temperature effectively. It became clear by comparing the panel power outputs that this cooling might boost daily electricity efficiency by up to 29.11% (Elminshawy et al., 2019).

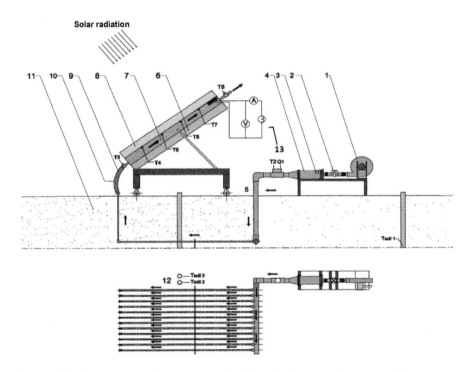

Figure 2.24 Diagrammatic representation of a PV system's air-cooling system (Elminshawy et al., 2019).

2.5.2.2 Liquid-based cooling

Air cooling is a standard method to cool PV systems. It involves using air as a cooling medium to remove excess heat from the solar panels. There are a few different approaches to implementing air cooling for PV systems:

1. Natural convection: This method uses natural airflow around the PV panels to dissipate heat. By mounting the panels with a small gap between them and allowing for adequate airflow, heat is naturally transferred away from the panels. This method is relatively simple and cost-effective but may need to be improved in areas with high ambient temperatures or limited natural airflow.
2. Forced convection: In this method, fans or blowers actively move air over the PV panels' surface. This increases the rate of heat transfer and enhances cooling efficiency. Forced convection systems can be designed with air ducts or channels to direct the airflow more effectively. They are commonly used in larger PV installations and areas with limited natural airflow.
3. Liquid cooling: This approach involves circulating a liquid coolant, such as water or a glycol mixture, through a closed-loop system that is in contact with the back surface of the PV panels. The liquid absorbs the heat from the panels and carries it to a heat exchanger, cooled before recirculating. Liquid cooling is highly efficient and effective at removing heat, but it requires additional infrastructure, including pump pipes and a heat exchanger.

The choice of air-cooling method depends on several factors, including the size of the PV system, available space ambient temperature, and cost considerations. In general, forced convection and liquid cooling methods offer better cooling performance and are suitable for larger PV installations or those operating in hot climates. On the other hand, natural convection is a more straightforward and cost-effective solution for smaller PV systems or areas with moderate ambient temperatures. To achieve a more consistent temperature in the cells, several novel approaches for liquid cooling of PV modules are being researched; one such proposal is the application of a converging channel, which was tried in a study (Baloch et al., 2015). The 2 degrees converging angle offered the best results for temperature homogeneity. On a typical hot day in June and a typical cold day in December, respectively, the PV temperature can be decreased by using converging channels from 71.2°C to 45.1°C and from 48.3°C to 36.4°C. The amount by which the temperature is reduced in PV panels that use liquid cooling is influenced by additional parameters, such as the type of coolant used (Khanjari et al., 2016).

It is important to note that proper maintenance and regular cleaning of the PV panels are essential for maximizing the effectiveness of any cooling method. Accumulated dirt, dust, or debris can reduce airflow and impede heat transfer, compromising the PV system's overall performance.

2.5.2.3 Heat pipe-based cooling

Liquid-based cooling methods are often deployed in PV systems to enhance efficiency and mitigate thermal stress. These cooling methods involve circulating a liquid such as water or a specialized coolant across the surface of the solar panels to remove excess heat. This helps to maintain optimal operating temperatures and prevents performance degradation. There are various types of liquid-based cooling techniques commonly used in PV systems:

1. Direct liquid cooling: In this method, a liquid coolant directly flows through channels or pipes attached to the backside or front surface of the solar panels. The heat absorbed from the panels is transferred to the liquid and then circulated to a heat exchanger, where the heat is dissipated into the surrounding environment. This method is efficient in removing heat directly from the PV panels.
2. Indirect liquid cooling: This technique uses a heat transfer fluid such as water or glycol to absorb heat from the PV modules. The heat transfer fluid circulates through a closed-loop system, transferring the absorbed heat to a heat exchanger. The heat is then dissipated using cooling towers' ambient air or other cooling methods while the cooled fluid is circulated back to the panels. Indirect liquid cooling is helpful in cases where the cooling system needs to be isolated from the PV modules.
3. Passive cooling: Passive cooling systems rely on natural convection and radiation to dissipate heat from the PV panels. This method involves incorporating heat sinks, heat pipes, or fin-like structures on the back surface of the modules. These structures increase the surface area for heat dissipation, allowing air or other gases to carry the heat passively. Passive cooling systems do not require additional power for operation, but their effectiveness depends on ambient conditions and air circulation.

The implementation of liquid-based cooling methods in PV systems offers several advantages. It enhances the power output of the panels by reducing temperature-induced performance losses. The life span and reliability of the PV modules can be extended by maintaining lower operating temperatures. This is particularly beneficial in hot climates or installations with densely packed solar panels, such as rooftops or ground-mounted PV arrays. Moreover, liquid-based cooling can optimize the overall system efficiency by allowing panels to operate closer to their maximum power point. This can result in higher energy yields and improved return on investment. Additionally, liquid cooling can contribute to maintaining stable temperatures, reducing the risk of hotspots and thermal stress on the PV modules. Despite the advantages, liquid-based cooling methods may require additional infrastructure such as pumps, heat exchangers, and tubing. These components add complexity and cost to the PV system. Furthermore, the choice of cooling liquid and system design must consider compatibility, durability, and maintenance requirements.

A suitable replacement for solid heat sinks or, in some situations, a viable alternative to the pumped liquid cooling system is heat pipes. They are long-lasting, have a high thermal conductivity, and are simple to bend and shape. A heat pipe is a two-sided, sealed pipe made of aluminum or copper that uses ammonia or water as the working fluid. Pulsating heat pipes (PHPs) are among the most often employed heat pipes because they are less sensitive to tilt angles than their gravity-assisted counterparts (Alhuyi Nazari et al., 2019). PHPs are a practical substitute for conductive fins for cooling PV modules because of their unique qualities. In a published study (Alizadeh et al., 2018), a single-turn PHP was used to control the thermal behavior of a PV panel. The PHP was positioned toward the module's rear in this study, as shown in Fig. 2.25. The heat was spread through radiation and convection, as depicted in Fig. 2.26. At a heat flow of 1000 W/m^2 and an immediate surrounding temperature of 291K, the transient heat equation on monocrystalline silicon solar cells was solved using a finite difference method. The numerical results showed that, while keeping the same size and geometry, utilizing PHP instead of copper resulted in a higher cooling of the PV modules.

Overall, liquid-based cooling techniques are an effective way to address thermal issues and optimize the performance of PV systems. With ongoing advancements in technology and increasing awareness of cooling strategies, the solar industry's adoption of liquid-based cooling is expected to grow.

2.5.2.4 Heat sink-based cooling

Heat sink-based cooling systems for PV panels are becoming increasingly popular for improving solar energy systems' overall efficiency and performance. PV panels are susceptible to degradation and efficiency losses when exposed to high temperatures, a

Figure 2.25 PV panel with a heat pipe model (Alizadeh et al., 2018).

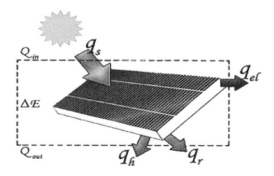

Figure 2.26 PV cell energy conservation (Alizadeh et al., 2018).

common issue in sunny regions or during periods of intense sunlight. Heat sinks are devices designed to dissipate excess heat by conducting it away from the source. They are commonly used in electronics and industries where thermal management is crucial. In the context of PV panels, heat sinks are used to reduce the temperature of the solar cells, thereby improving their electrical output and prolonging their life span. A heat sink-based cooling system for PV panels typically consists of three main components: heat sinks, a coolant circulation system, and a control system.

1. Heat sinks: These are generally made of thermally conductive materials such as aluminum or copper. The heat sinks are attached to the backside of the PV panels to absorb the excess heat generated by sunlight. The design of the heat sink is crucial for effective heat dissipation as it maximizes the surface area in contact with the air. Some heat sinks feature fins or other structures to increase their cooling efficiency.
2. Coolant circulation system: The heat absorbed by the heat sinks needs to be dissipated into the surrounding environment. This is usually achieved through a coolant circulation system. The system circulates a fluid such as water or a mixture of water and glycol through pipes or channels within the heat sinks. As the fluid absorbs heat from the heat sinks, it is pumped to a radiator or a heat exchanger, transferring the heat to the ambient air.
3. Control system: A control system is often employed to optimize the cooling process. This system monitors the temperature of the PV panels and regulates the flow rate and temperature of the coolant accordingly. It may be equipped with sensors such as thermocouples to provide accurate measurements and can be automated to adjust the cooling parameters based on real-time data.

The benefits of heat sink-based cooling systems for PV panels include the following:

1. Improved efficiency: By reducing the operating temperature of the solar cells, heat sinks prevent thermal losses that can negatively impact the electrical output of PV panels. More relaxed cells operate more efficiently and produce higher energy yields.
2. Extended life span: Excessive heat can lead to PV panel degradation and accelerated aging. Cooling systems help maintain lower temperatures, prolonging the panels' life span and reducing the need for replacements.
3. Enhanced reliability: By controlling and reducing operating temperatures, heat sink-based cooling systems contribute to the overall reliability and stability of PV systems, especially in high-temperature environments.
4. Increased energy harvest: The increased efficiency and extended life span of PV panels provided by heat sink-based cooling systems can result in higher energy production over the lifetime of the solar panels.

It has a lot of potential for cooling PV panels because it is straightforward and affordable. The PV panel's heat sink cooling system is depicted in Fig. 2.27. One study applied thermal conductive paste to the rear of the PV panel and attached aluminum fins in the form of an "L." The investigation revealed that randomly spaced fins with openings on the rear gave the PV panel the best cooling possible since air could pass through the interior of the structure at a speed of 1 m/s (Firoozzadeh et al., 2019).

In summary, heat sink-based cooling systems for PV panels offer effective thermal management solutions that improve solar energy systems' performance and reliability. As the demand for renewable energy continues to grow, such cooling systems can contribute to PV installations' overall efficiency and sustainability.

Figure 2.27 The PV panel's backside acts as a heat sink (Grubišić-Čabo et al., 2018).

2.5.2.5 Phase-change material-based cooling

Phase-change materials offer an innovative solution for cooling PV panels to improve efficiency and overall performance. PCMs can absorb, store, and release thermal energy during the phase change, typically between solid and liquid states. In a PCM-based cooling system for PV panels, the PCM is integrated into the panel structure or placed near it. When the panel surface temperature increases due to solar radiation, the PCM undergoes a phase change from solid to liquid, absorbing significant heat in the process. This phase change helps maintain the panel's temperature within an optimal range, preventing overheating and reducing the risk of performance degradation. One advantage of PCM-based cooling systems is their ability to store thermal energy. During periods of lower solar radiation, when the PV panel surface temperature decreases, the stored heat in the PCM is slowly released as it solidifies. This helps to maintain a stable temperature for the PV panels, allowing them to operate more efficiently even in varying weather conditions. PCMs can be selected based on their specific phase-change temperature, ensuring they absorb thermal energy within the desired temperature range. Additionally, PCMs with high thermal conductivity can facilitate heat transfer between the PV panel and the PCM, enhancing the cooling effect.

Implementation of PCM-based cooling systems for PV panels offers several benefits. First, it helps maintain the panels' operating temperature within an optimal range, which can significantly improve their electrical efficiency and power output. By reducing the panels' temperature, the risk of hotspots and potential damage to the PV cells is also minimized. Furthermore, PCM-based cooling systems require minimal maintenance and have relatively long life spans, making them a cost-effective solution for long-term energy production. Using PCMs can also extend the PV panels' life span by reducing thermal stress and degradation over time. A PV panel cooling system based on PCM was created in a study. Paraffin-based PCM with a melting range

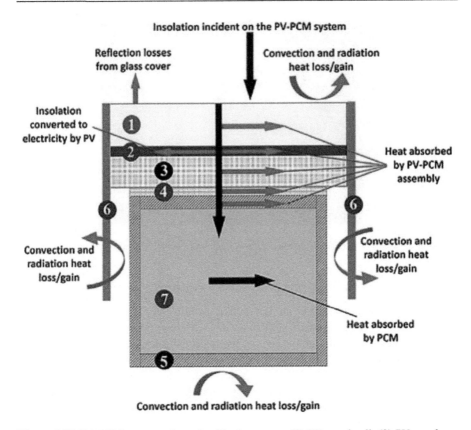

Figure 2.28 PV-PCM system schematic: (1) glass cover; (2) PV panel cell; (3) PV panel back sheet; (4) epoxy glue layer; (5) PCM container wall; (6) PV panel frame; and (7) layer of PCM (Hasan et al., 2017).

of 38°C−43°C was inserted at the back of a panel, as shown in Fig. 2.28. It was found that using PCM for cooling resulted in a 5.9% increase in electricity generation when comparing the annual electricity generated by the panel in hot temperature conditions. Additionally, less cooling was observed during extreme heat and cold periods, which was explained by partial solidification and melting, respectively (Hasan et al., 2017).

In conclusion, incorporating phase-change materials into cooling systems for PV panels effectively manages temperature, improves energy efficiency, and extends the panels' life span. With ongoing advancements in PCM technology, this innovative cooling solution holds promise for maximizing the performance and longevity of PV systems.

2.5.2.6 Nanofluid-based cooling

Nanofluid-based cooling of PV panels is an innovative approach that aims to enhance the efficiency and longevity of PV panels. Traditional PV panels often experience a decrease in efficiency as the panel temperature increases due to excess heat generation. This heat can be efficiently dissipated by implementing nanofluid-based cooling techniques, leading to improved panel performance. Nanofluids are suspensions of nanoparticles in a base fluid, typically water or oil. These nanoparticles possess unique thermal properties, such as high thermal conductivity and increased heat transfer capabilities. By adding these nanoparticles to the cooling medium flow across the PV panels, the heat generated by the panels can be more effectively transferred away, reducing the panel temperature.

One common technique is to use a microchannel heat exchanger integrated into the PV panel structure. The nanofluid flows through these microchannels in close contact with the PV cells, absorbing excess heat and transferring it away. The increased thermal conductivity of the nanofluid promotes efficient heat dissipation and ensures that the panels operate at lower temperatures, enhancing their overall electrical performance. The benefits of nanofluid-based cooling for PV panels are numerous. Firstly, the reduced panel temperature increases electrical efficiency as PV cells perform better at lower temperatures. This enhanced efficiency can translate into higher energy output and improved financial returns for solar power systems.

Additionally, the overall life span and reliability of the PV panels can be extended by maintaining lower operating temperatures. Elevated temperatures can accelerate the degradation of solar cells over time, leading to a decline in performance. With adequate cooling, the panels experience less thermal stress and wear, leading to longer life spans and reduced maintenance costs. Furthermore, nanofluid-based cooling methods can be easily integrated into new or existing PV panel systems. The design of the microchannel heat exchangers can be tailored to fit different panel sizes, shapes, and configurations. This flexibility allows for customization and optimization based on specific project requirements. However, it's important to note that implementing nanofluid-based cooling systems does come with some challenges. For instance, ensuring a consistent and reliable nanofluid flow across the panel surface requires careful design and attention to detail. Additionally, the choice of nanoparticles' concentration and stability in the fluid must be considered to prevent the formation of clogs or degradation over time. To determine the advantages of a PV/T system on the PV module, an experimental investigation was undertaken. The report acknowledged Hong Kong's potential for PV/T technology. The experiment's maximum electrical efficiency was 16%, but the scientists didn't look into the thermal performance. The setup of the experiment is displayed in Fig. 2.29 (Al-Shamani et al., 2016).

Overall, nanofluid-based cooling presents a promising solution for improving the efficiency and reliability of PV panels. As research and development in this field continue to advance, it is expected that nanofluid cooling will play an increasingly important role in maximizing the performance and longevity of solar power systems.

2.5.2.7 Hybrid cooling

Hybrid cooling refers to combining different techniques to enhance the efficiency and performance of PV panels. PV panels are sensitive to temperature, and their efficiency decreases as the temperature rises. Therefore, the panels can

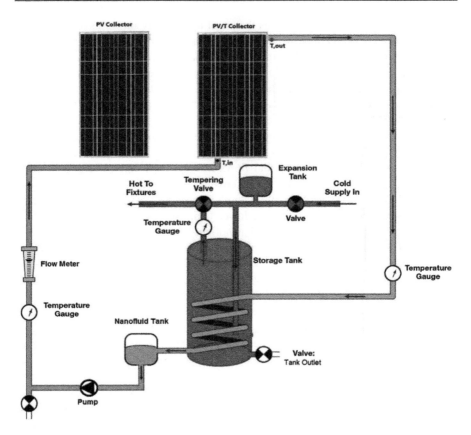

Figure 2.29 An illustration of the experimental setup (Al-Shamani et al., 2016).

maintain lower temperatures by implementing cooling systems, resulting in higher energy production. Various methods of hybrid cooling can be employed for PV panels:

1. Passive cooling: Passive cooling techniques involve design elements that enhance natural airflow and heat dissipation. This can include using light-colored or reflective materials for panel surfaces to reduce heat absorption and incorporating ventilation and spacing between panels to allow for air circulation.
2. Active cooling: Active cooling techniques involve external systems to remove heat from the PV panels. This can include the use of air-based or liquid-based cooling systems.
   a. Air-based cooling: Air-based cooling systems can utilize fans or blowers to circulate air around the PV panels, dissipating heat and reducing temperature. This technique is simple and cost-effective but may be less efficient in extremely hot or humid climates.
   b. Liquid-based cooling: Liquid-based cooling systems involve circulating a coolant such as water or a mixture of water and glycol through channels or pipes in direct contact with the PV panel surfaces. The coolant absorbs the heat generated by the panels and is then circulated to a heat exchanger or radiator for dissipation. Liquid-based cooling is more efficient than air-based cooling and is particularly effective in high-temperature environments.

3. Hybrid natural cooling combines passive and active cooling methods. It integrates natural cooling methods like ventilation and spacing with active cooling methods like fans or blowers. This combined approach maximizes heat dissipation while minimizing energy consumption.

Implementing hybrid cooling for PV panels offers several benefits, including the following:

1. Increased energy production: By maintaining lower temperatures, hybrid cooling techniques help PV panels operate at higher efficiencies, increasing energy production.
2. Extended life span: High temperatures can accelerate the degradation of PV panels over time. The panels' life span can be extended by reducing panel temperatures through cooling.
3. Enhanced reliability: Cooler operating temperatures can improve the overall reliability and performance of PV panels, reducing the risk of overheating-related failures.
4. Improved return on investment: Increasing the energy output and life span of PV panels through hybrid cooling can enhance the return on investment for solar installations.

A PV module and a thermal absorber are connected in a conventional PV/T setup. Technically, it is possible to design solar collectors for hybrid PV/T systems to operate with a cumulative efficiency of more than 80%. The electrical and thermal results in the PV/T system cannot be significantly increased simultaneously. Additionally, it is rare to find research on hybrid cooling systems that combine the following configurations: PV/T with PCM, PCM with nanofluids, a heat pipe with a heat sink, and PCM with a heat sink. Fig. 2.30 depicts the experimental

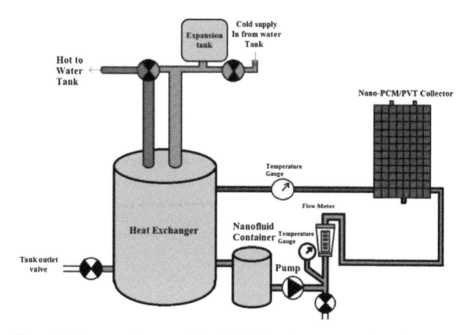

Figure 2.30 Diagrammatic representation of a hybrid cooling system experimental setup (Al-Waeli et al., 2017).

configuration for a hybrid cooling system. Examined and contrasted with the usage of a conventional PV module is the use of a nanofluid in a PCM system based on PV/T (Hassan et al., 2020).

It's important to note that selecting a specific hybrid cooling method depends on various factors such as climatic conditions, installation space budget, and maintenance requirements. Engineering expertise and analysis are typically required to determine a specific PV system's most suitable cooling approach.

2.6 Recommendations and future challenges

Solar PV systems have seen significant advancements in recent years, increasing efficiency, reliability, and cost-effectiveness. These advancements have made solar PV an increasingly attractive option for renewable energy generation worldwide. This comprehensive review will discuss various aspects of solar PV systems, including their performance influencing factors and mitigation techniques for potential challenges.

2.6.1 Performance analysis

Evaluating the performance of solar PV systems is crucial to ensure optimal energy generation. Parameters such as energy yield efficiency and reliability are critical indicators of system performance. Various factors such as solar irradiance, temperature, shading, soiling, and module degradation can impact the performance of PV systems. Understanding these factors and their effects on performance is essential for system optimization.

2.6.2 Influencing factors

2.6.2.1 Solar irradiance

Solar irradiance levels vary depending on location, time of day, and season. Accurate estimation of solar irradiance is crucial for predicting system performance.

2.6.2.2 Temperature

Higher operating temperatures can negatively impact the efficiency of PV modules. Managing temperature effectively through proper system design, cooling techniques, and material choices can improve performance.

2.6.2.3 Shading

Partial shading caused by nearby objects or vegetation can significantly reduce the overall energy yield of a PV system. Advanced power optimization techniques like bypass diodes and maximum power point tracking (MPPT) can mitigate shading effects.

2.6.2.4 Soiling

Accumulation of dust, dirt, or pollutants on PV modules can reduce their efficiency. Regular cleaning and maintenance are necessary to maintain optimal performance.

Module degradation: PV modules may experience degradation over time due to temperature, humidity, and UV exposure. Effective quality control measures and proper module selection can minimize degradation and ensure system longevity.

2.6.3 Mitigation techniques

To address the challenges associated with solar PV systems, several mitigation techniques have been developed:

1. Advanced monitoring systems: Real-time monitoring of system performance allows for early detection of issues, enabling prompt maintenance and maximizing energy production.
2. Optimal system design: Considering factors like tilt angle, azimuth angle, and shading analysis during the design phase can significantly improve system performance.
3. Module-level power electronics: Technologies like microinverters and DC optimizers enable the management of module-level power, mitigating the effects of shading and module mismatch.
4. Antisoiling coatings: Applying specialized coatings to PV modules can reduce dirt accumulation, improving system performance and reducing maintenance efforts.
5. Module inspection and maintenance: Regular inspections, cleaning, and maintenance are essential to prevent performance degradation.

2.6.4 Future challenges

Although solar PV systems have made substantial progress, several challenges remain:

1. Integration with the electrical grid: The intermittent nature of solar energy poses challenges for grid integration and stability. Developing advanced energy storage solutions and innovative grid technologies will be crucial for optimal utilization of solar PV systems.
2. Material sustainability: Ensuring the sustainability of materials used in PV modules, such as rare earth elements and heavy metals, is crucial to minimize environmental impact. Research and development into alternative materials and recycling technologies are needed.
3. Cost reduction: While solar PV costs have significantly decreased, further reductions are necessary for widespread adoption. Continued research and technological advancements can help drive down costs and improve affordability.

Solar PV systems have become integral to the global renewable energy landscape. Understanding the performance characteristics influencing factors and mitigation techniques is vital for optimizing system performance and addressing challenges. Continued research and technological advancements will drive the further development and deployment of solar PV systems as a sustainable solution for energy generation.

2.7 Conclusions

In conclusion, the advances in solar PV systems have significantly improved their performance and efficiency, making them a viable and sustainable alternative to conventional energy sources. This comprehensive review has discussed various factors that influence the performance of solar PV systems and highlighted different mitigation techniques to enhance their efficiency. One of the critical factors influencing PV performance is solar irradiance, which can be optimized through advanced tracking and concentration techniques. Additionally, the angle and orientation of PV modules play a crucial role in capturing maximum sunlight, and various strategies, such as tilt and azimuth angle optimization, have been proposed to improve system performance. The materials used in PV modules also significantly impact their efficiency. Recent advancements in PV technology have led to the development of new materials, such as perovskite and multijunction cells, which have shown promising results in higher efficiency and lower production costs. Moreover, various degradation mechanisms that affect PV performance over time have been identified, including temperature-induced stress, humidity, and soiling. These issues can be mitigated through proper system design, module encapsulation techniques, and regular maintenance.

Researchers have focused on developing innovative cooling techniques, such as thermal management systems and passive cooling strategies, to improve PV performance and reduce temperature-induced losses. Additionally, integrating energy storage systems and adopting innovative grid technologies can help optimize solar energy utilization and ensure a reliable power supply. Furthermore, using predictive modeling tools and advanced monitoring systems has enabled real-time analysis of PV performance and early detection of system faults, improving maintenance and overall efficiency. In conclusion, the advances in solar PV systems have addressed various challenges and limitations, ultimately enhancing their performance and efficiency. The continuous research and development in this field will further contribute to the widespread adoption and integration of solar energy as a clean and sustainable solution for meeting the world's growing energy demands.

References

Agrawal, M., Chhajed, P., & Chowdhury, A. (2022). Performance analysis of photovoltaic module with reflector: Optimizing orientation with different tilt scenarios. *Renewable Energy, 186*, 10−25.

Ahmed, N., et al. (2021). Techno-economic potential assessment of mega scale grid-connected PV power plant in five climate zones of Pakistan. *Energy Conversion and Management, 237*, 114097.

Al-Bashir, A., Al-Dweri, M., Al−Ghandoor, A., Hammad, B., & Al−Kouz, W. (2019). Analysis of effects of solar irradiance, cell temperature and wind speed on photovoltaic systems performance. *International Journal of Energy Economics and Policy, 10*(1), 353−359.

Alhuyi Nazari, M., Ghasempour, R., & Ahmadi, M. H. (2019). A review on using nanofluids in heat pipes. *Journal of Thermal Analysis and Calorimetry, 137*, 1847−1855.

Alizadeh, H., Ghasempour, R., Shafii, M. B., Ahmadi, M. H., Yan, W.-M., & Nazari, M. A. (2018). Numerical simulation of PV cooling by using single turn pulsating heat pipe. *International Journal of Heat and Mass Transfer*, *127*, 203−208.

Allouhi, A., Saadani, R., Kousksou, T., Saidur, R., Jamil, A., & Rahmoune, M. (2016). Grid-connected PV systems installed on institutional buildings: Technology comparison, energy analysis and economic performance. *Energy and Buildings*, *130*, 188−201.

Alonso-Garcia, M., Ruiz, J., & Chenlo, F. (2006). Experimental study of mismatch and shading effects in the I−V characteristic of a photovoltaic module. *Solar Energy Materials and Solar Cells*, *90*(3), 329−340.

Al-Shamani, A. N., Sopian, K., Mat, S., Hasan, H. A., Abed, A. M., & Ruslan, M. (2016). Experimental studies of rectangular tube absorber photovoltaic thermal collector with various types of nanofluids under the tropical climate conditions. *Energy Conversion and Management*, *124*, 528−542.

Al-Waeli, A. H., et al. (2017). Evaluation of the nanofluid and nano-PCM based photovoltaic thermal (PVT) system: An experimental study. *Energy Conversion and Management*, *151*, 693−708.

Ameur, A., Berrada, A., Bouaichi, A., & Loudiyi, K. (2022). Long-term performance and degradation analysis of different PV modules under temperate climate. *Renewable Energy*, *188*, 37−51.

Ameur, A., Sekkat, A., Loudiyi, K., & Aggour, M. (2019). Performance evaluation of different photovoltaic technologies in the region of Ifrane, Morocco. *Energy for Sustainable Development*, *52*, 96−103.

Anderson, M., et al. (2010). *Robotic device for cleaning photovoltaic panel arrays. Mobile robotics: Solutions and challenges* (pp. 367−377). World Scientific.

Aslam, A., Ahmed, N., Qureshi, S. A., Assadi, M., & Ahmed, N. (2022). Advances in solar PV systems; A comprehensive review of PV performance, influencing factors, and mitigation techniques. *Energies*, *15*, 7595. Available from https://doi.org/10.3390/en15207595.

Aste, N., Del Pero, C., & Leonforte, F. (2014). PV technologies performance comparison in temperate climates. *Solar Energy*, *109*, 1−10.

Atsu, D., Seres, I., & Farkas, I. (2021). The state of solar PV and performance analysis of different PV technologies grid-connected installations in Hungary. *Renewable and Sustainable Energy Reviews*, *141*, 110808.

Azouzoute, A., Zitouni, H., El Ydrissi, M., Hajjaj, C., Garoum, M., & Ghennioui, A. (2021). Developing a cleaning strategy for hybrid solar plants PV/CSP: Case study for semi-arid climate. *Energy*, *228*, 120565.

Baloch, A. A., Bahaidarah, H. M., Gandhidasan, P., & Al-Sulaiman, F. A. (2015). Experimental and numerical performance analysis of a converging channel heat exchanger for PV cooling. *Energy Conversion and Management*, *103*, 14−27.

do Nascimento, L. R., Braga, M., Campos, R. A., Naspolini, H. F., & Rüther, R. (2020). Performance assessment of solar photovoltaic technologies under different climatic conditions in Brazil. *Renewable Energy*, *146*, 1070−1082.

Elminshawy, N. A., El Ghandour, M., Gad, H., El-Damhogi, D., El-Nahhas, K., & Addas, M. F. (2019). The performance of a buried heat exchanger system for PV panel cooling under elevated air temperatures. *Geothermics*, *82*, 7−15.

Emziane, M., & Al Ali, M. (2015). Performance assessment of rooftop PV systems in Abu Dhabi. *Energy and Buildings*, *108*, 101−105.

Farahmand, M. Z., Nazari, M., Shamlou, S., & Shafie-khah, M. (2021). The simultaneous impacts of seasonal weather and solar conditions on PV panels electrical characteristics. *Energies*, *14*(4), 845.

Feldman, D., Dummit, K., Zuboy, J., & Margolis, R. (2023). *Spring 2023 solar industry update*. Golden, CO: National Renewable Energy Lab (NREL).

Firoozzadeh, M., Shiravi, A., & Shafiee, M. (2019). An experimental study on cooling the photovoltaic modules by fins to improve power generation: Economic assessment. *Iranian (Iranica) Journal of Energy & Environment, 10*(2), 80–84.

Fouad, M., Shihata, L. A., & Morgan, E. I. (2017). An integrated review of factors influencing the performance of photovoltaic panels. *Renewable and Sustainable Energy Reviews, 80*, 1499–1511.

Green, M. A., Emery, K., King, D. L., Hisikawa, Y., & Warta, W. (2006). Solar cell efficiency tables (version 27). *Progress in Photovoltaics, 14*(1), 45–52.

Grubišić-Čabo, F., Nižetić, S., Čoko, D., Kragić, I. M., & Papadopoulos, A. (2018). Experimental investigation of the passive cooled free-standing photovoltaic panel with fixed aluminum fins on the backside surface. *Journal of Cleaner Production, 176*, 119–129.

Hassan, A., et al. (2020). Thermal management and uniform temperature regulation of photovoltaic modules using hybrid phase change materials-nanofluids system. *Renewable Energy, 145*, 282–293.

Hasan, A., Sarwar, J., Alnoman, H., & Abdelbaqi, S. (2017). Yearly energy performance of a photovoltaic-phase change material (PV-PCM) system in hot climate. *Solar Energy, 146*, 417–429.

Hassan, M. U., Nawaz, M. I., & Iqbal, J. (2017). *Towards autonomous cleaning of photovoltaic modules: Design and realization of a robotic cleaner. 2017 First international conference on latest trends in electrical engineering and computing technologies (INTELLECT)* (pp. 1–6). IEEE.

He, G., Zhou, C., & Li, Z. (2011). Review of self-cleaning method for solar cell array. *Procedia Engineering, 16*, 640–645.

Høiaas, I., Grujic, K., Imenes, A. G., Burud, I., Olsen, E., & Belbachir, N. (2022). Inspection and condition monitoring of large-scale photovoltaic power plants: A review of imaging technologies. *Renewable and Sustainable Energy Reviews, 161*, 112353.

Jaradat, M. A., et al. (2015). *A fully portable robot system for cleaning solar panels. 2015 10th International symposium on mechatronics and its applications (ISMA)* (pp. 1–6). IEEE.

Kaddoura, T. O., Ramli, M. A., & Al-Turki, Y. A. (2016). On the estimation of the optimum tilt angle of PV panel in Saudi Arabia. *Renewable and Sustainable Energy Reviews, 65*, 626–634.

Kawamura, H., et al. (2003). Simulation of I−V characteristics of a PV module with shaded PV cells. *Solar Energy Materials and Solar Cells, 75*(3–4), 613–621.

Khanjari, Y., Pourfayaz, F., & Kasaeian, A. (2016). Numerical investigation on using of nanofluid in a water-cooled photovoltaic thermal system. *Energy Conversion and Management, 122*, 263–278.

Lamont, L. A., & El Chaar, L. (2011). Enhancement of a stand-alone photovoltaic system's performance: Reduction of soft and hard shading. *Renewable Energy, 36*(4), 1306–1310.

Mazumder, M., et al. (2006). Solar panel obscuration by dust and dust mitigation in the Martian atmosphere. *Particles on Surfaces: Detection, Adhesion and Removal, 9*, 167–195.

Mazumder, M., et al. (2011). *Electrostatic removal of particles and its applications to self-cleaning solar panels and solar concentrators. Developments in surface contamination and cleaning* (pp. 149–199). Elsevier.

Mohamed, A. O., & Hasan, A. (2012). Effect of dust accumulation on performance of photovoltaic solar modules in Sahara environment. *Journal of Basic and Applied Scientific Research*, 2(11), 11030–11036.

Moharram, K., Abd-Elhady, M., Kandil, H., & El-Sherif, H. (2013). Influence of cleaning using water and surfactants on the performance of photovoltaic panels. *Energy Conversion and Management*, 68, 266–272.

Pavan, A. M., Mellit, A., & De Pieri, D. (2011). The effect of soiling on energy production for large-scale photovoltaic plants. *Solar Energy*, 85(5), 1128–1136.

Prăvălie, R., Patriche, C., & Bandoc, G. (2019). Spatial assessment of solar energy potential at global scale. A geographical approach. *Journal of Cleaner Production*, 209, 692–721.

Rathinadurai Louis, J., Shanmugham, S., Gunasekar, K., Atla, N. R., & Murugesan, K. (2016). Effective utilisation and efficient maximum power extraction in partially shaded photovoltaic systems using minimum-distance-average-based clustering algorithm. *IET Renewable Power Generation*, 10(3), 319–326.

Sharma, M. K., Kumar, D., Dhundhara, S., Gaur, D., & Verma, Y. P. (2020). Optimal tilt angle determination for PV panels using real time data acquisition. *Global Challenges*, 4(8), 1900109.

Smith, M. K., et al. (2014). Water cooling method to improve the performance of field-mounted, insulated, and concentrating photovoltaic modules. *Journal of Solar Energy Engineering*, 136(3), 034503.

Srivastava, R., Tiwari, A., & Giri, V. (2020). An overview on performance of PV plants commissioned at different places in the world. *Energy for Sustainable Development*, 54, 51–59.

Tejwani, R., & Solanki, C. S. (2010). *360 sun tracking with automated cleaning system for solar PV modules. 2010 35th IEEE photovoltaic specialists conference* (pp. 002895–002898). IEEE.

Vasiljev, P., Borodinas, S., Bareikis, R., & Struckas, A. (2013). Ultrasonic system for solar panel cleaning. *Sensors and Actuators A: Physical*, 200, 74–78.

Technologies review for solar thermal integrated photovoltaic desalination

Yashar Aryanfar[1], Mamdouh El Haj Assad[2], Jorge Luis García Alcaraz[3], Julio Blanco Fernandez[4], Mohamed M. Awad[5], Shabbir Ahmad[6,7], Abdallah Bouabidi[8,9] and Ali Keçebaş[10]

[1]Department of Electric Engineering and Computation, Autonomous University of Ciudad Juárez, Ciudad Juárez, Chihuahua, México, [2]Department of Sustainable and Renewable Energy Engineering, University of Sharjah, Sharjah, United Arab Emirates, [3]Department of Industrial Engineering and Manufacturing, Autonomous University of Ciudad Juárez, Ciudad Juárez, Chihuahua, México, [4]Department of Mechanical Engineering, University of La Rioja, Logroño, La Rioja, Spain, [5]Mechanical Power Engineering Department, Faculty of Engineering, Mansoura University, Mansoura, Egypt, [6]Institute of Geophysics and Geomatics, China University of Geosciences, Wuhan, P.R. China, [7]Department of Basic Sciences and Humanities, Muhammad Nawaz Sharif University of Engineering and Technology, Multan, Pakistan, [8]Mechanical Modeling, Energy and Material (MEM), National School of Engineering of Gabes (ENIG), University of Gabes, Gabes, Tunisia, [9]Higher Institute of Industrial Systems of Gabes (ISSIG), University of Gabes, Gabes, Tunisia, [10]Department of Energy Systems Engineering, Technology Faculty, Muğla Sıtkı Koçman University, Muğla, Turkey

3.1 Introduction

Solar thermal integrated PV desalination is an innovative and sustainable solution that combines both solar thermal and PV technologies to produce freshwater from seawater or brackish water sources. This integrated system harnesses sunlight's power to generate electricity and thermal energy, simultaneously enabling the efficient desalination of water. In this system, PV panels generate electricity by converting sunlight directly into electrical energy. The PV panels absorb solar radiation and convert it into direct current (DC) electricity, which can power the desalination process and other electrical loads in the system.

At the same time, solar thermal collectors are employed to capture the heat from the sunlight and transfer it to a heat transfer fluid (HTF) such as water or oil. This thermal energy is then utilized in the desalination process to heat the seawater or brackish water, resulting in its evaporation and subsequent condensation into fresh purified water.

Performance Enhancement and Control of Photovoltaic Systems. DOI: https://doi.org/10.1016/B978-0-443-13392-3.00003-7

The integrated system maximizes the utilization of solar energy resources by combining electrical and thermal energy production. It offers several advantages over conventional desalination methods, including higher energy efficiency, reduced operational costs, and decreased environmental impact. One key advantage of solar thermal integrated PV desalination is its ability to operate in stand-alone mode, i.e., without relying on grid electricity. This makes it particularly suitable for remote coastal areas or islands where access to electricity may be limited or unreliable. Furthermore, the system can be designed with thermal energy storage capabilities allowing for continuous operation even when sunlight is unavailable. The stored thermal energy can be used during cloudy periods or at night to maintain the desalination process, ensuring a reliable and uninterrupted freshwater supply. Another benefit is the potential for cogeneration. The waste heat generated from the desalination process can be utilized for various applications such as space heating, agricultural processes, or other industrial processes, further enhancing the system's overall energy efficiency and sustainability. Solar thermal integrated PV desalination provides a promising solution for addressing water scarcity and increasing freshwater demand in regions with abundant solar resources. It offers a clean and sustainable alternative to conventional desalination technologies contributing to climate change mitigation and preserving precious water resources.

Different desalination methods are currently commercialized, and all methods require energy. Various desalination technologies' current installed capacities and their contributions to overall installed capacity are shown in Fig. 3.1 (Esmaeilion, 2020; Jones et al., 2019). Desalination technology is always improving to increase

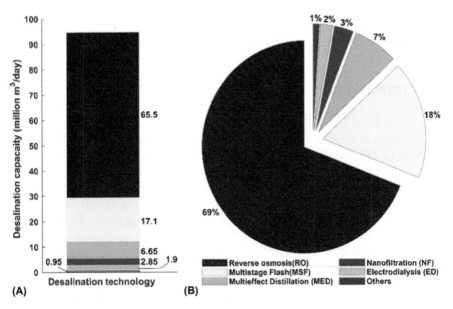

Figure 3.1 (A) Current installed desalination technology capacity and (B) its share of the installed capacity overall (Anand et al., 2021).

productivity, save energy, and lower costs. Freshwater costs are currently less than 0.6 US\$/m^3, down from around 10 US\$/m^3 in the 1960s (Al-Karaghouli & Kazmerski, 2013). Desalination systems, however, need substantial energy, whether through direct use of fossil fuels or electricity. In general, various factors may affect energy use and the choice of desalination technique (Fig. 3.2).

Early photovoltaic thermal (PVT) research dates back to the middle of the 1970s. Based on the configuration, PVT collectors are broadly categorized into flat plate and concentrated PVT (CPVT) collectors. The PVT technology has advanced over time, and current work includes developing electrical and thermal models and applying various techniques to improve performance (Chow, 2010; Joshi & Dhoble, 2018a; Sharaf & Orhan, 2015). Fig. 3.3 illustrates specific information on the PVT panel type, setups, and general applications. Solar PVT technologies are currently highly developed and widely adopted (Zenhäusern et al., 2017). The goal is to

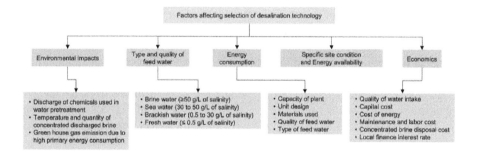

Figure 3.2 Technologies that affect desalination (Anand et al., 2021).

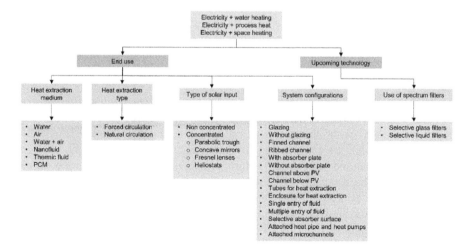

Figure 3.3 Photovoltaic thermal collector classifications (Anand et al., 2021; Joshi & Dhoble, 2018a).

reduce costs, primary energy consumption, and space requirements by effectively utilizing electricity and heat from the coolant in other applications.

The current work offers a thorough analysis of the technical advancements in PVT coupled with various desalination systems. Studying and outlining the specifics of PVT-integrated desalination technologies is the primary goal, along with investigating ways to use PVT panels as energy sources in other desalination technologies. PVT desalination systems' economic viability and environmental friendliness are discussed and contrasted with other desalination systems powered by renewable energy.

3.2 Direct desalination methods integrated with a photovoltaic thermal collector

Photovoltaic thermal collectors integrated with direct desalination technologies offer an innovative solution to address the pressing challenges of renewable energy generation and clean water production. This integrated system combines PV technology, which converts sunlight into electricity, with thermal collectors that harness solar heat for water desalination. The basic principle of a PVT system is to simultaneously generate electricity and utilize the waste heat from the PV cells for thermal applications. The system can further increase its efficiency and self-sustainability by integrating direct desalination technologies such as membrane distillation or multieffect distillation (MED) (Anand et al., 2021).

The PV cells in the PVT panels convert sunlight into electricity that can be used to power the desalination process. Any excess electricity can be stored in batteries or fed into the grid contributing to the overall energy supply. Meanwhile, the waste heat generated by the PV cells is captured and utilized for desalination. Direct desalination technologies use the captured solar heat to evaporate seawater or brackish water. The vapor produced is then condensed into freshwater, leaving the concentrated salts and impurities behind. These salts can be further processed and utilized, ensuring minimal waste and potentially providing additional value (Khalil et al., 2023).

Integration of PVT collectors with direct desalination technologies offers several advantages. Firstly, it maximizes the utilization of solar energy by combining electricity generation with thermal applications, resulting in higher efficiency. Secondly, it reduces the reliance on external energy sources, making the system more self-sufficient and independent. Moreover, as the thermal collectors generate heat, it reduces the need for additional heating systems, making the desalination process more cost-effective. This integrated system can be particularly beneficial in areas with abundant sunlight and limited access to clean water, such as arid regions or coastal communities. It can provide a sustainable and decentralized solution for freshwater production, reducing the burden on centralized water infrastructure and mitigating the environmental impact of traditional desalination methods (Chen et al., 2021).

However, there are challenges to overcome in developing and implementing such systems. Ensuring the long-term reliability and durability of PVT collectors and optimizing the efficiency of direct desalination technologies require further research and development. Additionally, cost considerations and scalability must be addressed to make the technology more accessible and economically viable on a larger scale (Tyagi et al., 2012). In conclusion, integrating PVT collectors with direct desalination technologies offers a promising renewable energy generation and clean water production solution. This integrated approach can contribute to the sustainability and resilience of water and energy systems, ultimately improving the quality of life for communities in need.

3.2.1 Combined photovoltaic thermal collector and solar still

Photovoltaic thermal collector integrated with solar stills combines the benefits of both technologies to maximize energy output and increase efficiency. This innovative system not only generates electricity but also produces clean drinking water. A PVT collector is a hybrid solar energy system that combines PV panels and thermal collectors into one unit. PV panels convert sunlight into electricity, while thermal collectors capture the sun's heat. A PVT collector can simultaneously provide electricity and thermal energy by combining these two technologies.

On the other hand, solar stills use solar energy to separate water from impurities and contaminants through evaporation and condensation. They are commonly used in areas with limited access to clean drinking water. Integrating a PVT collector with a solar still allows the system to harness solar electricity and thermal energy to power the still's operation. The PV panels generate electricity to power the still's pumps, fans, and other components, while the thermal collector provides the necessary heat for evaporation.

The integration of these technologies has numerous advantages. Firstly, it increases the overall efficiency of the system. The excess heat produced by the PV panels is utilized to enhance the evaporation process, reducing energy consumption and improving water production. Secondly, the integration allows for simultaneous electricity production and clean drinking water. This is particularly beneficial in remote areas where access to both resources may be limited. The system can provide sustainable and reliable power and water source without relying on external energy sources or infrastructure. Integrating a PVT collector with a solar still promotes sustainability and environmental stewardship. By utilizing solar energy, the system reduces dependence on fossil fuels and helps mitigate greenhouse gas emissions. Furthermore, the integrated system can be designed to be modular and scalable, allowing for flexibility in installation and adaptation to varying energy and water demands. It can be implemented in various settings, such as residential buildings, community facilities, or even larger scale projects for humanitarian purposes.

In conclusion, integrating a PVT collector with solar stills offers a promising solution for sustainable energy and clean water production. By combining the benefits of both technologies, this innovative system can maximize energy output, increase efficiency, and provide a valuable resource for communities in need.

One of the hybrid technologies that produce both energy and freshwater concurrently is a photovoltaic/thermal (PV/T) integrated active solar still. It has two crucial parts: a solar still and a PVT collector. A conceptual representation of a typical PVT-integrated solar still is shown in Fig. 3.4. A straightforward hybrid PVT integration with a single-slope active solar still was started by Kumar and Tiwari (Kumar & Tiwari, 2008). In this, a solar flat plate collector (FPC) with a surface area of 2 m^2 and a PV panel of 0.66 m^2 are partially integrated. The PV panel is cooled by water running behind it, but it is also heated by the heat the PV panel generates. Due to its self-sustainability, this hybrid active solar still can be employed in any isolated area and produces 3.2−5.5 times more energy than traditional passive solar still (Singh et al., 2011).

A CPC-PVT combined with a single-slope active solar still was examined by Mishra et al. (2019) in natural flow mode. Compared to conventional solar stills, CPC-PVT stills maintain a 28% higher temperature variation between the basin and the glass cover while having a 5.2% bigger intercepting surface. Xinxin et al. (2019) created a low-concentration ratio CPC-PVT-driven single-slope solar still. The CPC-PVT was parallel and connected to three solar stills with varied glass cover angles (24, 35, and 45) (Fig. 3.5).

A PVT single-slope stepped solar still with a bottom channel in the basin was created by Xiao et al. (2019) (Fig. 3.6). The bottom channel, which was created between the insulation and absorber plate, was used to channel preheated salt/saline water from the PVT panel. Saline water was heated inside the bottom channel even more by absorbing heat from the absorber plate and then dispersed throughout the solar still's steps. The average heat transfer rate between the absorber plate and salt water was raised by 44% at an optimal depth of the bottom channel (0.01 m). As a result, there was an

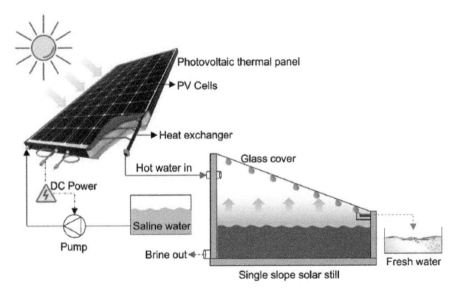

Figure 3.4 A typical PVT integrated solar still's conceptual diagram (Anand et al., 2021).

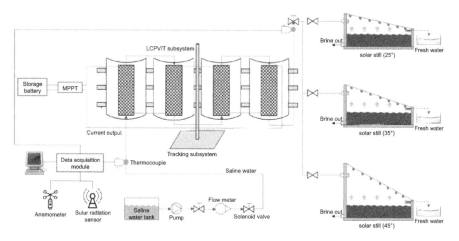

Figure 3.5 A low-concentration ratio CPC-PVT-coupled solar still's schematic diagram (Xinxin et al., 2019).

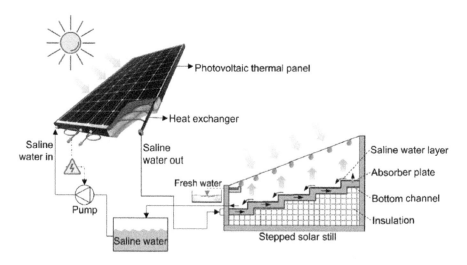

Figure 3.6 A PVT-linked stepped solar still's schematic diagram (Xiao et al., 2019).

improvement in average thermal efficiency, exergy efficiency, daily freshwater production, and saline water temperature of 16.4%, 51.7%, 17%, and 3%, respectively.

3.2.2 Combined desalination, humidification, and dehumidification photovoltaic thermal collector

The integrated system of a PVT collector with humidification and dehumidification desalination combines various technologies to generate electricity efficiently, produce heat, and desalinate water using renewable energy sources. The PVT collector is a device that combines PV cells with a thermal absorber. It utilizes solar radiation to generate electricity through the PV effect while also harnessing the waste heat for other applications. The PV cells convert sunlight directly into electricity, while the thermal absorber captures the excess heat for desalination (Gabrielli et al., 2019).

Humidification and dehumidification (HDH) desalination is a sustainable and energy-efficient method to convert saltwater into freshwater. The process involves two main stages: humidification and dehumidification. In the humidification stage, the saltwater is heated, causing evaporation and separation of the water vapor from the brine. The water vapor is then condensed and collected as freshwater. In the dehumidification stage, the remaining brine is heated further to remove residual moisture before discharge. By integrating these two technologies, the excess heat generated by the PV cells can be directly utilized in the desalination process. The heat can provide the necessary energy for the evaporation and condensation stages of the HDH desalination system. This integration improves the overall energy efficiency of the combined system and reduces the dependency on external energy sources (Elsafi, 2017).

The integration of the PV thermal collector with HDH desalination has several advantages. Firstly, it maximizes the use of solar energy as the same setup can generate electricity and desalinate water simultaneously. Secondly, it reduces the cost and space requirement of separate PV and desalination systems. Lastly, it promotes sustainable and environmentally friendly water production by utilizing renewable energy sources and reducing greenhouse gas emissions. Integrating a PVT collector with humidification and dehumidification desalination is a promising solution for sustainable water production. Utilizing solar energy efficiently, it contributes to the development of more environmentally friendly and cost-effective desalination technologies (Pourafshar et al., 2020).

Lawal and Qasem (2020) investigated various HDH desalination systems linked with geothermal, renewable, and waste heat from power plants and heat pumps. HDH desalination powered by heat pumps and waste heat from power plants is more productive and effective. However, isolated areas need access to these waste heat sources. Therefore, it is recommended that solar thermal collectors, geothermal energy, and PVT panels be used in conjunction with HDH desalination in remote areas. However, most of the HDH systems powered by geothermal energy and solar thermal collectors need extra electrical power (Alnaimat & Klausner, 2012; Li & Zhang, 2016). As a result, remote places can benefit more from PVT-combined HDH desalination. The HDH desalination and its accompanying equipment require both thermal and electrical energy, which the PVT technology provides. The conceptual diagram of a typical PVT-integrated HDH desalination system is shown in Fig. 3.7.

Anand and Murugavelh also created and investigated a thorough economic model of CPVT-HDH desalination (Anand & Murugavelh, 2019, 2022). By integrating the CPVT collector with a two-stage HDH desalination and cooling plant, Ananda et al. (2019) further optimized the system. The two-stage HDH desalination

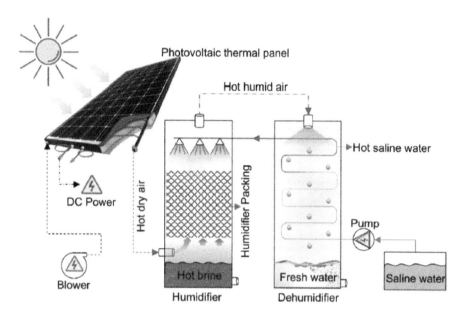

Figure 3.7 Process flow schematic of a PVT-integrated HDH desalination plant using air cooling.

plant controlled by CPVT uses 76% less electricity than a solar flat plate collector while still getting the necessary thermal energy from the CPVT collector. The two-stage HDH desalination plant controlled by CPVT produced 8.3% more specific water and utilized 52.1% more energy than the plant run by a solar flat plate collector, respectively.

A new PVT open water closed-air HDH desalination coupled with heat pump was created by Pourafshar et al. (2020) (Fig. 3.8). A double-pass dual-fluid heat exchanger (DP-DF-HE), which serves as a humidifier for HDH desalination, was installed beneath the PVT panel (Fig. 3.9). Saline water and air from the heat pump condenser were interacting when they entered the DP-DF-HE. The resulting humid air was cooled and dehumidified using the cooling load in the heat pump's evaporator. In that order, the maximum PVT-humidifier evaporation rate, average freshwater production rate, and cost of freshwater were 1.06 L/h-m^2 of humidifier, 0.99 L/h-m^2 of humidifier, and 0.018 US$/L.

3.2.3 Observations and prospects of PVT-coupled solar still and HDH desalination exploration in the future

Photovoltaic thermal-coupled solar still and HDH desalination is an innovative and promising approach to address two significant challenges simultaneously: clean water scarcity and renewable energy generation (Elsafi, 2017; Pourafshar et al., 2020).

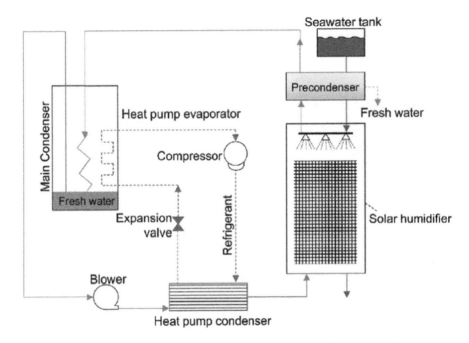

Figure 3.8 Diagram of the HDH desalination system's process, which is helped by a heat pump and a PVT humidifier (Pourafshar et al., 2020).

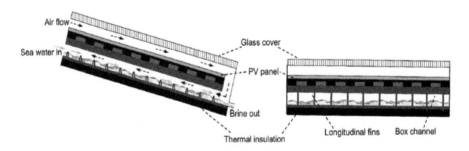

Figure 3.9 Solar PVT humidifier in cross section (Pourafshar et al., 2020).

Here are some observations and future opportunities to explore in PVT-coupled solar still and HDH desalination:

1. Enhanced energy efficiency: Integrating PV and thermal collectors in a PVT system improves overall energy efficiency. By capturing both electricity and heat from sunlight, the system can generate more energy for the desalination process, thereby increasing the overall system's efficiency. Future research can focus on optimizing the PVT system design and operational parameters to maximize energy yield.
2. Performance optimization: The performance of the PVT-coupled solar still and HDH desalination system can be further enhanced by studying various parameters such as water

flow rate, solar radiation intensity, wind speed, and ambient temperature. Understanding the effect of these parameters on the system's performance can lead to better control strategies and operational guidelines.

3. Salt concentration management: Seawater contains a high concentration of salts that can accumulate in the system and reduce its efficiency over time. Developing effective salt concentration management techniques such as periodic cleaning or antiscaling measures is essential to ensure sustained performance and longevity of the system.

4. Integration with energy storage: Incorporating energy storage technologies such as batteries or thermal storage systems can address the intermittent nature of solar radiation. This allows for continuous operation of the desalination system even during periods of low solar availability. Future research can explore the optimal sizing and control strategies for integrating energy storage into the PVT-coupled solar still and HDH desalination system.

5. Techno-economic analysis: Conducting a comprehensive techno-economic analysis of the PVT-coupled solar still and HDH desalination system is crucial to evaluate its feasibility and cost-effectiveness. This analysis should consider factors such as capital costs, operational costs, maintenance requirements, and the potential for revenue generation from excess electricity production.

6. Scale-up and commercialization: While the concept of PVT-coupled solar still and HDH desalination has shown promising results in small-scale experimental setups, scaling up the technology for a commercial application requires further investigation. This includes the development of large-scale prototype testing under real-world conditions and assessing the system's performance and economic viability on a larger scale.

7. Environmental impact assessment: It is crucial to ensure their sustainability to evaluate the environmental impact of PVT-coupled solar still and HDH desalination systems. This assessment should consider factors such as energy consumption, greenhouse gas emissions, water usage, and the potential for ecosystem disruption. Future research should optimize the system's design and operation to minimize its environmental footprint.

In conclusion, PVT-coupled solar still and HDH desalination is a promising technology that can simultaneously address water scarcity and renewable energy generation. Continued research and development in areas such as energy efficiency performance optimization integration with energy storage and scale-up will be essential to realize the full potential of this technology and make it commercially viable.

3.3   Combined nonmembrane indirect desalination technology and photovoltaic thermal collector

Photovoltaic thermal collectors integrated with nonmembrane indirect desalination technologies offer a promising solution for sustainable energy generation and clean water production. PVT collectors combine the benefits of PV panels and solar thermal collectors (Anand et al., 2021). They capture solar energy to generate electricity while utilizing the excess heat produced for various applications, including desalination. This integrated approach maximizes the efficiency of solar energy utilization. Nonmembrane indirect desalination technologies such as MED and multistage flash (MSF) distillation are commonly used. These technologies utilize the waste heat generated by the PVT collectors to vaporize seawater and condense it to produce freshwater. Combining PVT collectors with nonmembrane indirect desalination technologies makes the process more energy-efficient and cost-effective (Sharon & Reddy, 2015).

The PVT collectors absorb solar radiation converting it into electricity that can be used to power the desalination plant's electrical components and pumps. The excess heat generated by the PVT collectors, which would otherwise be wasted, is utilized in the distillation process. This integration reduces the overall energy consumption of the desalination plant and improves its sustainability (Zewdie et al., 2021).

Furthermore, using nonmembrane indirect desalination technologies eliminates the need for expensive membrane materials and energy-intensive processes. MED and MSF distillation rely on heat transfer and evaporation principles, making them suitable for integration with PVT collectors. The combined PVT and nonmembrane indirect desalination system offers several advantages. Firstly, it utilizes renewable energy sources, reducing dependence on fossil fuels and minimizing greenhouse gas emissions.

Secondly, the integration of PVT collectors with desalination technologies improves energy efficiency and reduces operational costs (Zheng, 2017). Additionally, the system provides a reliable source of clean water, addressing water scarcity issues in coastal areas. However, it is essential to note that implementing such systems requires careful planning and optimization. Factors such as solar radiation levels, climatic conditions, and water demand must be considered to ensure the efficiency and feasibility of the system in a specific location.

In conclusion, integrating PVT collectors with nonmembrane indirect desalination technologies offers a sustainable solution for simultaneous electricity generation and water desalination. This integrated approach maximizes solar energy utilization, reduces energy consumption, and provides a cost-effective means of producing clean water (Rabiee et al., 2019).

3.3.1   Multieffect distillation powered by photovoltaic thermal energy

Photovoltaic thermal-driven multieffect distillation (PVT-MED) is an innovative technology that combines PV solar cells and thermal energy to power the multieffect distillation process. Distillation is widely used to separate and purify liquids by exploiting their different boiling points. Multieffect distillation is a variant that uses multiple stages or "effects" to increase overall efficiency and reduce energy consumption. In traditional multieffect distillation systems, heat is typically supplied by fossil fuels or electricity (Zhang et al., 2019).

Photovoltaic thermal-driven multieffect distillation takes a sustainable and efficient approach to powering multieffect distillation by utilizing PV and thermal energy sources. Photovoltaic cells convert sunlight directly into electricity, while the thermal energy component captures and utilizes the excess heat generated by the PV panels. The PVT-MED system consists of three main components: the PV panels, a thermal collector, and the multieffect distillation unit. The PV panels generate electricity to power the distillation process, and any excess energy is used to heat the thermal collector. The thermal collector, typically a heat exchanger or a solar thermal collector, efficiently captures and transfers the excess heat to the distillation unit. The multieffect distillation unit utilizes the combined electrical and thermal energy to heat the feed water and create the necessary temperature gradient for the distillation process. The feed water is evaporated in several stages, each utilizing the vapor generated by the previous stage as a heat source. This cascading effect significantly improves the overall energy efficiency of the distillation process (Chen et al., 2021; Prajapati et al., 2022).

The advantages of PVT-MED are numerous. Firstly, it reduces the dependency on fossil fuels, making it a more sustainable and environmentally friendly option. Secondly, it maximizes solar energy utilization by combining both PV and thermal approaches. The excess thermal energy generated by the PV panels is harnessed, eliminating waste and increasing overall system efficiency. Additionally, PVT-MED systems can be designed to be modular and scalable, allowing for easy installation and expansion as per the required capacity. This makes it suitable for various applications such as water desalination, wastewater treatment, and purification of industrial or agricultural streams (Ong et al., 2012).

In conclusion, PVT-MED is an innovative technology that combines PV and thermal energy sources to power the distillation process. It provides a sustainable and efficient approach to water purification and other liquid separation applications reducing dependence on fossil fuels and maximizing solar energy utilization.

3.3.2 Process for multistage flash desalination using thermal photovoltaic energy

Photovoltaic thermal-driven multistage flash (PT-MSF) desalination is an advanced technology that combines solar and thermal energy to operate a multistage flash (MSF) desalination process. This process converts sunlight into electricity through PV panels. The generated electricity is then used to power the desalination system and drive the components. At the heart of the PT-MSF desalination system is the multistage flash process. It is a widely used method for desalination where seawater is heated and converted into steam in multiple stages at successively lower pressures. The steam is then condensed to produce distilled water (Khoshrou et al., 2017).

The PV panels provide the required electricity to power the process and generate excess thermal energy. This excess thermal energy is used to heat the seawater before entering the multistage flash unit. By preheating the seawater, the system's overall energy efficiency is improved, reducing the amount of electricity required from the PV panels. The multistage flash desalination process typically consists of several stages, each operating at progressively lower pressures. As the seawater enters each stage, it is heated by flashing steam from the previous stage. The flashing steam condenses, releasing heat energy transferred to the incoming seawater. This process is repeated in subsequent stages, further increasing the temperature and maximizing the efficiency of the thermal energy utilization (Sharaf et al., 2013).

By integrating PV and thermal technologies, PT-MSF desalination systems can achieve high energy efficiency and reduce operating costs. Renewable solar energy

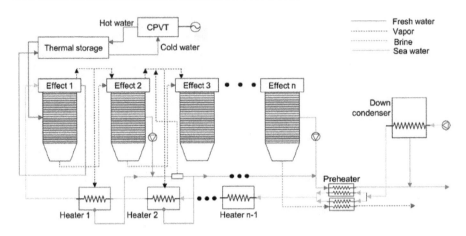

Figure 3.10 Desalination process using CPVT coupled with MED diagram.

also makes them more environmentally friendly than conventional desalination processes that rely on fossil fuels. PT-MSF desalination systems have the potential to provide clean and sustainable freshwater for regions facing water scarcity. They can be used in coastal areas with abundant sunlight and access to seawater. The technology is scalable and can be deployed for small-scale applications such as individual households and large-scale desalination plants. Overall, the PT-MSF desalination system offers a promising solution for addressing water scarcity challenges while minimizing the environmental impact of traditional desalination processes (Shaobo et al., 2008).

The economic viability of large-scale CPVT-driven MED desalination with thermal energy storage was examined by Mittelman et al. (2009) (Fig. 3.10). Electrical energy was delivered to the grid. In contrast, the thermal energy produced by the CPVT collector was used in MED desalination. Using thermal storage technology, the desalination plant was sized and constructed to run continuously for 24 hours daily throughout the summer. When the storage tank was empty during the winter, the desalination machine was intended to be turned off and restarted when sunshine was available. Additionally, the traditional RO, PV-RO, and solar thermal-driven MED plants were evaluated and compared to the CPVT-MED desalination plant under a range of installation costs (1−6 US\$/W) and energy prices. When the energy price was high, and the cost of the installed solar collector was low, it was claimed that the CPVT-MED plant was competitive with other alternative technologies.

3.4 Methods for membrane desalination powered by photovoltaic thermal energy

Photovoltaic thermal-powered membrane desalination is a technology that utilizes solar energy, specifically PV and thermal energy, to power the desalination process. This technology combines two different systems − PV panels and a thermal desalination unit − to achieve both electricity generation and freshwater production.

Photovoltaic panels are used to convert sunlight into electricity through the PV effect. These panels consist of multiple solar cells made of semiconductor materials such as silicon that generate DC when exposed to sunlight. The generated electricity can power various components of the desalination system (Anand et al., 2021).

In addition to electricity generation, the PV system also produces excess thermal energy. This excess thermal energy can be integrated with desalination to enhance efficiency. One way to utilize this thermal energy is through a thermal desalination unit such as a MED or an adsorption desalination (AD) system. A multieffect distillation system utilizes multiple evaporation and condensation stages to extract freshwater from seawater. In this system, the excess thermal energy generated by the PV system is used to heat the seawater and initiate the evaporation process. As the vapor rises, it passes through a series of condensation stages where it gradually cools and condenses to form freshwater. The condensed freshwater is collected and separated from the remaining concentrated brine. On the other hand, adsorption desalination systems utilize excess thermal energy to drive adsorption and desorption processes. These systems typically consist of adsorbent beds that absorb water vapor from the air or a saline solution during adsorption. The adsorbed water vapor is then desorbed by applying heat provided by the thermal energy produced by the PV system. The desorbed water vapor is condensed to produce freshwater (He et al., 2023; Lotfy et al., 2022).

This integration of PV panels and thermal desalination units allows for the simultaneous utilization of solar energy for electricity generation and thermal energy for desalination. This technology offers several advantages, including reduced reliance on grid electricity, decreased operational costs, and lower carbon emissions than conventional desalination methods. However, it is essential to note that PVT-powered membrane desalination technologies are still developing and under research. Further advancements and optimizations are required to improve efficiency, cost-effectiveness, and scalability. Nonetheless, these technologies hold great promise in addressing the increasing global demand for freshwater in a sustainable and environmentally friendly manner (Giwa et al., 2020; Ong et al., 2012).

3.4.1 Reverse osmosis desalination powered by photovoltaic and thermal energy

Photovoltaic/thermal-powered reverse osmosis (PV/T-RO) desalination is an innovative approach to address the challenges of water scarcity and energy sustainability. It combines PV panels and thermal energy generated from the sun to power the RO desalination process. PV/T-RO systems utilize solar energy in two ways. Firstly, solar panels capture sunlight and convert it into electricity through the PV effect. This electricity powers RO, removing salt and other impurities from seawater or brackish water. It suits various applications, including drinking water irrigation and industrial processes. Secondly, the excess heat generated by the PV panels is harnessed through a thermal energy collection system. This heat energy can be used in the desalination process specifically for heating the feed water before it enters the RO unit. Solar thermal energy PV/T-RO systems improve the overall energy efficiency of the desalination process.

The integration of PV and thermal technologies offers several advantages. First and foremost, it eliminates the reliance on grid electricity or fossil fuels, making the desalination process more environmentally friendly and reducing operational costs in the long run. Additionally, it provides a sustainable solution in areas with limited access to electricity or where power outages are expected. Moreover, PV/T-RO systems can be designed with modular configurations allowing for scalability based on the water demand. This flexibility makes them suitable for both small-scale and large-scale applications. Additionally, the systems can be easily integrated into existing desalination plants or implemented as stand-alone units.

Despite their numerous benefits, PV/T-RO systems face some challenges. One significant challenge is the intermittency of solar energy. The availability of sunlight varies throughout the day and across different seasons, affecting the system's performance. However, with advancements in solar energy storage technologies such as batteries, the energy generated during peak sunlight hours can be stored and used during low sunlight, ensuring a continuous water supply.

Therefore, integrating PV thermal panels with a RO desalination system is preferable to integrating RO with a separate PV panel and solar thermal collector. The conceptual diagram of a typical PVT-driven RO desalination system is shown in Fig. 3.11. The autonomous PVT-RO desalination plant should be appropriately designed to deliver freshwater on cloudy days and meet daily water needs.

In conclusion, PV/T-RO systems offer a sustainable and efficient solution for desalinating water using solar energy. By harnessing electricity and thermal energy from the sun, these systems reduce reliance on traditional energy sources, decrease operating costs, and contribute to environmental sustainability. Continued research and development in this field will further improve the efficiency and affordability of PV/T-RO systems, making them more accessible for water-scarce regions worldwide.

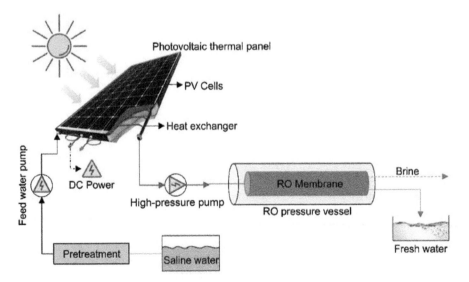

Figure 3.11 Process flow diagram of the integration of RO desalination and PVT.

3.4.2   Photovoltaic/thermal-powered membrane distillation

Photovoltaic/thermal-powered membrane distillation is an innovative technology that combines two renewable energy sources, PV and thermal energy, to drive the membrane distillation process. This technology offers a sustainable and efficient solution for water treatment and purification. Membrane distillation is a process that uses a hydrophobic membrane to separate hot saline water from a colder purified water stream. By maintaining a temperature gradient across the membrane, water vapor selectively passes through the membrane leaving behind impurities and contaminants. The purified water vapor condenses on the membrane's cold side, resulting in clean water. In traditional membrane distillation, the process is typically powered by conventional energy sources such as electricity or natural gas. However, by integrating PV and thermal technologies, the dependence on fossil fuels can be reduced, making the process more environmentally friendly (Hughes et al., 2014).

The PV component uses solar panels to convert sunlight directly into electricity. These solar panels generate the necessary electrical power to operate the pump valves and control systems involved in the membrane distillation process. The excess electricity generated can also be used to power other equipment or stored for later use. The thermal component of the system utilizes solar thermal collectors or concentrated solar power (CSP) systems to capture and convert sunlight into heat energy. This thermal energy is utilized to heat the saline water on the membrane's hot side, creating the necessary temperature gradient required for membrane distillation (Chen et al., 2021; Santana et al., 2023).

Combining PV and thermal energy ensures a continuous and sustainable power supply for the membrane distillation process. Depending on the available solar resource, the system can operate independently or in hybrid mode, utilizing solar electricity and thermal energy to meet the demand (Han et al., 2023). The advantages of PV/T-powered membrane distillation include the following:

1. Renewable energy: The system significantly reduces reliance on conventional energy sources by harnessing solar energy, contributing to a more sustainable and environmentally friendly water treatment process.
2. Energy efficiency: Integrating solar PV and thermal energy allows for efficient utilization and conversion of solar radiation into electricity, maximizing overall system efficiency.
3. Cost-effectiveness: Photovoltaic/thermal-powered membrane distillation can be economically viable over the long term due to reduced operating costs and lower dependence on external energy sources.
4. Modular and scalable: The system can be designed and implemented in various capacities allowing for scalability and adaptation to different water treatment needs.
5. Water availability: Membrane distillation can treat various water sources, including brackish water, seawater, and wastewater, providing a reliable source of clean water.

This results in low-cost, effective, stand-alone MD desalination, considerably lowering the initial investment cost, area needs, and grid dependency. The conceptual diagram of a typical PVT-integrated MD desalination is shown in Fig. 3.12. However, because sun radiation varies during the day, it also impacts the quantity, quality, and temperature of freshwater produced through MD desalination.

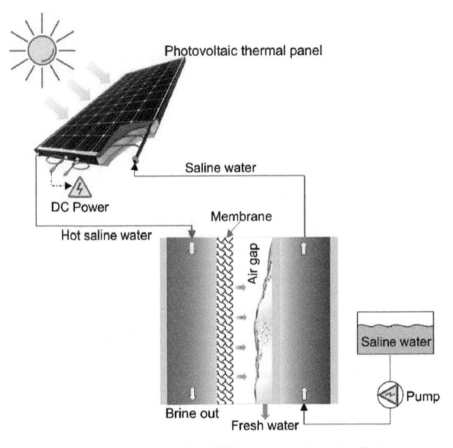

Figure 3.12 Process flow schematic of the PVT-integrated membrane desalination system.

Overall PV/T-powered membrane distillation is an innovative, sustainable, and efficient solution that addresses water scarcity and purification needs while reducing the reliance on fossil fuels. With further research and development, this technology has the potential to be widely deployed and make a significant impact on global water sustainability.

3.5 The photovoltaic thermal system's suitability and prospective chances for use with other desalination techniques

Photovoltaic thermal systems have the potential to be highly suitable for various desalination technologies. PVT systems combine the advantages of PV and solar thermal collectors, allowing the simultaneous generation of electricity and thermal energy from solar radiation. One desalination technology that can be coupled with PVT systems is solar-driven MED. MED is a widely used desalination method that utilizes the evaporation and condensation of seawater to produce freshwater. PVT systems can provide the necessary heat for the MED process, improving the overall system efficiency. With simultaneous electricity production, PVT systems can also power the auxiliary systems reducing external energy requirements (He et al., 2023).

Another desalination technology that can benefit from PVT systems is solar stills. Solar stills are simple distillation devices that use solar heat to evaporate water and condense it into freshwater. Solar stills can benefit from the generated electricity to power pumps and other components by integrating PVT systems, enhancing their efficiency and reliability (Joshi & Dhoble, 2018b).

Reverse osmosis is the most common desalination technology that can also be coupled with PVT systems. RO systems use high-pressure pumps to separate salt from water through a semipermeable membrane. The electricity produced by the PVT systems can power the pumps, reducing the electricity demand from the grid or other sources. The suitability and potential opportunities of PVT systems for other desalination technologies depend on various factors, such as the location of solar radiation water demand and system design. Determining the optimal configuration and sizing of the PVT systems requires careful evaluation of these factors. Additionally, integrating PVT systems with desalination technologies can offer several advantages, including reduced operational costs, decreased reliance on fossil fuels, and increased sustainability. Utilizing solar energy, PVT systems can mitigate greenhouse gas emissions and address the growing global water scarcity challenge (Huang et al., 2020).

In conclusion, PVT systems hold great potential for various desalination technologies, including MED solar stills and RO. Their ability to generate electricity and thermal energy makes them an attractive option for improving the efficiency and sustainability of desalination processes. Proper design and integration can lead to cost-effective and environmentally friendly solutions for addressing water scarcity issues (Sohani et al., 2023).

3.5.1 Electrodialysis thermal photovoltaic collector

Photovoltaic thermal collectors are advanced solar energy systems that combine the functions of solar thermal collectors and PV panels. These innovative devices are designed to harness electricity and heat from the sun's radiation, making them highly efficient and versatile. On the other hand, an electrodialysis system is a technology used for desalination and purification processes. It utilizes an electrochemical process to remove salts and impurities from the water, making it suitable for various applications such as drinking water production, industrial processes, and agriculture (Anderson et al., 2022; Salari et al., 2022; Shen et al., 2023). Combining a PV thermal collector with an electrodialysis system can offer several benefits. Here's how it might work:

1. Energy generation: The PV component of the PVT collector converts sunlight into electricity. This electricity can power the electrodialysis system, reducing the dependence on external power sources.

2. Thermal energy collection: The thermal component of the PVT collector absorbs the sun's heat and transfers it to a heat exchange system. This collected thermal energy can be directly utilized in the electrodialysis system for heating purposes improving its energy efficiency.

3. Desalination and purification: The electrodialysis system receives water input, which may contain impurities and high salt content. Using the generated electricity and thermal energy from the PVT collector, the electrodialysis system can perform the desalination and purification process more effectively.

4. Energy optimization: The PVT collector can act as a heat sink for the electrodialysis system, preventing excessive temperature rise during operation. Additionally, surplus thermal energy can be stored or used for other purposes, increasing overall system efficiency.

By integrating PV thermal collectors with an electrodialysis system, the combined system can benefit from increased energy efficiency, reduced energy costs, and improved water purification capabilities. Moreover, this technology can find applications in remote areas with limited access to clean water and electricity, contributing to sustainable development and environmental conservation. However, it's important to note that such a system has technical and economic challenges. Integrating PVT collectors and electrodialysis systems requires careful design control and optimization to ensure compatibility and maximize energy utilization. Additionally, cost considerations and maintenance requirements should be considered for this technology's broader adoption and scalability (Chandrasekar et al., 2022; Hu et al., 2023; Zheng et al., 2023).

Overall, combining PVT collectors and electrodialysis systems holds excellent potential for efficient and sustainable water purification in various settings, bringing together the benefits of solar energy and electrochemical processes.

3.5.2   Photovoltaic thermal collector for forward osmosis

A PVT collector for forward osmosis (FO) is a hybrid device that combines the principles of PV technology and thermal energy collection with the FO process. Forward osmosis is a membrane-based separation process that utilizes the natural osmotic pressure difference between two solutions to draw water through a semi-permeable membrane. This process favors reverse osmosis in specific applications due to its lower energy requirements and ability to handle feed solutions with high fouling potential or low salinity.

In a PVT collector for FO, PV cells are integrated directly into the membrane module. These cells convert sunlight into electricity that can be used to power the FO process and other components such as pumps or monitoring equipment. The thermal aspect of the collector is achieved by incorporating a heat exchanger within the collector design. This heat exchanger allows heat transfer from the PV cells to a working fluid such as water or a heat transfer fluid. The collected thermal energy can be used for various purposes, such as preheating the feed solution, maintaining desirable temperatures within the FO module or generating additional electricity through a thermal-electric conversion process.

Combining PV and thermal energy collection in a FO system offers several advantages. Firstly, it improves the overall energy efficiency of the process by utilizing both the electrical and thermal energy generated by the PV cells. Secondly, it reduces the reliance on external energy sources, making the system more self-sustainable and economically viable. Furthermore, the integration of PV technology allows for the direct use of sunlight as a power source, eliminating the need for grid electricity or other traditional energy sources. This makes the system particularly suitable for remote or off-grid applications such as desalination in rural areas or water treatment in disaster-stricken regions. The PVT collector for FO holds great potential for sustainable water treatment and desalination applications. Its ability to harness both electrical and thermal energy makes it an efficient and environmentally friendly option for simultaneously addressing water scarcity and energy requirements.

3.5.3 Photovoltaic thermal collector for vapor compression desalination

Photovoltaic thermal collectors are hybrid solar panels combining PV and thermal energy collection functionalities. These collectors are designed to convert solar radiation into electricity while harnessing the waste heat generated during the PV process. This waste heat can be utilized for various applications, including vapor compression desalination. Vapor compression desalination is a process that uses heat to produce freshwater from saltwater or brackish water sources. It employs the principle of evaporation and condensation to separate water vapor from saline water leaving behind salt or other impurities. Various sources can supply the heat required for this process, and one efficient option is to utilize the waste heat from PVT collectors (Gado et al., 2022).

When the PVT collector operates to generate electricity, it produces heat as a by-product. This excess heat can be redirected to a vapor compression desalination system helping meet the energy requirements of the desalination process. By integrating PVT collectors and desalination units, this technology enables the combined production of electricity and freshwater more efficiently and sustainably. The PVT collector incorporates PV cells on the front surface to convert solar radiation into electricity while heat-absorbing material and fluid channels collect the waste heat. The fluid channels allow a HTF to circulate, absorbing the heat and transferring it to a thermal energy storage system or the desalination unit. The heat from the PVT collector can be used in various stages of the vapor compression desalination process. In the evaporation stage, heat is applied to the saline water to convert it into water vapor, separating it from the salt and impurities. This vapor is then condensed back into freshwater, and the remaining heated HTF can be continuously recirculated to provide the required energy (Eisavi et al., 2021).

The integration of PVT collectors with vapor compression desalination offers several advantages. Firstly, it maximizes solar energy utilization by simultaneously generating electricity and producing freshwater. This synergy ensures more efficient use of available solar resources. Secondly, the waste heat from the PVT collectors contributes to the desalination process's thermal energy requirements, reducing the need for additional energy sources. This not only enhances the overall energy efficiency but also lowers the carbon footprint of the desalination system. Furthermore, PVT collectors have the potential for a compact design, making them suitable for decentralized desalination applications. This can provide access to clean water in remote areas where electricity and freshwater resources are limited (Chen et al., 2020; George et al., 2019).

Photovoltaic thermal collectors are thus perfectly suitable to integrate with FO desalination. Thermal energy from the collectors can be employed to draw solution recovery or used to warm the FO desalination solutions appropriately. Additionally, electrical energy from the collectors can be drawn for the solution recovery process or supplied to auxiliary equipment to create a freestanding, sustainable system. A conceptual or potential process flow diagram for PVT-integrated FO-MD hybrid desalination is shown in Fig. 3.13. Saline water is heated in a PVT panel in this process before going into the FO desalination, and the draw solution is heated in the MD desalination first before going into the FO unit. Saline water escaping the PVT panel further warms the diluted draw solution leaving the FO unit. The heat exchanger's dilute hot draw solution is transferred to the air gap membrane desalination unit's evaporator side to produce freshwater.

The solar cell temperature generally limits the coolant outlet temperature and thermal energy output from PVT/CPVT collectors (Sharaf & Orhan, 2015). Depending on the PV panel type, position, cooling method, configurations, etc., the electrical energy production from PVT/CPVT collectors varies. However, when it is correctly sized, the electrical energy generated is sufficient to meet the power needs of vapor compression desalination. To meet the energy requirements of the vapor compression desalination unit, the heat and thermal energy generated from PVT/CPVT collectors is thus adequate. A potential conceptual flow diagram of PVT-coupled mechanical vapor compression desalination can be seen in Fig. 3.14.

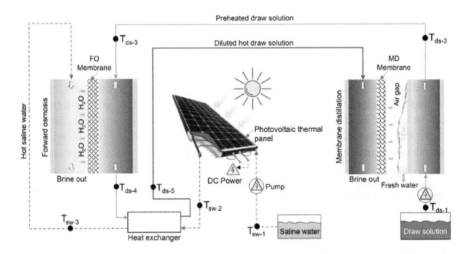

Figure 3.13 A conceptual flow chart of PVT-integrated FO-MD hybrid desalination.

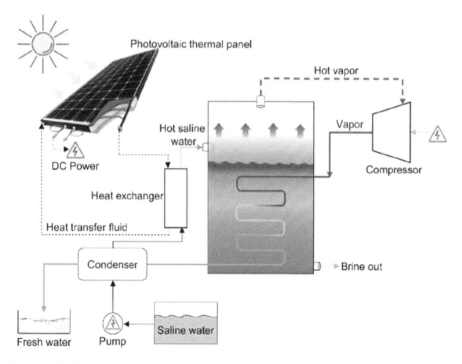

Figure 3.14 Potential conceptual representation of mechanical vapor compression desalination using PVT.

In summary, integrating PVT collectors and vapor compression desalination offers a promising solution for sustainable water production. By combining electricity generation with waste heat utilization, this technology efficiently harnesses solar energy while simultaneously meeting the freshwater demands of water-stressed regions.

3.5.4 Photovoltaic/thermal collector for adsorption desalination

Photovoltaic/thermal collectors are innovative devices that combine the benefits of PV solar cells and thermal collectors. These collectors can be effectively utilized in adsorption desalination systems to enhance their performance and energy efficiency. Adsorption desalination is a water desalination process that utilizes heat energy to drive the desalination process. It typically employs an adsorbent material such as silica gel to adsorb water vapor from the air, and then the adsorbed water is desorbed using heat. The desorbed water vapor is condensed to produce freshwater. Integrating PV/T collectors into adsorption desalination systems offers several advantages. Firstly, the PV cells capture solar radiation and convert it into electricity that can be used to power the system's pumps, fans, and control systems. This reduces the reliance on external electricity sources, making the system more autonomous and sustainable (Ghazy et al., 2022).

Secondly, the thermal component of the PV/T collectors can supply the required heat energy for the desorption process of the adsorption desalination system. The heat generated by the PV cells can be collected and used directly or stored in a thermal storage medium such as water or phase-change materials. This stored heat can be released on demand to drive the desalination process even during periods of low solar irradiance (Gado et al., 2022; Ghazy et al., 2022). The combination of PV and thermal technologies in a single collector reduces the system's overall footprint, making it more space efficient than separate PV and thermal collectors. This is particularly advantageous in desalination applications with limited space, such as remote or off-grid areas. Moreover, using PV/T collectors increases the overall system efficiency by utilizing both the electrical and thermal energy from solar radiation, which would otherwise be wasted in conventional PV or thermal systems (Buonomano et al., 2018).

In conclusion, integrating PV/T collectors in adsorption desalination systems provides an efficient and sustainable solution for water desalination. By harnessing solar energy in the form of electricity and heat, these collectors optimize the system's energy utilization, reduce its environmental impact, and contribute to the overall reliability and self-sufficiency of the desalination process.

3.6 Economics and other aspects of PVT desalination system

Private desalination systems, also known as PVT desalination systems, refer to the operation of desalination plants by private entities rather than government institutions. This approach has gained significant attention recently due to its potential to improve efficiency, promote innovation, and attract private investment in the desalination sector. One of the primary advantages of PVT desalination systems is the potential for increased efficiency. Private entities are often driven by profit and competition, which can lead to adopting innovative technologies and operational practices that enhance efficiency and reduce costs. This can result in higher production rates, lower energy consumption, and improved overall performance of the desalination plant. Additionally, private ownership of desalination plants can foster competition within the market, leading to further advancements in technology and operational practices. The competition encourages companies to find ways to differentiate themselves from their competitors, leading to a continuous push for innovation. This can result in more efficient desalination processes, lower costs, and improved environmental sustainability (Calise et al., 2014).

Moreover, private ownership can attract significant investment capital from both domestic and foreign sources. The desalination industry requires substantial financial resources for infrastructure development, technology upgrades, and operational costs. Private companies have the potential to tap into these resources, leading to increased investment in the sector and expanding access to clean water for various communities. Despite these potential benefits, PVT desalination systems are also associated with challenges. One concern is the risk of monopolistic practices if a single private company dominates the market. This may lead to higher consumer prices and reduced competition, potentially stalling innovation and limiting consumer choice. Proper regulatory frameworks and oversight are crucial in ensuring fair competition and preventing potential abuses of market power (Al Jabri et al., 2019).

Another challenge is that private companies may prioritize profitability over social welfare or environmental concerns. Public institutions often have the mandate to prioritize the community's needs and may consider factors such as affordability, equity, and sustainability. Therefore, careful monitoring and regulation are necessary to ensure that private desalination companies operate socially and environmentally sustainable (Kettani & Bandelier, 2020; Madani, 1990).

In conclusion, PVT desalination systems have the potential to improve efficiency, encourage innovation, and attract private investment in the desalination sector. However, careful oversight and regulation are necessary to address potential challenges such as monopolistic practices and ensure social and environmental considerations are considered. By striking a balance between private sector efficiency and public interest, PVT desalination systems can contribute to addressing global water scarcity challenges.

3.7 Conclusion

In conclusion, integrating solar thermal and PV technologies in desalination processes shows excellent potential for addressing global water scarcity. Combining solar thermal energy for seawater distillation and PV energy for powering desalination plants offers several advantages.

Firstly, using solar energy reduces dependence on fossil fuels, making the desalination process more sustainable and environmentally friendly. Solar energy is a renewable resource, and by harnessing its power, we can significantly decrease greenhouse gas emissions and mitigate the negative impacts of traditional desalination methods.

Secondly, integrating solar thermal and PV technologies improves overall energy efficiency. Solar thermal technology utilizes the heat from the sun to enhance the distillation process, reducing the amount of energy needed for evaporation. Meanwhile, PV systems convert sunlight directly into electricity, providing a clean and efficient power source for desalination.

Furthermore, this integrated approach can capitalize on the synergies between solar thermal and PV systems. For example, excess electricity generated by the PV panels during peak sunlight hours can heat the distillation process through solar thermal collectors, maximizing overall system efficiency.

Although solar thermal integrated PV desalination is a promising solution, it is essential to consider a few challenges. The cost of implementing these integrated systems can be higher initially, requiring significant investment in infrastructure and technology. However, as solar technologies continue to advance and become more affordable, these costs are expected to decrease over time.

Additionally, desalination plants using solar thermal integrated PV systems may face limitations due to weather conditions, such as lack of sunlight on cloudy days. To overcome this challenge, energy storage systems can be incorporated to store excess solar energy during sunny periods and use it during low-light conditions.

In summary, integrating solar thermal and PV technologies in desalination processes offers a sustainable and efficient solution to water scarcity. While there are initial challenges, the long-term benefits of reducing dependence on fossil fuels improving energy efficiency and utilizing renewable resources make this technology a promising option. Continued research and development in this field will further enhance the viability and effectiveness of solar thermal integrated PV desalination systems.

References

Al Jabri, S. A., Zekri, S., Zarzo, D., & Ahmed, M. (2019). Comparative analysis of economic and institutional aspects of desalination for agriculture in the Sultanate of Oman and Spain. *Desalination and Water Treatment, 156*, 1—6.

Al-Karaghouli, A., & Kazmerski, L. L. (2013). Energy consumption and water production cost of conventional and renewable-energy-powered desalination processes. *Renewable and Sustainable Energy Reviews, 24*, 343—356. Available from https://doi.org/10.1016/j.rser.2012.12.064.

Alnaimat, F., & Klausner, J. F. (2012). Solar diffusion driven desalination for decentralized water production. *Desalination, 289*, 35—44. Available from https://doi.org/10.1016/j.desal.2011.12.028.

Anand, B., & Murugavelh, S. (2019). Techno-economic analysis of solar trigeneration system. *IOP Conference Series: Earth and Environmental Science, 312*(1), 012030, IOP Publishing.

Anand, B., & Murugavelh, S. (2022). A hybrid system for power, desalination, and cooling using concentrated photovoltaic/thermal collector. *Energy Sources, Part A: Recovery, Utilization, and Environmental Effects, 44*(1), 1416—1436.

Anand, B., Shankar, R., Murugavelh, S., Rivera, W., Prasad, K. M., & Nagarajan, R. (2021). A review on solar photovoltaic thermal integrated desalination technologies. *Renewable and Sustainable Energy Reviews, 141*, 110787. Available from https://doi.org/10.1016/j.rser.2021.110787.

Ananda, B., Shankara, R., Srinivasb, T., & Murugavelha, S. (2019). Performance analysis of combined two stage desalination and cooling plant with different solar collectors. *Desalination and Water Treatment, 156*, 136—147.

Anderson, A., et al. (2022). Effects of nanofluids on the photovoltaic thermal system for hydrogen production via electrolysis process. *International Journal of Hydrogen Energy, 47*(88), 37183—37191.

Buonomano, A., Calise, F., & Palombo, A. (2018). Solar heating and cooling systems by absorption and adsorption chillers driven by stationary and concentrating photovoltaic/thermal solar collectors: Modelling and simulation. *Renewable and Sustainable Energy Reviews, 82*, 1874—1908.

Calise, F., d'Accadia, M. D., & Piacentino, A. (2014). A novel solar trigeneration system integrating PVT (photovoltaic/thermal collectors) and SW (seawater) desalination: Dynamic simulation and economic assessment. *Energy, 67*, 129—148.

Chandrasekar, M., Gopal, P., Kumar, C. R., & Geo, V. E. (2022). Effect of solar photovoltaic and various photovoltaic air thermal systems on hydrogen generation by water electrolysis. *International Journal of Hydrogen Energy, 47*(5), 3211−3223.

Chen, H., Li, Z., & Xu, Y. (2020). Assessment and parametric analysis of solar trigeneration system integrating photovoltaic thermal collectors with thermal energy storage under time-of-use electricity pricing. *Solar Energy, 206*, 875−899.

Chen, Q., et al. (2021). A decentralized water/electricity cogeneration system integrating concentrated photovoltaic/thermal collectors and vacuum multi-effect membrane distillation. *Energy, 230*, 120852.

Chow, T. T. (2010). A review on photovoltaic/thermal hybrid solar technology. *Applied Energy, 87*(2), 365−379. Available from https://doi.org/10.1016/j.apenergy.2009.06.037.

Eisavi, B., Nami, H., Yari, M., & Ranjbar, F. (2021). Solar-driven mechanical vapor compression desalination equipped with organic Rankine cycle to supply domestic distilled water and power−Thermodynamic and exergoeconomic implications. *Applied Thermal Engineering, 193*, 116997.

Elsafi, A. M. (2017). Integration of humidification-dehumidification desalination and concentrated photovoltaic-thermal collectors: Energy and exergy-costing analysis. *Desalination, 424*, 17−26.

Esmaeilion, F. (2020). Hybrid renewable energy systems for desalination. *Applied Water Science, 10*(3), 84. Available from https://doi.org/10.1007/s13201-020-1168-5.

Gabrielli, P., et al. (2019). Combined water desalination and electricity generation through a humidification-dehumidification process integrated with photovoltaic-thermal modules: Design, performance analysis and techno-economic assessment. *Energy Conversion and Management: X, 1*, 100004.

Gado, M. G., Megahed, T. F., Ookawara, S., Nada, S., & El-Sharkawy, I. I. (2022). Potential application of cascade adsorption-vapor compression refrigeration system powered by photovoltaic/thermal collectors. *Applied Thermal Engineering, 207*, 118075.

George, M., Pandey, A., Abd Rahim, N., Tyagi, V., Shahabuddin, S., & Saidur, R. (2019). Concentrated photovoltaic thermal systems: A component-by-component view on the developments in the design, heat transfer medium and applications. *Energy Conversion and Management, 186*, 15−41.

Ghazy, M., Ibrahim, E., Mohamed, A., & Askalany, A. A. (2022). Experimental investigation of hybrid photovoltaic solar thermal collector (PV/T)-adsorption desalination system in hot weather conditions. *Energy, 254*, 124370.

Giwa, A., Yusuf, A., Dindi, A., & Balogun, H. A. (2020). Polygeneration in desalination by photovoltaic thermal systems: A comprehensive review. *Renewable and Sustainable Energy Reviews, 130*, 109946.

Han, X., Ding, F., Huang, J., & Zhao, X. (2023). Hybrid nanofluid filtered concentrating photovoltaic/thermal-direct contact membrane distillation system for co-production of electricity and freshwater. *Energy, 263*, 125974.

He, W., Huang, G., & Markides, C. N. (2023). Synergies and potential of hybrid solar photovoltaic-thermal desalination technologies. *Desalination, 552*, 116424.

Hu, L., et al. (2023). Development and evaluation of an electro-Fenton-based integrated hydrogen production and wastewater treatment plant coupled with the solar and electrodialysis units. *Process Safety and Environmental Protection, 177*, 568−580.

Huang, G., Curt, S. R., Wang, K., & Markides, C. N. (2020). Challenges and opportunities for nanomaterials in spectral splitting for high-performance hybrid solar photovoltaic-thermal applications: A review. *Nano Materials Science, 2*(3), 183−203.

Hughes, A., O'Donovan, T., & Mallick, T. (2014). Experimental evaluation of a membrane distillation system for integration with concentrated photovoltaic/thermal (CPV/T) energy. *Energy Procedia, 54*, 725−733.

Jones, E., Qadir, M., van Vliet, M. T. H., Smakhtin, V., & Kang, S.-M. (2019). The state of desalination and brine production: A global outlook. *Science of the Total Environment, 657*, 1343−1356. Available from https://doi.org/10.1016/j.scitotenv.2018.12.076.

Joshi, S. S., & Dhoble, A. S. (2018a). Photovoltaic-thermal systems (PVT): Technology review and future trends. *Renewable and Sustainable Energy Reviews, 92*, 848−882. Available from https://doi.org/10.1016/j.rser.2018.04.067.

Joshi, S. S., & Dhoble, A. S. (2018b). Photovoltaic-thermal systems (PVT): Technology review and future trends. *Renewable and Sustainable Energy Reviews, 92*, 848−882.

Kettani, M., & Bandelier, P. (2020). Techno-economic assessment of solar energy coupling with large-scale desalination plant: The case of Morocco. *Desalination, 494*, 114627.

Khalil, A., Khaira, A. M., Abu-Shanab, R. H., & Abdelgaied, M. (2023). A comprehensive review of advanced hybrid technologies that improvement the performance of solar dryers: Photovoltaic/thermal panels, solar collectors, energy storage materials, biomass, and desalination units. *Solar Energy, 253*, 154−174.

Khoshrou, I., Nasr, M. J., & Bakhtari, K. (2017). New opportunities in mass and energy consumption of the multi-stage flash distillation type of brackish water desalination process. *Solar Energy, 153*, 115−125.

Kumar, S., & Tiwari, A. (2008). An experimental study of hybrid photovoltaic thermal (PV/T)-active solar still. *International Journal of Energy Research, 32*(9), 847−858. Available from https://doi.org/10.1002/er.1388.

Lawal, D. U., & Qasem, N. A. A. (2020). Humidification-dehumidification desalination systems driven by thermal-based renewable and low-grade energy sources: A critical review. *Renewable and Sustainable Energy Reviews, 125*, 109817. Available from https://doi.org/10.1016/j.rser.2020.109817.

Li, G.-P., & Zhang, L.-Z. (2016). Investigation of a solar energy driven and hollow fiber membrane-based humidification−dehumidification desalination system. *Applied Energy, 177*, 393−408. Available from https://doi.org/10.1016/j.apenergy.2016.05.113.

Lotfy, H. R., Staš, J., & Roubík, H. (2022). Renewable energy powered membrane desalination—Review of recent development. *Environmental Science and Pollution Research, 29*(31), 46552−46568.

Madani, A. (1990). Economics of desalination for three plant sizes. *Desalination, 78*(2), 187−200.

Mishra, K. N., Meraj, M., Tiwari, A. K., & Tiwari, G. (2019). Performance evaluation of PVT-CPC integrated solar still under natural circulation. *Desalination and Water Treatment, 156*, 117−125.

Mittelman, G., Kribus, A., Mouchtar, O., & Dayan, A. (2009). Water desalination with concentrating photovoltaic/thermal (CPVT) systems. *Solar Energy, 83*(8), 1322−1334. Available from https://doi.org/10.1016/j.solener.2009.04.003.

Ong, C. L., Escher, W., Paredes, S., Khalil, A., & Michel, B. (2012). A novel concept of energy reuse from high concentration photovoltaic thermal (HCPVT) system for desalination. *Desalination, 295*, 70−81.

Pourafshar, S. T., Jafarinaemi, K., & Mortezapour, H. (2020). Development of a photovoltaic-thermal solar humidifier for the humidification-dehumidification desalination system coupled with heat pump. *Solar Energy, 205*, 51−61.

Prajapati, M., Shah, M., & Soni, B. (2022). A comprehensive review of the geothermal integrated multi-effect distillation (MED) desalination and its advancements. *Groundwater for Sustainable Development*, 100808.

Rabiee, H., Khalilpour, K. R., Betts, J. M., & Tapper, N. (2019). *Energy-water nexus: Renewable-integrated hybridized desalination systems. Polygeneration with polystorage for chemical and energy hubs* (pp. 409−458). Elsevier.

Salari, A., Hakkaki-Fard, A., & Jalalidil, A. (2022). Hydrogen production performance of a photovoltaic thermal system coupled with a proton exchange membrane electrolysis cell. *International Journal of Hydrogen Energy, 47*(7), 4472−4488.

Santana, J. P., Rivera-Solorio, C. I., Chew, J. W., Tan, Y. Z., Gijón-Rivera, M., & Acosta-Pazmiño, I. (2023). Performance assessment of coupled concentrated photovoltaic-thermal and vacuum membrane distillation (CPVT-VMD) system for water desalination. *Energies, 16*(3), 1541.

Shaobo, H., Zhang, Z., Huang, Z., & Xie, A. (2008). Performance optimization of solar multistage flash desalination process using Pinch technology. *Desalination, 220*(1−3), 524−530.

Sharaf Eldean, M. A., & Fath, H. (2013). Exergy and thermo-economic analysis of solar thermal cycles powered multi-stage flash desalination process. *Desalination and Water Treatment, 51*(40−42), 7361−7378.

Sharaf, O. Z., & Orhan, M. F. (2015). Concentrated photovoltaic thermal (CPVT) solar collector systems: Part I − Fundamentals, design considerations and current technologies. *Renewable and Sustainable Energy Reviews, 50*, 1500−1565. Available from https://doi.org/10.1016/j.rser.2015.05.036.

Sharon, H., & Reddy, K. (2015). A review of solar energy driven desalination technologies. *Renewable and Sustainable Energy Reviews, 41*, 1080−1118.

Shen, T., et al. (2023). Experimental analysis of photovoltaic thermal system assisted with nanofluids for efficient electrical performance and hydrogen production through electrolysis. *International Journal of Hydrogen Energy, 48*(55), 21029−21037.

Singh, G., Kumar, S., & Tiwari, G. N. (2011). Design, fabrication and performance evaluation of a hybrid photovoltaic thermal (PVT) double slope active solar still. *Desalination, 277*(1), 399−406. Available from https://doi.org/10.1016/j.desal.2011.04.064.

Sohani, A., et al. (2023). Building integrated photovoltaic/thermal technologies in Middle Eastern and North African countries: Current trends and future perspectives. *Renewable and Sustainable Energy Reviews, 182*, 113370.

Tyagi, V., Kaushik, S., & Tyagi, S. (2012). Advancement in solar photovoltaic/thermal (PV/T) hybrid collector technology. *Renewable and Sustainable Energy Reviews, 16*(3), 1383−1398.

Xiao, L., Shi, R., Wu, S.-Y., & Chen, Z.-L. (2019). Performance study on a photovoltaic thermal (PV/T) stepped solar still with a bottom channel. *Desalination, 471*, 114129. Available from https://doi.org/10.1016/j.desal.2019.114129.

Xinxin, G., Heng, Z., Haiping, C., Kai, L., Jiguang, H., & Haowen, L. (2019). Experimental and theoretical investigation on a hybrid LCPV/T solar still system. *Desalination, 468*, 114063. Available from https://doi.org/10.1016/j.desal.2019.07.003.

Zenhäusern, D., Bamberger, E., Baggenstos, A., & Häberle, A. (2017). PVT Wrap-Up: Energy systems with photovoltaic thermal solar collectors. *Final Report, 31*.

Zewdie, T. M., Habtu, N. G., Dutta, A., & Van der Bruggen, B. (2021). Solar-assisted membrane technology for water purification: A review. *Water Reuse, 11*(1), 1−32.

Zhang, Z., et al. (2019). Theoretical analysis of a solar-powered multi-effect distillation integrated with concentrating photovoltaic/thermal system. *Desalination, 468*, 114074.

Zheng, H. (2017). *Solar energy desalination technology*. Elsevier.

Zheng, N., Zhang, H., Duan, L., Wang, Q., Bischi, A., & Desideri, U. (2023). Techno-economic analysis of a novel solar-driven PEMEC-SOFC-based multi-generation system coupled parabolic trough photovoltaic thermal collector and thermal energy storage. *Applied Energy, 331*, 120400.

Thermodynamic analysis of solar photovoltaic energy conversion systems

Laveet Kumar[1], Mohammad Waqas Chandio[1] and
Mamdouh El Haj Assad[2]
[1]Department of Mechanical Engineering, Mehran University of Engineering and
Technology, Jamshoro, Pakistan, [2]Department of Sustainable and Renewable Energy
Engineering, University of Sharjah, Sharjah, United Arab Emirates

4.1 Introduction

4.1.1 Solar photovoltaic energy conversion

Solar photovoltaic (PV) system converts electromagnetic radiations from the sun into electricity. The system consists of an array of solar cells. The commonly used semiconductor material in solar cells is silicon. The electric current can be obtained by separating the electrons and the holes by creating an electric field in a semiconductor, a p−n diode. Zone n and Zone p are rich in electrons and holes, respectively. The PV effect gives rise to an electric field separating the charges. The potential difference is established at the terminals of the cell as shown in Fig. 4.1. The standard PV module can be integrated with thermal unit for removal of heat generated. The removed heat can be utilized for heating applications. such modified PV module is termed as photovoltaic thermal (PVT) system.

The conversion efficiency of a solar cell, η_{cell} can be represented as follows Eq. 4.1 (Rawat et al., 2017):

$$\eta_{cell} = \frac{\dot{W}}{A \times G} \tag{4.1}$$

where $\dot{W}$ is the electrical power output, A (m^2) is the area of the solar cell, and G (W/m^2) is the solar irradiation incident to the solar cell.

The Carnot efficiency ($\eta_{cell,max}$) relation as given in Eq. 4.2 can be used to obtain an extreme limit for solar cell efficiency by using the effective surface temperature (T_H) of the sun (5780K) and an ambient temperature (T_L) of 298K (Mehmet Kanoglu & Cimbala, 2019).

$$\eta_{cell,max} = 1 - \frac{T_L}{T_H} = 1 - \frac{298}{5780} = 0.948 \tag{4.2}$$

Performance Enhancement and Control of Photovoltaic Systems. DOI: https://doi.org/10.1016/B978-0-443-13392-3.00004-9

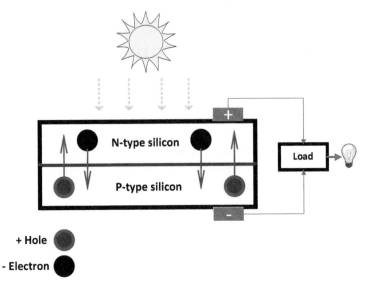

Figure 4.1 Schematic of a solar PV cell.

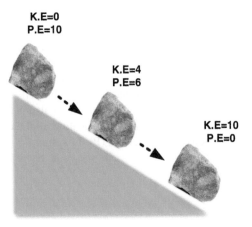

Figure 4.2 Illustration of the first law of thermodynamics.

4.1.2   Laws of thermodynamics

4.1.2.1   First law of thermodynamics (FLT)

This principle states that energy can neither be created nor destroyed, but it can change from one form to another. Fig. 4.2 shows an example that illustrates the conservation of energy of a stone rolling down the inclined surface where values are given for both kinetic energy (KE) as the function of velocity and potential energy (PE) as the function of height (elevation). The total energy remains constant at 10 units from the beginning to the end at every position. The FLT is a presentation of the conservation of energy principle. The performance of energy systems can be quantified using FLT through energy efficiency.

The conservation of energy principle is essentially derived from the FLT, which highlights the net change in the total energy of the system during a process is equal to the difference between incoming and outgoing total energies, which can be presented in terms of symbols as follows:

$$E_{in} - E_{out} = \Delta E_{system} \tag{4.3}$$

Energy can transport the closed system boundaries by heat and work, whereas it can transfer by heat, work, and mass flow in an open system.

The energy balance for a general steady flow system can be written as follows:

$$\dot{Q} - \dot{W} = \sum_{out} \dot{m} \times \theta - \sum_{in} \dot{m} \times \theta \tag{4.4}$$

where $\dot{Q}$ is heat transfer rate, $\dot{W}$ is power, $\dot{m}$ is mass flow rate, and θ is flow energy.

4.1.2.2 Second law of thermodynamics (SLT)

The key weakness of the FLT is that it does not account for losses. The SLT considers both quantity and quality of energy, and it hence states that actual processes occur in the direction of decreasing quality of energy. There are two key statements related to SLT. These are known as Kelvin–Plank statement and Clausius statement.

4.1.2.2.1 Kelvin–Plank statement
No system can produce a net amount of work while operating in a cycle and exchanging heat with a single thermal energy reservoir as shown in Fig. 4.3A.

4.1.2.2.2 Clausius statement
It is impossible to construct a refrigerator that transfers heat from a cold reservoir to a hot reservoir without the aid of work input or external agency, Moreover, a refrigeration system cannot operate without work input or external energy source to provide the desired cooling load as shown in Fig. 4.3A.

The SLT can be represented by exergy (Ex) balance as follows:

$$Ex_{in} - Ex_{out} - Ex_{dest} = \Delta Ex_{system} \tag{4.5}$$

Exergy can transport the closed system boundaries by heat and work, whereas it can transfer by heat, work, and mass flow in an open system.

The exergy rate balance for the general steady flow system can be written as follows:

$$\sum \left(1 - \frac{T_0}{T}\right) \times \dot{Q} - \dot{W} + \sum_{in} \dot{m} \times \psi - \sum_{out} \dot{m} \times \psi - \dot{X}_{destroyed} = 0 \tag{4.6}$$

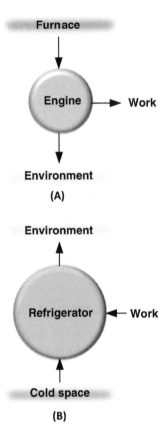

Figure 4.3 (A) Illustration of the second law of thermodynamics (Kelvin—Plank statement). (B) Illustration of the second law of thermodynamics (Clausius statement statement).

where $\dot{Q}$ is the heat transfer rate, $\dot{W}$ is the power, $\dot{m}$ is the mass flow rate, and ψ is the flow exergy.

4.2 Energy and exergy analysis of solar photovoltaic energy conversion systems

4.2.1 Energy analysis

This section presents energy analysis of solar photovoltaic systems. The conventional energy analysis consists of carrying out energy balances based on the first law of thermodynamics and determining energy efficiencies. The energy balance equation for all the systems under control volume in the equilibrium state is given as follows (Agrawal & Tiwari, 2010; Ma et al., 2020; Sardarabadi et al., 2017):

$$\sum \dot{E}_i = \sum \dot{E}_o + \sum \dot{E}_{loss} \qquad (4.7)$$

or

$$\dot{E}_s + \dot{E}_{m,i} = \dot{E}_{m,o} + \dot{E}_{el} + \dot{E}_{loss} \qquad (4.8)$$

where $\dot{E}_i$, $\dot{E}_o$, and $\dot{E}_{loss}$ are the energy input, output, and loss, respectively. $\dot{E}_{el}$ is the amount of energy transformed into electricity, and $\dot{E}_s$ is the amount of solar energy reaching the surface of collectors.

For a solar photovoltaic thermal (PVT) system, The total energy influx from the sun is the summation of the energy intercepted by the solar photovoltaic thermal (PVT) collector.

$$\dot{E}_{s(PVT)} = \tau \alpha A_{PVT} I \qquad (4.9)$$

where τ and α are the transmittance and absorptance coefficients, respectively.

Heat gain and thermal energy efficiency of PVT collector are calculated as follows:

(a) Heat gain

$$\dot{E}_{g(PVT)} = \dot{E}_{m,i(PVT)} = \dot{m} C_p (T_{o(PVT)} - T_{i(PVT)}) \qquad (4.10)$$

(b) Thermal efficiency

$$\eta_{Thermal(PVT)} = \frac{\dot{E}_g}{\dot{E}_{s(PVT)}} = \frac{\dot{m} C_p (T_o - T_i)}{\tau \alpha A_{PVT} I} \qquad (4.11)$$

where T_o and T_i are the inlet and outlet water temperature, respectively.

4.2.2 Exergy analysis

In this section, a general description of exergy analysis is presented, and then an exergy analysis of solar PV systems is given in detail taking into account the internal as well as the external exergy losses rate. Exergy is defined as the maximum available work that can be obtained from a process or system when the process or the system is brought reversibly into thermal, mechanical, and chemical equilibrium with its surroundings at temperature T_o and pressure p_o. The use of exergy analysis is very beneficial to locate and estimate the lost work of a process or system. An energy balance, as explained in the preceding section, does not focus on the degradation of energy during a process and provides no information on the quality of various energy types. Exergy is the amount of useful energy that is accessible to expense or convert into useful work. The basic theory of exergy analysis is based on the second law of thermodynamics, which states that conversion of thermal energy is not possible without the temperature difference, but electrical energy can be converted into other forms. The exergy balance equation for all the systems under control volume in the equilibrium state is given as follows (Fudholi et al., 2018; Hossain et al., 2019; Kumar et al., 2019, 2021; Park et al., 2014; Sardarabadi et al., 2017):

Exergy for open systems (control volume) is considered the summation of physical and chemical exergy.

The specific exergy of a system consisting of many components is expressed as follows:

$$e_i = (h_i - h_o) - T_o(s_i - s_o) + \sum_i x_i(\mu_i - \mu_{io}) \tag{4.12}$$

where h is the specific enthalpy, s is the specific entropy, T is the temperature, x is the mole fraction, and μ is the chemical potential. The subscript o refers to reference state at T_o and p_o.

For a steady-state open system, the useful work is

$$\dot{W}_{useful} = \sum_i \dot{m}_i e_i + \sum_j \left(1 - \frac{T_o}{T_j}\right)\dot{Q}_j - T_o\dot{S}_{gen} \tag{4.13}$$

where $\dot{m}_i$ is the mass flow rate, $\dot{Q}_i$ is the heat rate from or into the system, T_j is the temperature at which $\dot{Q}_i$ is transferred, and $\dot{S}_{gen}$ is the entropy generation rate.

$$\sum \dot{E}_{x_i} = \sum \dot{E}_{x_o} + \sum \dot{E}_{x_{dest}} \tag{4.14}$$

or

$$\dot{E}_{x_s} + \dot{E}_{x_{m,in}} = \dot{E}_{x_{m,o}} + \dot{E}_{x_{el}} + \dot{E}_{x_{dest}} \tag{4.15}$$

where $\dot{E}_{x_i}$, $\dot{E}_{x_o}$, and $\dot{E}_{x_{dest}}$ are the exergy input, output, and destruction, respectively. $\dot{E}_{x_{el}}$ is the electrical exergy that is the electrical generation of the PV panel. $\dot{E}_{x_s}$ is the exergy of solar radiations reaching the surface of PV. For PVT system, it can be written as follows:

$$\dot{E}_{x_{s(PVT)}} = A_{PVT}I_{rad}\left(1 - \frac{T_a}{T_s}\right) \tag{4.16}$$

where T_a and T_s are the ambient temperature and equivalent temperature of the sun, respectively. T_s is taken as 5780K.

The thermal and electrical exergy efficiency, exergy destruction, improvement potential, and entropy generation of integrated PVT system can be depicted as follows:

(c) Thermal exergy efficiency

$$\varepsilon_{th(PVT)} = \frac{\dot{E}_{x_{th}}}{\dot{E}_{x_{s(PVT)}}} = \frac{\dot{m}C_p\left(T_{o(PVT)} - T_{i(PVT)}\right)}{A_{PVT}I_{rad}\left(1 - \frac{T_a}{T_s}\right)} \tag{4.17}$$

(d) Electrical exergy efficiency

$$\varepsilon_{el(SAPH)} = \frac{\dot{E}_{x_{el}}}{\dot{E}_{x_{s(PVT)}}} \tag{4.18}$$

(e) Overall exergy efficiency

$$\varepsilon_{ov(PVT)} = \frac{\dot{E}_{x_{th}}}{\dot{E}_{x_{s(PVT)}}} + \frac{\dot{E}_{x_{el}}}{\dot{E}_{x_{s(PVT)}}} \tag{4.19}$$

Furthermore, simple exergy analysis has been used to assess only the quantitative inefficiencies throughout the system, whereas in advanced-level exergy analysis, the irreversibility factors are taken to compute the actual change. The maximum improvement in the system exergy performance is achieved when the rate of exergy destruction is decreased.

(f) Exergy destruction

$$\dot{E}_{x_{dest(PVT)}} = \dot{E}_{x_s} - \dot{E}_{x_{th}} - \dot{E}_{x_{el}} \tag{4.20}$$

(g) Improvement potential

$$IP = \left(1 - \frac{\varepsilon_{ov(PVT)}}{100}\right) \times \dot{E}_{x_{dest(PVT)}} \tag{4.21}$$

(h) Entropy generation

$$\dot{S} = \frac{\dot{E}_{x_{dest(PVT)}}}{T_a} \tag{4.22}$$

4.3 Concluding remarks

This chapter presented an energy and exergy analysis of a PV systems based on the first and second law of thermodynamics. The chapter presented an overview of all the thermodynamic equations used to determine the energy and exergy efficiency of the PV systems. The chapter aims to provide a summary of the thermodynamics of the solar PV energy conversion process through energy and exergy balance equations for the design and performance improvement of the system. This chapter will serve as a guideline for researchers in the field of solar PV systems.

References

Agrawal, B., & Tiwari, G. N. (2010). Optimizing the energy and exergy of building integrated photovoltaic thermal (BIPVT) systems under cold climatic conditions. *Applied Energy*, *87*(2), 417−426. Available from https://doi.org/10.1016/j.apenergy.2009.06.011.

Fudholi, A., Zohri, M., Jin, G. L., Ibrahim, A., Yen, C. H., Othman, M. Y., ... Sopian, K. (2018). Energy and exergy analyses of photovoltaic thermal collector with V-groove. *Solar Energy*, *159*, 742−750. Available from https://doi.org/10.1016/j.solener.2017.11.056.

Hossain, M. S., Pandey, A. K., Selvaraj, J., Rahim, N. A., Islam, M. M., & Tyagi, V. V. (2019). Two side serpentine flow based photovoltaic-thermal-phase change materials (PVT-PCM) system: Energy, exergy and economic analysis. *Renewable Energy*, *136*, 1320−1336. Available from https://doi.org/10.1016/j.renene.2018.10.097.

Kumar, L., Hasanuzzaman, M., & Rahim, N. (2019). Global advancement of solar thermal energy technologies for industrial process heat and its future prospects: A review. *Energy Conversion and Management*, *195*, 885−908.

Kumar, L., Hasanuzzaman, M., Rahim, N., & Islam, M. (2021). Modeling, simulation and outdoor experimental performance analysis of a solar-assisted process heating system for industrial process heat. *Renewable Energy*, *164*, 656−673.

Ma, T., Li, M., & Kazemian, A. (2020). Photovoltaic thermal module and solar thermal collector connected in series to produce electricity and high-grade heat simultaneously. *Applied Energy*, *261*, 114380. Available from https://doi.org/10.1016/j.apenergy.2019.114380.

Mehmet Kanoglu, Y. C., & Cimbala, J. (2019). *Fundamentals and applications of renewable energy*. McGraw-Hill Education.

Park, S. R., Pandey, A. K., Tyagi, V. V., & Tyagi, S. K. (2014). Energy and exergy analysis of typical renewable energy systems. *Renewable and Sustainable Energy Reviews*, *30*, 105−123. Available from https://doi.org/10.1016/j.rser.2013.09.011.

Rawat, R., Lamba, R., & Kaushik, S. C. (2017). Thermodynamic study of solar photovoltaic energy conversion: An overview. *Renewable and Sustainable Energy Reviews*, *71*, 630−638. Available from https://doi.org/10.1016/j.rser.2016.12.089.

Sardarabadi, M., Hosseinzadeh, M., Kazemian, A., & Passandideh-Fard, M. (2017). Experimental investigation of the effects of using metal-oxides/water nanofluids on a photovoltaic thermal system (PVT) from energy and exergy viewpoints. *Energy*, *138*, 682−695.

Performance assessment of three photovoltaic systems

Adar Mustapha, Babay Mohamed-Amine and Mabrouki Mustapha
Department of Physics, Laboratory of Industrial Engineering, Sultan Moulay Slimane University, Beni Mellal, Morocco

5.1 Introduction

Recently, the photovoltaic (PV) industry has experienced enormous progress and, as a result, PV cells have ascended to the throne of low-cost green power generation technologies (Renewable Energy Market Update: Outlook for 2022 and 2023, 2022). This is due to the improved efficiency of the conversion of solar energy into electrical energy by different PV cell technologies. The record lab monocrystalline cell efficiency is 26.7% and 24.4% for the multicrystalline silicon one. In comparison with wafer-based technology, the highest lab efficiency in thin-film technology is 23.4% for CIGS and 21.0% for CdTe solar cells. The record lab cell efficiency for perovskite is 23.7%. This evolution is accompanied by another evolution in the conversion efficiency of the inverters, which has reached 98% or higher for state-of-the-art brand products. As is well known, the efficiency of these systems under real operating conditions differs from that measured in the laboratory. Environmental factors and inappropriate manipulation when transporting and installing PV systems can lead to performance degradation. That is why the control and analysis of the performance of the PV systems from the first moment of their implementation are required.

Generally, the performance analysis of PV systems is based on the evaluation of the parameters described by the IEC 61215 standard (Adar et al., 2017). The most commonly used parameter to analyze and compare the performance of different PV systems is the performance ratio (PR) (Adar, Khaouch, et al., 2018). Based on this parameter, several studies have analyzed and compared the performance of different PV systems installed around the world. On the Asian continent, the performance ratio reaches 88.7% for a PV system installed in the city of Yazid in Iran (Ghodusinejad et al., 2022). The PR of a mc-si-based PV system installed in Malaysia reaches 85.4% (Saleheen et al., 2021). In Europe, it reaches 81.15% in Turkey (Cubukcu & Gumus, 2020) and is confined between 62% and 72% in Spain (Drif et al., 2007). In Africa, it is reported that it reaches 85.5% in Algeria (Ihaddadene et al., 2022) and 70% in Mauritania (Elhadj Sidi et al., 2016), and it was found to be about 83% in the case of Brazil (de Lima et al., 2017). Other studies are devoted to analyzing the performance of PV systems with respect to the prevailing climatic conditions. Under tropical meteorological conditions, a study

Performance Enhancement and Control of Photovoltaic Systems. DOI: https://doi.org/10.1016/B978-0-443-13392-3.00005-0

has revealed that polycrystalline silicon (pc-si) PV systems outperform amorphous thin-film silicon (a-si) in terms of energy efficiency, annual yield, and overall losses (Tripathi et al., 2014). The same technology had lower overall system losses than both mc-Si and thin-film a-si (Akhter et al., 2020). In the climatic conditions of the Sahara, the thin-film modules perform better than mc-si and pc-si (Al-Otaibi et al., 2015).

As with every product, consumers of PV modules are continually looking for more affordable items that offer the highest level of reliability and a durable lifetime of service. For this reason, manufacturers include a warranty on the data sheet to reassure customers (Adar et al., 2022). They promise that the performance does not fall below 80% over the first 20 years of operation (Schlothauer et al., 2012). However, real-world operating conditions differ from normal test conditions (STC). Because of the varying behavior of PV modules based on weather conditions, quantifying degradation rates allows for the circumvention of certain budgetary and technological obstacles. Financial risks are involved with long-term energy efficiency estimation, notably due to assessments of the rate of deterioration and volatility of solar resources, as represented in levelized electricity cost calculations for PV systems (Ascencio-Vásquez et al., 2019). Several degradation determinants and mechanisms are described and investigated in the literature. Temperature, humidity, precipitation, dust, snow, and sun irradiation are the most common elements associated with PV module performance degradation during field operation (Park et al., 2013; Phinikarides et al., 2014). The presence of broken cells, glass breakage, and mechanical stress damage to the frame causes moisture to permeate the interior of the PV module, resulting in deterioration processes such as corrosion. When PV modules are exposed to UV light, the encapsulant discolors and experiences photochemical deterioration (Badiee et al., 2014). Environmental factors frequently lead to encapsulant delamination. Front glass encapsulant delamination, cells, interconnecting ribbons, and backsheets are all examples of encapsulant delamination. All of these types of delamination contribute to various types of PV module degradation (Wohlgemuth et al., 2017). All these types of delamination contribute to different types of degradation of the PV module. Potential-induced degradation (Luo et al., 2017; Naumann et al., 2014; Papargyri et al., 2020; Pingel et al., 2010; Schwark et al., 2013) and light-induced degradation (Lindroos & Savin, 2016) are other degradation mechanisms for which the encapsulant is responsible. The effect of these performance degradation factors is evaluated for the different photovoltaic technologies in several studies. In Egypt, the average yield degradation value was −1.19%/year, −1.17%/year, and −1.67%/year for pc-si, mc-si, and thin-film CdTe, respectively (Othman & Hatem, 2022). In Italy, a study showed that the degradation of a mc-si PV system's performance was −1.12%/year (Malvoni et al., 2020). A performance degradation study of 834 fielded PV modules representing 13 module types in three climates has shown that the degradation rates were highly nonlinear over time, and seasonal variations were present in some module types. The mean and median degradation rate values of −0.62%/year and −0.58%/year, respectively, were consistent with rates measured for older modules. Of the 23 systems studied, 6 have degradation rates that will exceed the warranty limits in the future, whereas 13 systems demonstrate the potential to achieve a lifetime beyond 30 years (Theristis et al., 2023).

In this work, we analyze the performance of three PV systems using parameters such as conversion efficiency, capacity factor, and performance ratio. PV modules based on three silicon technologies are evaluated in the field for performance decline using their ratio. From January 2015 until the last month of 2019, the three PV plants' performance was tracked. The 60 performance ratio values were used to create a monthly time series. For each technology, seasonal adjustment operations have been performed on their time series, such as classical seasonal decomposition (CSD), to extract performance trends.

5.2 Materials and methods

5.2.1 Description of PV grid-connected plants

The first and second PV grid-connected systems are based on eight monocrystalline and eight polycrystalline silicon PV modules, respectively. Each module has a nominal power of 255 Wp. The modules are wired in series to create a 2 kWp peak power for each technology. They are injected into the grid through two SMA 2000 inverters. The third PV plant is made up of 12 NEXPOWER NT_155AF amorphous silicon modules with a total output of 155 Wp. The modules form two strings connected in parallel. Every string consists of six PV modules that are then grid-connected through the same type of inverter as the first and second PV plants. PV modules are cleaned regularly. Two minipolycrystalline silicon PV modules, calibrated with a Kipp and Zonen pyranometer, were used to monitor solar irradiation (Adar et al., 2022; Babay et al., 2022). The temperatures of the three photovoltaic technologies and the surrounding air are measured using four PT100 temperature sensors. The research (Adar et al., 2017, 2020; Adar, Bazine, et al., 2018; Adar, Khaouch, et al., 2018; Bahanni et al., 2020, 2022; (Babay et al., 2023)) discusses and offers further details about the technologies that have been installed, the collection of meteorological data, and some early findings in the investigation of the performance of these PV systems.

5.2.2 Performance metrics

Photovoltaic performance metrics are used to evaluate the efficiency and effectiveness of a solar energy system. These metrics can help to identify potential issues, monitor system performance over time, and optimize the system to improve energy production.

5.2.2.1 Reference yield

The reference yield (Yr) is the ratio of the reference radiation quantity G (1 kW/m^2) to the total solar radiation H (kWh/m^2) reaching the surface of PV solar panels (Adar et al., 2020). In Eq. (5.1), the reference yield is given as follows:

$$Yr = H/G \qquad (5.1)$$

5.2.2.2   Final yield

The final yield (Yf) is the total energy produced by the PV system, E_{AC} (kWh), with respect to the nominal installed power P (kWp). This quantity represents the number of hours during which the PV field operates at its nominal power (Adar et al., 2020). The final yield is given by Eq. (5.2).

$$Yf = E_{AC}/P \qquad (5.2)$$

5.2.2.3   Performance ratio

The performance ratio (PR) values show how a PV system performs close to its ideal performance in real-world situations (Adar et al., 2017). PR is defined as the ratio between the final yield and the reference yield as shown in Eq. (5.3).

$$PR = Yf/Yr \qquad (5.3)$$

5.2.2.4   Conversion efficiency of PV modules

The link between the electrical energy generated by the module and the solar energy that is incident on the plane of the solar modules is known as the conversion efficiency of solar PV modules (Adar et al., 2020). The Eq. (5.4) is used to compute the PV system's monthly energy yield:

$$n = \sum_{i=1}^{n}(E_D)i/S\sum_{i=1}^{n}(G_{opt})i \qquad (5.4)$$

where n is the number of days in a month, E_D is the total amount of electrical energy produced by the PV system and transmitted to the power grid during the day (Wh), G_{opt} is the total amount of overall solar energy falling on the plane of the solar modules (Wh/m^2) during the day, and S is the total area of the solar modules.

5.2.2.5   Capacity factor of the solar PV system

The capacity factor (CF) is the relationship between the actual annual electrical energy produced by the PV system and the electrical energy that could be produced if the solar PV system is operated at its full installed capacity 24 hours a day for 1 year. According to the reference (Adar et al., 2020), the capacity factor of the solar PV system is calculated using Eq. (5.5).

$$CF = Yf/8760 \qquad (5.5)$$

5.2.3   Classical seasonal decomposition (CSD)

This method divides the time series into three parts: the trend, the seasonal components, and any residual random parts. A moving average centered on two steps is used to extract the trend from the time series. The centered average at time t is derived using Eq. (5.6) for a moving average of $2k$, where k represents the order of the moving average (Makrides et al., 2014; Malvoni et al., 2020; Singh et al., 2020; Solís-Alemán et al., 2019).

$$T_t = 1/2k \left(\sum_{i=t-m}^{t+m-1} Y_i + \sum_{i=t-m+1}^{t+m} Y_i \right) \tag{5.6}$$

where T_t represents the trend at time t, $(t > m)$, and m is defined as half the width of a moving average, $m = k/2$. As a result, the calculated moving averages are used to build the trend for each time series. After that, the trend is subtracted from the initial time series data to calculate the seasonal component. Using Eq. (5.7), the gross seasonal component was recovered (Adar et al., 2022; Lindig et al., 2018).

$$S_t = Y_t - T_t \tag{5.7}$$

where seasonality S_t represents the difference between the original data T_t and the trend at time t. The linear regression is reapplied after extracting the trend for each technology to get the performance degradation rate (R_D). The degradation rate is calculated using Eqs. (5.5)−(5.8):

$$R_D = 12 \times a \tag{5.8}$$

where a represents the slope of the linear regression curve.

5.3   Results and discussions

5.3.1   Performance analysis

The performance of any PV system depends on the environmental conditions at the location of the PV plant to provide insight into the weather conditions under which the PV system performance evolves. Fig. 5.1 illustrates the total monthly insolation measured in the PV module plane between January 2015 and December 2019. The monthly insolation varied from 139 kWh/m^2 in November 2016 to 205.6 kWh/m^2 in May 2019, averaging 173.3 kWh/m^2/m over 60 months. The total annual solar radiation measured is 2051.7 kWh/m^2, 2091.3 kWh/m^2, 2125.5 kWh/m^2, 1659.5 kWh/m^2, and 2157.6 kWh/m^2 for the years 2015, 2016, 2017, 2018, and 2019, respectively, making an average of 2017.14 kWh/m^2/year.

The same figure shows the average monthly ambient temperature measured over the same monitoring period. The average monthly ambient temperature ranged

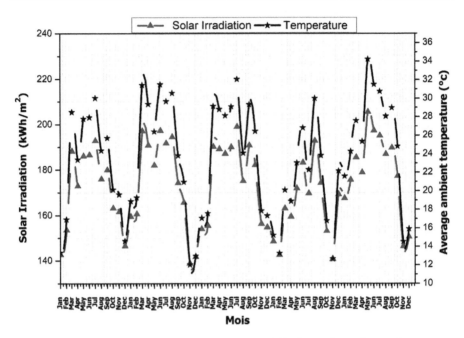

Figure 5.1 Solar irradiation and ambient temperature. This figure shows the weather conditions such as the evolution of the ambient temperature and solar irradiance on the photovoltaic array formed by the three silicon-based photovoltaic technologies.

from 12.06°C in November 2016 to 34.22°C in May 2019, while the average value was 23.5°C/m. The year 2019 was the warmest with an average of 25.63°C, and 2018 was the coldest with an average of 20.41°C.

The capacity factor indicates the fraction of a year in which the PV system operates at its rated power. Although the capacity factor is known for its strong dependence on solar irradiance, Fig. 5.2 shows that the capacity factors of the three PV systems follow the temperature variation. The maximum values of FCs are 24.95%, 24.06%, and 23.90% for a-si, pc-si, and mc-si, respectively, which were recorded in July 2017 for amorphous and May 2019 for crystalline. While the minimum values were 14.43% and 15.61% in January 2018 for a-si and pc-si, respectively, for mc-si, the minimum value is 14.93%, recorded in January 2015. On average, the PV modules of the mc-si, pc-si, and a-si types have operated, respectively, in their nominal operation for the fractions of 19.8%, 20.24%, and 19.84% of the 5-year period (1826 days). Therefore, the PV systems produced electricity at their maximum capacity for about 361.54 days for mc-si, about 369.58 days for pc-si, and about 362.27 days for a-si.

Another factor that can be used to evaluate the performance of PV modules is the conversion efficiency of light into electricity. Fig. 5.3 shows that the variation in efficiency of a-si PV modules follows almost the same pattern of temperature variation compared to that of c-si. The yield of mc-si PV panels varies from 11.55% in November 2017 to 13.39% in March 2018, with higher values during the

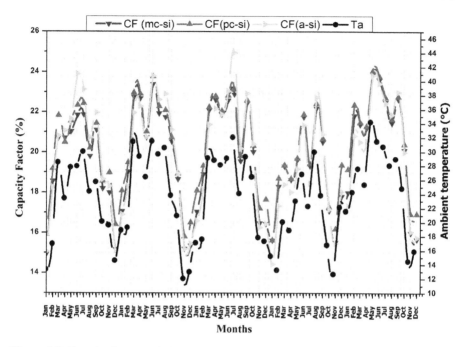

Figure 5.2 Capacity factor evolution. This figure shows the evolution and the effect of the ambient temperature on the capacity factor of the three silicon-based photovoltaic technologies.

most irradiated months. The efficiency of pc-si-type PV panels took values between 11.9% and 13.43%, recorded in January 2018 and March 2018, respectively, while the efficiency of a-si PV panels was found to be confined between 7.14% in December 2017 and 9.21% in July 2017. The average efficiency of pc-si modules is 12.97%, making a difference of 2.24% with their efficiency at STC conditions, which is quantified at 15.21%. The average yield of the mc-si modules is 12.67%; this average value deviates from the laboratory yield of 2.54%. Under STC conditions, the yield of a-si PV modules is 9.87%, and under real operating conditions, this yield is an average of 8.22%. The difference in efficiency between the two conditions is 1.65%. This shows that amorphous silicon PV technology experiences reduced efficiency losses under real operating conditions compared to crystalline silicon.

In comparison with other performance parameters, the performance ratio is the only parameter that can effectively make a comparison between PV systems regardless of their capacity, technology, or location. The performance ratio depicts the overall losses of the PV installation's nominal power that may be brought on by factors like module temperature, wiring, inverter inefficiencies, component failures, etc. Fig. 5.4 depicts the average monthly ambient temperature and the performance ratio of the three PV technologies. With a deviation of 10.05%, the PR of pc-si cell-based PV modules experience the least amount of variation. It ranges from 78.19% in January 2018 to 88.94% in March 2018. The mc-si technology has slightly larger monthly performance fluctuations quantified at 12.1%, ranging from

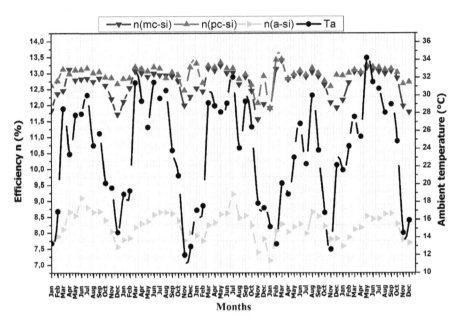

Figure 5.3 Conversion efficiency evolution. This figure shows the evolution and the effect of the ambient temperature on the conversion efficiency of the three silicon-based photovoltaic technologies.

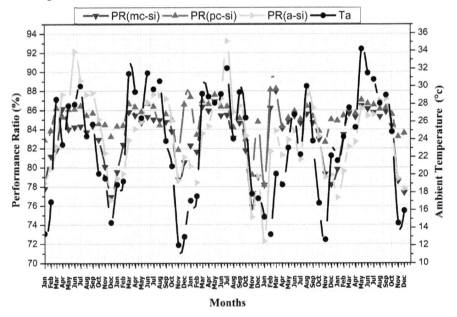

Figure 5.4 Performance ratio and ambient temperature. This figure shows the evolution and the effect of the ambient temperature on the performance of the three silicon-based photovoltaic technologies.

75.89% in November 2017 to 87.99% in March 2018. The PR of a-si modules fluctuated more widely, ranging from 72.28% in January 2018 to 93.24% in July 2017, a difference of around 21%. During 5 years of operation, the average monthly performance ratio of the pc-si, mc-si, and a-si technologies was obtained at 85.2%, 83.23%, and 83.28%, respectively.

According to the power temperature coefficients of each photovoltaic module technology, as the ambient temperature rises, the PR of pc-si, mc-si, and a-si modules should indeed decrease. As a result, and taking solar irradiation into account, the performance of the mc-si and pc-si technologies reaches its maximum values during the period from February to May each year. This period is characterized by an almost constant low temperature and good solar irradiation. On the contrary, the performance of a-si modules increases during the warmer months, exceeding the performance of crystalline technology and reaching high performance during the period between May and September. This makes sense given that amorphous silicon (a-si) cells can enhance electrical performance at high temperatures. Solar is the term for this effect. It enables the restoration of some of the initial nominal power lost as a result of the degradation brought on by prolonged exposure to high temperatures and light (Adar et al., 2020; Amin et al., 2009; Makrides et al., 2012).

In Fig. 5.5, it can be clearly seen that the effect of thermal annealing became more evident for a-si technology from average temperatures above 23°C, which correspond to average temperature values recorded in the period between May and September of each year. Amorphous solar modules are cooler than c-si solar modules

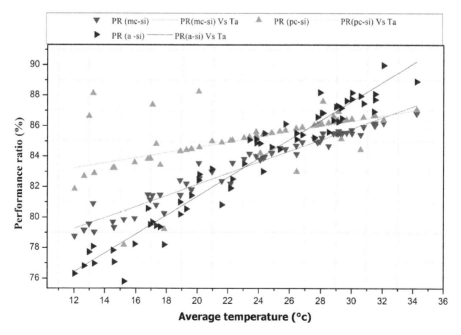

Figure 5.5 The linear regression of performance against temperature. This figure shows the effect of thermal annealing on the performance of the three silicon-based photovoltaic technologies.

for a second reason, which is their bigger size, that allows for more heat dissipation (Amin et al., 2009). As a result, throughout the summer, it performs better than other technologies. The months of January and February are when the temperatures are at their lowest levels; the c-si technologies, unlike the amorphous silicon technology, operate with high performance; the pc-si shows higher performance than the mc-si.

5.3.2 Annual degradation rate assessment using classical seasonal decomposition method

A five-year time series of monthly average PR values, comprising 60 data sets, was elaborated for the period June 2015 to December 2019. The graphs in Figs. 5.6–5.8

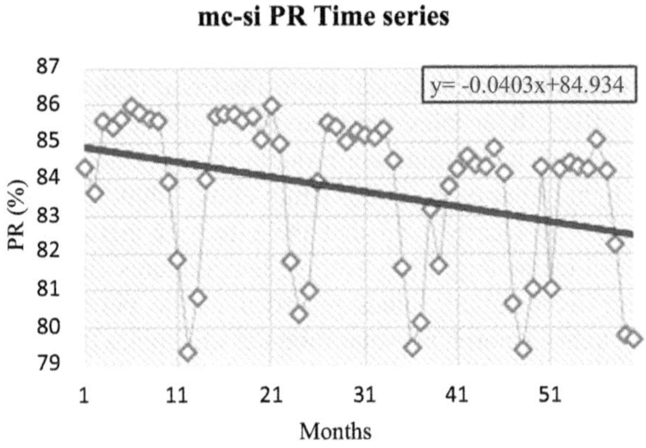

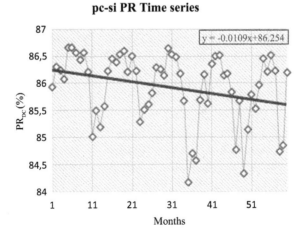

Figure 5.6 Time series of mc-si PR values. This figure shows the constructed time series of monthly performance ratio values of monocrystalline silicon-based photovoltaic cells.

Figure 5.7 Time series of pc-si PR values. This figure shows the constructed time series of monthly performance ratio values of polycrystalline silicon photovoltaic cells.

a-si PR Time series

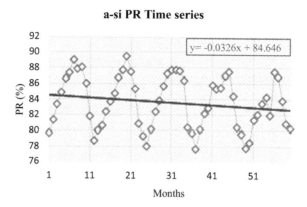

Figure 5.8 Time series of a-si PR. This figure shows the constructed time series of monthly performance ratio values of amorphous silicon-based photovoltaic cells.

mc-si PR Trend component

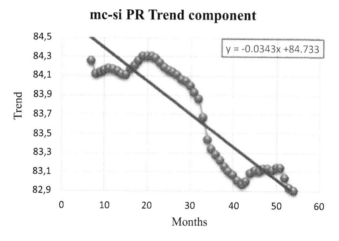

Figure 5.9 mc-si trend component. This figure shows the decreasing trend in the performance of monocrystalline silicon-based photovoltaic cells.

illustrate the monthly PR time series formed for the mc-si, pc-si, and a-si PV technologies, respectively. These figures demonstrate that seasonal behavior is present across all technologies, while all silicon PV technologies show a pattern of gradually declining performance. The trend of decreasing performance is visible from the latest months of the first year (2015) of field operation. This is attributed to potential-induced degradation (PID) for crystalline silicon (c-si) technologies (Amin et al., 2009; Naumann et al., 2014; Pingel et al., 2010; Wohlgemuth, 2020) and to the Staebler-Wronski impact and thermal annealing cycles that are attributed to amorphous silicon PV technology (Adar et al., 2020; Lotfi et al., 2021; Nofuentes et al., 2017; Silvestre et al., 2016; Tahri et al., 2017).

The linear regression of the PR values is used to quantify this decreasing trend. In terms of performance, mc-si degrades at a rate of roughly -0.48%/year, while a-si degrades at a rate of roughly -0.40%/year. The pc-si has the lowest degradation rate, with an R_D of -0.13% per year. The values found by the linear regression are not correct because the variation in performance is affected by the effect of seasonality.

The classical seasonal decomposition, which was used to split the time series by extracting the trend and seasonality for each PV technology, was also used to evaluate the degradation rates (Adar et al., 2022).

Figs. 5.9−5.11 represent the trend components of the three PV systems, extracted from their performance time series. The linear regression applied to the trend components

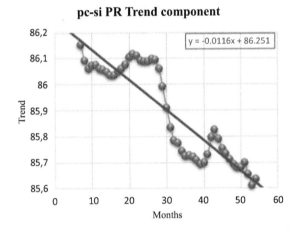

pc-si PR Trend component

$y = -0.0116x + 86.251$

Figure 5.10 pc-si trend component. This figure shows the decreasing trend in the performance of polycrystalline silicon-based photovoltaic cells.

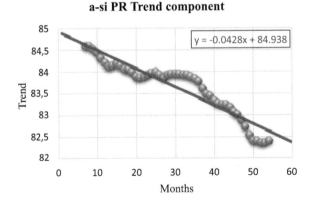

a-si PR Trend component

$y = -0.0428x + 84.938$

Figure 5.11 a-si trend component. This figure shows the decreasing trend in the performance of amorphous silicon-based photovoltaic cells.

shows that the degradation rate of the a-si PV system is the highest at -0.51%/year, while the mc-si system has a quantified annual performance degradation rate of -0.41%/year. Over the same 5-year period, pc-si technology demonstrated the lowest annual performance degradation rate of about -0.14%/year.

5.4   Conclusion

The exposure during 5 years to the real conditions of functioning characterized by solar irradiation, temperature, and humidity of the three PV systems based on three technologies of silicon induced the degradation of their initial performance. Degradation is potential-induced (PID) in the case of crystalline silicon (c-si) technologies and by the Staebler-Wronski effect and thermal annealing cycles for amorphous silicon PV technology. The mc-si system demonstrated a defined yearly performance degradation rate of -0.41%/year, whereas the degradation rate of the a-si PV system was the greatest, at -0.51%/year. A lower annual performance degradation rate of roughly -0.14%/year has been observed for pc-si technology.

References

Adar, M., Bazine, H., Najih, Y., Bahanni, C., Mabrouki, M., Chebak, A. (2018). Institute of Electrical and Electronics Engineers Inc. Morocco Simulation study of three PV systems. Proceedings of 2018 6th International Renewable and Sustainable Energy Conference, IRSEC 2018, Available from https://doi.org/10.1109/IRSEC.2018.8702827, http://ieeexplore.ieee.org/xpl/mostRecentIssue.jsp?punumber = 8694926, 9781728111827.

Adar, M., Khaouch, Z., Mabrouki, M., Benouna, A., Chebak, A. (2018). Performance Analysis of PV Grid-Connected in Fours Special Months of the Year. Proceedings of 2017 International Renewable and Sustainable Energy Conference, IRSEC 2017, 10.1109/IRSEC.2017.8477373, 9781538628478, Institute of Electrical and Electronics Engineers Inc. Morocco. http://ieeexplore.ieee.org/xpl/mostRecentIssue.jsp?punumber = 8467300.

Adar, M., Mabrouki, M., Bennouna, A., & Chebak, A. (2017). *Production study of a grid connected PV plant. Proceedings of 2016 International Renewable and Sustainable Energy Conference, IRSEC 2016* (pp. 116−120). Morocco: Institute of Electrical and Electronics Engineers Inc. Available from https://doi.org/10.1109/IRSEC.2016.7983963.

Adar, M., Najih, Y., Chebak, A., Mabrouki, M., & Bennouna, A. (2022). Performance degradation assessment of the three silicon PV technologies. *Progress in Photovoltaics: Research and Applications, 30*(10), 1149−1165. Available from https://doi.org/10.1002/pip.3532, http://onlinelibrary.wiley.com/journal/10.1002/(ISSN)1099-159X.

Adar, M., Najih, Y., Gouskir, M., Chebak, A., Mabrouki, M., & Bennouna, A. (2020). Three PV plants performance analysis using the principal component analysis method. *Energy, 207*. Available from https://doi.org/10.1016/j.energy.2020.118315, https://www.journals.elsevier.com/energy.

Akhter, M. N., Mekhilef, S., Mokhlis, H., Olatomiwa, L., & Muhammad, M. A. (2020). Performance assessment of three grid-connected photovoltaic systems with combined capacity of 6.575 kWp in Malaysia. *Journal of Cleaner Production*, *277*, 123242. Available from https://doi.org/10.1016/j.jclepro.2020.123242.

Al-Otaibi, A., Al-Qattan, A., Fairouz, F., & Al-Mulla, A. (2015). Performance evaluation of photovoltaic systems on Kuwaiti schools' rooftop. *Energy Conversion and Management*, *95*, 110−119. Available from https://doi.org/10.1016/j.enconman.2015.02.039.

Amin, N., Lung, C. W., & Sopian, K. (2009). A practical field study of various solar cells on their performance in Malaysia. *Renewable Energy*, *34*(8), 1939−1946. Available from https://doi.org/10.1016/j.renene.2008.12.005.

Ascencio-Vásquez, J., Kaaya, I., Brecl, K., Weiss, K. A., & Topič, M. (2019). Global climate data processing and mapping of degradation mechanisms and degradation rates of PV modules. *Energies*, *12*(24). Available from https://doi.org/10.3390/en12244749, https://www.mdpi.com/1996-1073/12/24.

Badiee, A., Wildman, R., Ashcroft, I. (2014). Effect of UV aging on degradation of Ethylene-vinyl Acetate (EVA) as encapsulant in photovoltaic (PV) modules. Proceedings of SPIE—The International Society for Optical Engineering. Available from https://doi.org/10.1117/12.2062007. 1996756X SPIE United Kingdom, http://spie.org/x1848.xml 9179.

Bahanni, C., Adar, M., Boulmrharj, S., Khaidar, M., Mabrouki, M. (2020). 5th International Conference on Renewable Energies for Developing Countries, Institute of Electrical and Electronics Engineers Inc. Morocco Analysis of weather impact on the yield of PV plants installed in two antagonists cities in Morocco. REDEC 2020 https://doi.org/10.1109/REDEC49234.2020.9163841, http://ieeexplore.ieee.org/xpl/mostRecentIssue.jsp?punumber = 9162003, 9781728155951.

Babay, M.-A., Adar, M., & Mabrouki, M. (2022). Modeling and simulation of a PEMFC using three-dimensional multi-phase computational fluid dynamics model. *IEEE*, 1−6. Available from https://doi.org/10.1109/IRSEC53969.2021.9741144.

Babay, M.-A., M. Adar, A. Chebak., & M. Mabrouki (2023). Dynamics of gas generation in porous electrode alkaline electrolysis cells: An investigation and optimization using machine learning. *Energies*, *16*(14), 1-21. Available from https://doi.org/10.3390/en16145365.

Bahanni, C., Adar, M., Boulmrharj, S., Khaidar, M., & Mabrouki, M. (2022). Performance comparison and impact of weather conditions on different photovoltaic modules in two different cities. *Indonesian Journal of Electrical Engineering and Computer Science*, *25*(3), 1275−1286. Available from https://doi.org/10.11591/ijeecs.v25.i3.pp1275-1286, http://ijeecs.iaescore.com/index.php/IJEECS/article/view/27210.

Cubukcu, M., & Gumus, H. (2020). Performance analysis of a grid-connected photovoltaic plant in eastern Turkey. *Sustainable Energy Technologies and Assessments*, *39*, 100724. Available from https://doi.org/10.1016/j.seta.2020.100724.

de Lima, L. C., de Araújo Ferreira, L., & de Lima Morais, F. H. B. (2017). Performance analysis of a grid connected photovoltaic system in northeastern Brazil. *Energy for Sustainable Development*, *37*, 79−85. Available from https://doi.org/10.1016/j.esd.2017.01.004, http://www.elsevier.com.

Drif, M., Pérez, P. J., Aguilera, J., Almonacid, G., Gomez, P., de la Casa, J., Aguilar, J. D., & Univer Project. (2007). A grid connected photovoltaic system of 200 kWp at Jaén University. Overview and performance analysis. *Solar Energy Materials and Solar Cells*, *91*(8), 670−683. Available from https://doi.org/10.1016/j.solmat.2006.12.006.

Elhadj, C. E. B., Sidi, M. L., Ndiaye, M., El Bah, A., Mbodji, A., & Ndiaye, P. A. (2016). Performance analysis of the first large-scale (15 MWp) grid-connected photovoltaic plant in Mauritania. *Energy Conversion and Management*, *119*, 411−421. Available from https://doi.org/10.1016/j.enconman.2016.04.070.

Ghodusinejad, M. H., Ghodrati, A., Zahedi, R., & Yousefi, H. (2022). Multi-criteria modeling and assessment of PV system performance in different climate areas of Iran. *Sustainable Energy Technologies and Assessments*, *53*, 102520. Available from https://doi.org/10.1016/j.seta.2022.102520.

Ihaddadene, R., El Hassen Jed, M., Ihaddadene, N., & De Souza, A. (2022). Analytical assessment of Ain Skhouna PV plant performance connected to the grid under a semi-arid climate in Algeria. *Solar Energy*, *232*, 52−62. Available from https://doi.org/10.1016/j.solener.2021.12.055, http://www.elsevier.com/inca/publications/store/3/2/9/index.htt.

Lindig, S., Kaaya, I., Weis, K. A., Moser, D., & Topic, M. (2018). Review of statistical and analytical degradation models for photovoltaic modules and systems as well as related improvements. *IEEE Journal of Photovoltaics*, *8*(6), 1773−1786. Available from https://doi.org/10.1109/JPHOTOV.2018.2870532, http://eds.ieee.org/jpv.html.

Lindroos, J., & Savin, H. (2016). Review of light-induced degradation in crystalline silicon solar cells. *Solar Energy Materials and Solar Cells*, *147*, 115−126. Available from https://doi.org/10.1016/j.solmat.2015.11.047, http://www.sciencedirect.com/science/journal/09270248/100.

Lotfi, H., Adar, M., Bennouna, A., Izbaim, D., Oum'Bark, F., & Ouacha, E. H. (2021). Silicon photovoltaic systems performance assessment using the principal component analysis technique. *Materials Today: Proceedings*, *51*, 1966−1974. Available from https://doi.org/10.1016/j.matpr.2021.04.374, https://www.sciencedirect.com/journal/materials-today-proceedings.

Luo, W., Khoo, Y. S., Hacke, P., Naumann, V., Lausch, D., Harvey, S. P., Singh, J. P., Chai, J., Wang, Y., Aberle, A. G., & Ramakrishna, S. (2017). Potential-induced degradation in photovoltaic modules: A critical review. *Energy and Environmental Science*, *10*(1) 43−68. Available from https://doi.org/10.1039/c6ee02271e, http://www.rsc.org/Publishing/Journals/EE/About.asp.

Makrides, G., Zinsser, B., Phinikarides, A., Schubert, M., & Georghiou, G. E. (2012). Temperature and thermal annealing effects on different photovoltaic technologies. *Renewable Energy*, *43*, 407−417. Available from https://doi.org/10.1016/j.renene.2011.11.046.

Makrides, G., Zinsser, B., Schubert, M., & Georghiou, G. E. (2014). Performance loss rate of twelve photovoltaic technologies under field conditions using statistical techniques. *Solar Energy*, *103*, 28−42. Available from https://doi.org/10.1016/j.solener.2014.02.011, http://www.elsevier.com/inca/publications/store/3/2/9/index.htt.

Malvoni, M., Kumar, N. M., Chopra, S. S., & Hatziargyriou, N. (2020). Performance and degradation assessment of large-scale grid-connected solar photovoltaic power plant in tropical semi-arid environment of India. *Solar Energy*, *203*, 101−113. Available from https://doi.org/10.1016/j.solener.2020.04.011, http://www.elsevier.com/inca/publications/store/3/2/9/index.htt.

Naumann, V., Lausch, D., Hähnel, A., Bauer, J., Breitenstein, O., Graff, A., Werner, M., Swatek, S., Großer, S., Bagdahn, J., & Hagendorf, C. (2014). Explanation of potential-induced degradation of the shunting type by Na decoration of stacking faults in Si solar cells. *Solar Energy Materials and Solar Cells*, *120*, 383−389. Available from https://doi.org/10.1016/j.solmat.2013.06.015.

Nofuentes, G., de la Casa, J., Solís-Alemán, E. M., & Fernández, E. F. (2017). Spectral impact on PV performance in mid-latitude sunny inland sites: Experimental vs. modelled results. *Energy*, *141*, 1857−1868. Available from https://doi.org/10.1016/j.energy.2017.11.078, http://www.elsevier.com/inca/publications/store/4/8/3/.

Othman, R., & Hatem, T. M. (2022). Assessment of PV technologies outdoor performance and commercial software estimation in hot and dry climates. *Journal of Cleaner Production, 340*, 130819. Available from https://doi.org/10.1016/j.jclepro.2022.130819.

Papargyri, L., Theristis, M., Kubicek, B., Krametz, T., Mayr, C., Papanastasiou, P., & Georghiou, G. E. (2020). Modelling and experimental investigations of microcracks in crystalline silicon photovoltaics: A review. *Renewable Energy, 145*, 2387−2408. Available from https://doi.org/10.1016/j.renene.2019.07.138, http://www.journals.elsevier.com/renewable-and-sustainable-energy-reviews/.

Park, N. C., Oh, W. W., & Kim, D. H. (2013). Effect of temperature and humidity on the degradation rate of multicrystalline silicon photovoltaic module. *International Journal of Photoenergy, 2013*. Available from https://doi.org/10.1155/2013/925280, http://www.hindawi.com/journals/ijp/contents/.

Phinikarides, A., Kindyni, N., Makrides, G., & Georghiou, G. E. (2014). Review of photovoltaic degradation rate methodologies. *Renewable and Sustainable Energy Reviews, 40*, 143−152. Available from https://doi.org/10.1016/j.rser.2014.07.155, https://www.journals.elsevier.com/renewable-and-sustainable-energy-reviews.

Pingel, S., Frank, O., Winkler, M., Oaryan, S., Geipel, T., Hoehne, H., Berghold, J. (2010). Germany Potential induced degradation of solar cells and panels. Conference Record of the IEEE Photovoltaic Specialists Conference. Available from https://doi.org/10.1109/PVSC.2010.5616823. 01608371, 2817−2822.

Renewable Energy Market Update: Outlook for 2022 and 2023. (2022) OECD, Available from https://doi.org/10.1787/faf30e5a-en.

Saleheen, M. Z., Salema, A. A., Mominul Islam, S. M., Sarimuthu, C. R., & Hasan, M. Z. (2021). A target-oriented performance assessment and model development of a grid-connected solar PV (GCPV) system for a commercial building in Malaysia. *Renewable Energy, 171*, 371−382. Available from https://doi.org/10.1016/j.renene.2021.02.108, http://www.journals.elsevier.com/renewable-and-sustainable-energy-reviews/.

Schlothauer, J., Jungwirth, S., Köhl, M., & Röder, B. (2012). Degradation of the encapsulant polymer in outdoor weathered photovoltaic modules: Spatially resolved inspection of EVA ageing by fluorescence and correlation to electroluminescence. *Solar Energy Materials and Solar Cells, 102*, 75−85. Available from https://doi.org/10.1016/j.solmat.2012.03.022.

Schwark, M., Berger, K., Ebner, R., Ujvari, G., Hirschl, C., Neumaier, L., Muhleisen, W. (2013). Austria Investigation of potential induced degradation (PID) of solar modules from different manufacturers. IECON Proceedings (Industrial Electronics Conference). Available from https://doi.org/10.1109/IECON.2013.6700486. 8090−8097.

Silvestre, S., Kichou, S., Guglielminotti, L., Nofuentes, G., & Alonso-Abella, M. (2016). Degradation analysis of thin film photovoltaic modules under outdoor long term exposure in Spanish continental climate conditions. *Solar Energy, 139*, 599−607. Available from https://doi.org/10.1016/j.solener.2016.10.030, http://www.elsevier.com/inca/publications/store/3/2/9/index.htt.

Singh, R., Sharma, M., Rawat, R., & Banerjee, C. (2020). Field analysis of three different silicon-based technologies in composite climate condition—Part II—Seasonal assessment and performance degradation rates using statistical tools. *Renewable Energy, 147*, 2102−2117. Available from https://doi.org/10.1016/j.renene.2019.10.015, http://www.journals.elsevier.com/renewable-and-sustainable-energy-reviews/.

Solís-Alemán, E. M., de la Casa, J., Romero-Fiances, I., Silva, J. P., & Nofuentes, G. (2019). A study on the degradation rates and the linearity of the performance decline of various thin film PV technologies. *Solar Energy, 188*, 813−824. Available from https://doi.org/10.1016/j.solener.2019.06.067, http://www.elsevier.com/inca/publications/store/3/2/9/index.htt.

Tahri, A., Silvestre, S., Tahri, F., Benlebna, S., & Chouder, A. (2017). Analysis of thin film photovoltaic modules under outdoor long term exposure in semi-arid climate conditions. *Solar Energy*, *157*, 587−595. Available from https://doi.org/10.1016/j.solener.2017.08.048, http://www.elsevier.com/inca/publications/store/3/2/9/index.htt.

Theristis, M., Stein, J. S., Deline, C., Jordan, D., Robinson, C., Sekulic, W., Anderberg, A., Colvin, D. J., Walters, J., Seigneur, H., & King, B. H. (2023). Onymous early-life performance degradation analysis of recent photovoltaic module technologies. *Progress in Photovoltaics: Research and Applications*, *31*(2), 149−160. Available from https://doi.org/10.1002/pip.3615, http://onlinelibrary.wiley.com/journal/10.1002/(ISSN)1099-159X.

Tripathi, B., Yadav, P., Rathod, S., & Kumar, M. (2014). Performance analysis and comparison of two silicon material based photovoltaic technologies under actual climatic conditions in Western India. *Energy Conversion and Management*, *80*, 97−102. Available from https://doi.org/10.1016/j.enconman.2014.01.013, https://www.journals.elsevier.com/energy-conversion-and-management.

Wohlgemuth, J. H. (2020). *Photovoltaic module reliability*. Wiley. Available from http://doi.org/10.1002/9781119459019.

Wohlgemuth, J. H., Hacke, P., Bosco, N., Miller, D. C., Kempe, M. D., & Kurtz, S. R. (2017). *Assessing the causes of encapsulant delamination in PV modules. IEEE 44th photovoltaic specialist conference, PVSC 2017* (pp. 301−304). United States: Institute of Electrical and Electronics Engineers Inc. Available from https://doi.org/10.1109/PVSC.2017.8366601, http://ieeexplore.ieee.org/xpl/mostRecentIssue.jsp?punumber = 8360188, 9781509056057.

Electrical models of photovoltaic modules

Hadeed Ahmed Sher[1] and Ali Faisal Murtaza[2]
[1]Faculty of Electrical Engineering, Ghulam Ishaq Khan Institute of Engineering Sciences and Technology, Topi, Pakistan, [2]Director Research, University of Central Punjab, Lahore, Pakistan

Nomenclature

α	ideality factor of diode
AM	air mass ratio
C_{bd}	breakdown capacitance
C_d	diffusion capacitance
C_j	junction capacitance
C_p	parallel capacitance
G	irradiance in W/m^2
G_{STC}	irradiance at STC, i.e., 1000 W/m^2
I_{actual}	current (A) read from the curve
I_{Cp}	current through the capacitor
I_D	current through diode
I_{Df}	current through forward diode
I_{Dr}	current through reverse diode
$I_{estimated}$	current (A) estimated through simulation
I_{mp}	current at maximum power
I_0	saturation current of diode
I_{ph}	photocurrent
$I_{ph\text{-}STC}$	photocurrent
I_{pv}	output current of solar cell/module
I_{sc}	short circuit current at STC
I_{sh}	current through the shunt resistance
J_{sc}	current density
k	Boltzmann constant (1.38065×10^{-23})
K_i	temperature coefficient of current (A/°C)
K_r	reverse breakdown scalar coefficient
K_v	temperature coefficient of voltage (V/°C)
L_s	series inductance
N_s	number of series connected cells in one PV module
P_{mp}	maximum power
q	charge on electron (1.6021×10^{-19})
R_D	forward resistance of the diode
R_s	series resistance

Performance Enhancement and Control of Photovoltaic Systems. DOI: https://doi.org/10.1016/B978-0-443-13392-3.00006-2

R_{sh}	shunt resistance
STC	standard testing condition 1000 W/m^2, 25°C, 1.5 AM
T	temperature in °C
T_{STC}	temperature at STC, i.e., 25°C or 298.15K
V_D	forward voltage drop of diode
V_t	thermal voltage
$V_{oc\text{-}STC}$	open circuit voltage at STC
V_{pv}	output voltage of solar cell/module
V_{mp}	voltage at maximum power
V_{oc}	open-circuit voltage at STC
V_{th}	Thevenin voltage

6.1 Introduction

As the world progresses toward globalization, the challenge to achieve sustainable and global economic security has become more and more challenging. The pollution index of the entire globe is rising by consuming resources like petroleum, coal, nuclear fuel, etc. (Eugene & Richard, 2014). These sources are depleting, which calls for an opportunity to discover new energy sources. The developed and developing countries have installed renewable energy power plants mostly for harnessing wind and solar energy (Sher et al., 2015). Among all the renewable energy technologies, solar PV is most feasible for rooftop installations and for electrifying remotely located villages and installations. Since sunlight is intermittent and the solar PV module is a current source, it is imperative to study the dynamics of the PV power system. This is usually done through computer-aided simulations (Sher et al., 2015). The design of a solar PV system requires the modeling of a practical solar PV module in terms of its equivalent electrical circuits (Nadeem et al., 2021). This procedure is known as the electrical modeling of the solar modules. Precise electrical modeling of solar PV modules accurately forecasts the system performance under uniform, partial shading conditions and under aging. Several researchers have developed models of PV modules to address related issues like accuracy, unknown parameters, computation burden, etc., to develop an insight into the actual characteristics of solar modules (Ahmad et al., 2017). In literature, there is no such study that encompasses the work done on the modeling of solar PV modules. In this chapter, the existing PV models are classified under two main categories, i.e., static and dynamic PV models. These are further divided into various types as shown in Fig. 6.1. Several authors have introduced variations in these models, which are also explored in this chapter.

6.1.1 Solar cell operation

A solar cell is a simple semiconductor device that utilizes sunlight to generate electric current using electron-hole flow. Essentially, it operates like a semiconducting diode, carefully designed to separate and collect the carriers (electrons and holes)

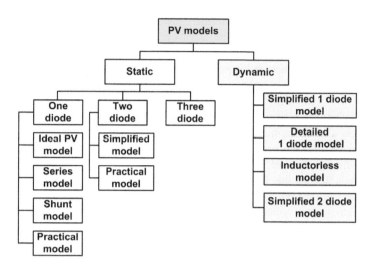

Figure 6.1 Classification. Classification of practical PV device models.

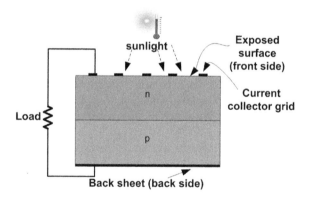

Figure 6.2 PV cell cross-sectional view. Structure of a solar cell.

and route them to the output terminals as shown in Fig. 6.2 (Jeffery & Luque, 2011).

The crystalline silicon (c-Si) solar cell under the standard testing conditions produces a voltage between 0.5 and 0.7 V and current according to the current density J_{sc}, which is between 25 and 35 mA/cm². This means that for a cell size of 12.5 cm² the short circuit current I_{sc} would be between 4 and 5.5 A. Note that while the voltage is not dependent on the surface area, the output current is proportional to the surface area of the PV cell. However, almost all kinds of domestic and commercial applications require a power rating much higher than a single solar cell. To achieve a reasonable output power, individual cells are connected in series and parallel configurations. The output power of a solar cell is also dependent on the bandgap of the semiconductor material. Nowadays, solar cells are manufactured using various materials. Most of the advanced materials with high efficiency are still under experimentation at a laboratory scale. The commercially available PV modules are either monocrystalline or polycrystalline forms of silicon cells because they require a relatively simpler manufacturing process for large-scale production. A thin layer of bulk Si or a thin Si film linked to electric terminals makes up PV cells. Thereafter, the process of doping is performed to create a p−n junction. The p−n junction can be homogeneous (made of the same semiconducting material), heterogeneous (different materials), or Ms junction (made of metal and semiconductor) forming a Schottky's barrier (Bashahu & Nkundabakura, 2007). A thin metal mesh is set on the front side, which is exposed to sunlight, to collect the charges. The sunlight falling on the front side surface is absorbed; the subsequently produced charges are collected through the metallic grid that is connected to the output terminals. Since solar radiation consists of photons with nonuniform energy levels, only those photons generating energy greater than the bandgap are used; the rest is wasted as heat. Semiconductors with a lower bandgap have a wide spectrum, but they produce lower output voltages. This process is electrically modeled in various forms, which are reviewed in this chapter (Villalva et al., 2009).

6.1.2   Datasheet of a solar module

The performance and characteristics of a solar cell and hence a module are largely based on the parameters provided by the manufacturer in the form of a datasheet usually affixed on the backside of a module. These parameters are short-circuit current (I_{sc}), open-circuit voltage (V_{oc}), maximum voltage (V_{mp}), and maximum current (I_{mp}) corresponding to the maximum power (P_{mp}), the temperature coefficient of voltage (K_v), and current (K_i), all measured in a controlled environment with irradiance set at 1000 W/m^2, temperature at 25°C, and air mass ratio (AM) = 1.5. These environmental conditions are referred to as standard operating conditions (STCs). In general, a PV module is characterized at STCs, although the datasheets of recently manufactured solar panels contain performance curves at nominal operating cell temperature (NOCT) as well. To understand these parameters, consider the characteristics of an arbitrary PV module shown in Fig. 6.3 (Ahmad et al., 2019).

Fig. 6.3 has several key points that are of interest.

- Short-circuit current I_{sc} is the maximum current that a solar module can offer and is obtained by shorting the solar cell output terminals.
- Open-circuit voltage V_{oc} is the maximum voltage generated by the cell and is obtained when the terminals are disconnected.
- Knee point holds the maximum available power P_{mp}. Corresponding to the knee point, the values of voltage and current are termed V_{mp} and I_{sc}, respectively.

With this information, one crude way of quantifying the quality of a PV module is to calculate the fill factor (FF). Fill factor is the ratio of maximum power generated by the solar cell to the theoretical maximum power, which is the product of

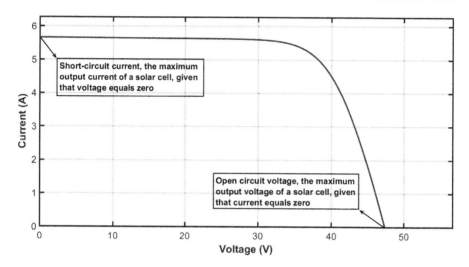

Figure 6.3 I−V characteristics. Current versus voltage characteristics of a solar module.

Table 6.1 Module datasheet.

Parameters	Description	Values
P_{mp}	Maximum power	245.328 W
V_{oc}	Open circuit voltage	38.1 V
I_{sc}	Short circuit current	8.59 A
N_s	Number of series connected cells	60
V_{mp}	Voltage at MPP	30.4 V
I_{mp}	Current at MPP	8.07 A
K_v	Temperature coefficient of V_{oc}	−0.0032 V/°C
K_i	Temperature coefficient of I_{sc}	0.00044 A/°C

Commercially available datasheet of PV Module Yingli YL245C-30B.

V_{oc} and I_{sc} (Nadeem et al., 2020). Mathematically, it can be expressed as follows:

$$FF = \frac{I_{mp}V_{mp}}{I_{sc}V_{oc}} \qquad (6.1)$$

With this information in view, a typical datasheet of a commercially available PV module is shown in Table 6.1.

6.2 Classification of electrical models of PV cell/module

An electrical model of a solar is crucial to simulate, design, evaluate, control, and optimize a PV system. It is also essential in calculating the efficiency, maximum power point tracking (MPPT), life expectation, and aging of a PV module (Caracciolo et al., 2012; Jordehi, 2016). The electrical modeling also makes it easier to use simulation software such as MATLAB to not only study the output behavior but also to estimate the design parameters essential for understanding a practical PV device. The electrical models of solar modules can be broadly classified into two types as follows:

1. Static models
2. Dynamic models

Each of these two classes is further divided into various subcategories. Strictly speaking, the main difference between static and dynamic modeling is the inclusion of nonlinear dynamic behavior. This is accomplished by the inclusion of dynamic electrical elements like capacitors and inductors.

6.3 Static models of PV module

Static models are general-purpose models that are widely adopted to assess the characteristics of a PV module under different environmental conditions. Several models have been presented in the literature to characterize the static features of solar modules. Among them, the following models are reviewed in this chapter.

- Ideal PV model
- Single-diode series resistance model (series model)
- Single-diode shunt resistance model (shunt model)
- Single-diode model with series and shunt resistance (practical 1D model)
- Two-diode model
- Simplified two-diode model
- Three-diode model

6.3.1 Ideal 1-D PV model

As stated in Section 6.1, a solar cell is a semiconductor device and its characteristics resemble those of a diode. An ideal PV model thus consists of a diode and a current source as shown in Fig. 6.4 (Masters, 2013). In the absence of any solar irradiation, the solar cell acts like a P-N junction diode expressed by Shockley equation as

$$I_D = I_0 \left[\exp\left(\frac{V_{pv}}{\alpha V_t}\right) - 1 \right] \tag{6.2}$$

where I_0 is the reverse saturation current of the diode, V_{pv} is the output terminal voltage, V_t is the thermal voltage as shown in the equation below, q is the charge on the electron, α is the ideality factor of the diode, K is the Boltzmann constant,

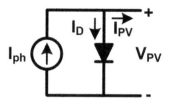

Figure 6.4 Ideal model of PV cell. Ideal model of a PV cell/module.

and T is the cell temperature in kelvin

$$V_t = \frac{KT}{q} \tag{6.3}$$

Referring to Fig. 6.4, the generated photocurrent I_{ph} is modeled as an ideal current source (Adeel et al., 2021; Chin et al., 2015). The I_{ph} depends on environmental conditions such as temperature and solar irradiation and is given as follows:

$$I_{ph} = \left(I_{ph_{STC}} + K_i(T - T_{STC})\right) \frac{G}{G_{STC}} \tag{6.4}$$

where $I_{ph_{STC}}$, T_{STC}, and G_{STC} represent photocurrent, temperature, and solar irradiance at standard testing conditions (1000 W/m^2 25°C). K_i is the temperature coefficient of current. An ideal model is lossless, and hence it does not have any resistance. The diode current (I_D), during a short circuit, is fairly negligible compared to the photocurrent generated; the short circuit current I_{sc} is usually taken approximately equal to the photocurrent. The output current I_{pv}, using Kirchhoff's current law, is given as follows:

$$I_{pv} = I_{ph} - I_0 \left[\exp\left(\frac{V_{pv}}{\alpha V_t}\right) - 1\right] \quad I_D \tag{6.5}$$

Although the datasheet of a PV panel contains several important features, the change in temperature can alter these values. To obtain these values as a function of temperature and irradiance, the following equations can be used:

$$V_{oc} = V_{oc_{STC}} + K_v(T - T_{STC}) + V_t \ln\left(\frac{G}{G_{STC}}\right) \tag{6.6}$$

$$V_{mp} = V_{mp_{STC}} + K_v(T - T_{STC}) + V_t \ln\left(\frac{G}{G_{STC}}\right) \tag{6.7}$$

$$V_{mp} = I_{mp_{STC}} + K_v(T - T_{STC}) + V_t \ln\left(\frac{G}{G_{STC}}\right) \tag{6.8}$$

Here, in Eqs. (6.6−6.8), K_v stands for the temperature coefficient of voltage, T is the temperature, G is the irradiance, and $V_{oc_{STC}}$ is the open circuit voltage at standard testing conditions (STC), which corresponds to $G_{STC} = 1000$ W/m^2 and $T_{STC} = 25°$C.

To implement this model in any computer-aided software, three design parameters, namely I_{ph}, I_0, and α, are required. Note that for all practical and simulation purposes, other models are preferred because of the apparent simplicity and lack of detail in this model. In addition, for series-connected PV cells, this model is not able to behave like a real model under partial shading. Nevertheless, an ideal model is used to theoretically describe the operation of a solar cell.

6.3.2 1-D model with series resistance (series model)

A more realistic and practical solar cell model should contain the losses in a module. A resistance between the electrodes and silicon surface represents losses, which are modeled using a resistor connected in series with the output terminals of an ideal PV model as shown in Fig. 6.5 (Gray, 2003). The calculation of photocurrent is based on Eq. (6.4). The diode current is given as follows:

$$I_D = I_0 \left[\exp\left(\frac{V_{pv} + I_{pv}R_s}{\alpha V_t}\right) - 1 \right] \tag{6.9}$$

where V_{pv} is the output voltage and I_{pv} is the output current. The output current is given as:

$$I_{PV} = I_{ph} - I_0 \left[\exp\left(\frac{V_{pv} + I_{pv}R_s}{\alpha V_t}\right) - 1 \right] \tag{6.10}$$

Modeling a PV module using this model requires four design parameters, namely I_{ph}, I_0, α, and R_s. It describes the behavior of a practical solar cell with a limited factor of nonideality, and hence it is not very accurate. In addition, this model is also not suitable for studying the partial shading phenomenon.

6.3.3 1-D model with shunt resistance (shunt model)

The shunt model of a PV module is presented in Fig. 6.6. The shunt resistance represents the losses due to the electron-hole recombination before it is delivered to the output.

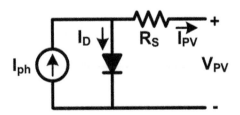

Figure 6.5 1-D series model. Series model of a PV module.

The expression for current and voltage is given as

$$I_{pv} = I_{ph} - I_0 \left[\exp\left(\frac{qV}{N_s KT\alpha} \right) - 1 \right] - \frac{V}{R_{sh}} \tag{6.11}$$

This model has been tested thoroughly in Mahmoud and Xiao (2018) and has been proven effective for large-scale PV array simulation.

6.3.4 Practical 1-D model

The circuit, as shown in Fig. 6.7, has series and parallel resistance in addition to an ideal PV model. Shunt resistance is used to model leakage current, while series resistance is used to model contact resistance (Villalva et al., 2009).

The output current is expressed as

$$I_{pv} = I_{ph} - I_{D1} - \left(\frac{V_{pv} + I_{pv}R_s}{R_{sh}} \right) \tag{6.12}$$

The current through the diode is the same as Eq. (6.9), is rewritten in terms of D_1 as I_{D1}, and is given as

$$I_{D1} = I_0 \left[\exp\left(\frac{V_{pv} + I_{pv}R_s}{V_t\alpha} \right) - 1 \right] \tag{6.13}$$

The thermal voltage V_t is the same as expressed in Eq. (6.3), and N_s is the number of cells. The photogenerated current I_{ph} is calculated using Eq. (6.4); $I_{ph\text{-}STC}$

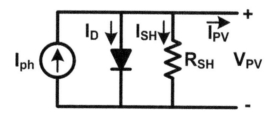

Figure 6.6 Shunt model. Shunt model of a PV module.

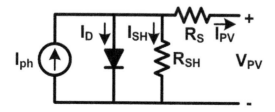

Figure 6.7 Practical 1-D model. Single-diode model with series and shunt resistance.

can be found as

$$I_{ph_{STC}} = I_{sc_{STC}} \left(\frac{R_s + R_{sh}}{R_{sh}} \right)$$ (6.14)

The leakage current of the diode can be calculated by

$$I_0 = \frac{I_{ph_{STC}} - \left(V_{oc_{STC}}/R_{sh} \right)}{\exp\left(\frac{V_{oc_{STC}}}{\alpha V_{t_{STC}}} \right) - 1}$$ (6.15)

The modeling of a PV module using a one-diode practical model requires five parameters that are not included in the datasheet. These parameters are the photo-generated current I_{ph}, diode leakage current I_0, diode ideality factor α, series resistance R_s, and shunt resistance R_{sh}.

6.3.5 Ideal 2-D model

A two-diode model shown in Fig. 6.8 was proposed by Babu and Gurjar (2014), and it does not include the series and shunt resistance. The additional diode in this model provides a more detailed performance by incorporating the recombination losses in the PN junction. A simplified two-diode model is seldom used in studies because the intention of creating a two-diode model is largely based on getting the finer details in the system optimization and performance. With the omission of losses and an additional diode, it does not add much to the detailing, but it adds to the unknown parameters. Moreover, like the 1-D ideal model, this model is also not able to work in partial shading conditions for a PV string.

6.3.6 Practical 2-D model

The practical two-diode model, shown in Fig. 6.9, includes the resistive losses that were explained in the single-diode PV model.

The output current is expressed using the following equation:

$$I_{pv} = I_{ph} - I_{D1} - I_{D2} - \left(\frac{V_{pv} + I_{pv}R_s}{R_{sh}} \right)$$ (6.16)

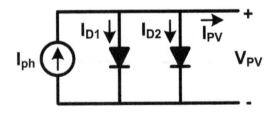

Figure 6.8 Ideal 2 D model. Ideal 2-D model of a PV module.

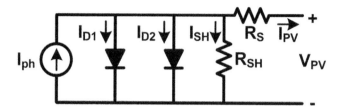

Figure 6.9 Practical 2 D model. Practical 2-D model of PV module.

where the photogenerated current I_{ph} is estimated by adopting the same method as discussed for the single-diode model. In addition, the current flowing through diodes D_1 and D_2 is based on Eq. (6.16) while having a modified nomenclature in terms of I_{D1} and I_{D2}. The resultant equations are as follows:

$$I_{D1} = I_{01}\left[\exp\left(\frac{V_{pv} + I_{pv}R_s}{V_t\alpha_1}\right) - 1\right] \tag{6.17}$$

$$I_{D2} = I_{02}\left[\exp\left(\frac{V_{pv} + I_{pv}R_s}{V_t\alpha_2}\right) - 1\right] \tag{6.18}$$

V_t is the same as in Eq. (6.3).
The reverse saturation currents can be calculated as

$$I_{01} = \frac{I_{ph_{STC}} - \left(V_{oc_{STC}}/R_{sh}\right)}{\exp\left(\frac{V_{oc_{STC}}}{\alpha_1 V_{t_{STC}}}\right) - 1} \tag{6.19}$$

$$I_{02} = \frac{I_{ph_{STC}} - \left(V_{oc_{STC}}/R_{sh}\right)}{\exp\left(\frac{V_{oc_{STC}}}{\alpha_2 V_{t_{STC}}}\right) - 1} \tag{6.20}$$

The total required number of parameters is seven, i.e., I_{ph}, I_{01}, I_{02}, $\alpha 1$, α_2, R_s, and R_{sh}. In Salam et al. (2010), a new computational approach for a 2D model is introduced. This model is structurally the same as Fig. 6.9; however, the model presented by Salam et al. (2010) is computationally easier with four unknown variables, and thus it converges fast compared to the conventional 2-D practical model.

6.3.7 Practical 3-D model

A three-diode model is shown in Fig. 6.10. The diodes contribute diode currents I_{D1}, I_{D2}, and I_{D3}, which are currents due to diffusion and recombination in the p−n junction region, space charge region, defected region, etc. The reasons for the inclusion of series resistance R_s and shunt resistance R_{sh} are the same as discussed

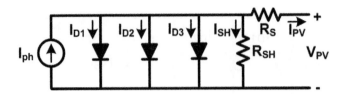

Figure 6.10 PV equivalent circuit for three-diode model. Practical 3-D model of a PV module.

in the practical 1-D model.

$$I_{pv} = I_{ph} - I_{D1} - I_{D2} - I_{D3} - \left(\frac{V_{pv} + I_{pv}R_s}{R_{sh}}\right) \tag{6.21}$$

where I_{ph} is the photocurrent, R_s is the series, R_{sh} is the shunt resistance, and V_{pv} is the output voltage. The current through the diodes is given as follows:

$$I_{D1} = I_{01}\left[\exp\left(\frac{V_{pv} + I_{pv}R_s}{V_t\alpha_1}\right) - 1\right] \tag{6.22}$$

$$I_{D2} = I_{02}\left[\exp\left(\frac{V_{pv} + I_{pv}R_s}{V_t\alpha_2}\right) - 1\right] \tag{6.23}$$

$$I_{D3} = I_{03}\left[\exp\left(\frac{V_{pv} + I_{pv}R_s}{V_t\alpha_3}\right) - 1\right] \tag{6.24}$$

I_0 is the reverse saturation current and V_t represents the thermal voltage of diode as in Eq. (6.3). The photocurrent I_{ph} comes from the expression given in Eq. (6.4). The reverse saturation current of any diode is given by

$$I_{01} = \frac{I_{ph_{STC}} - \left(V_{oc_{STC}}/R_{sh}\right)}{\exp\left(\frac{V_{oc_{STC}}}{\alpha_1 V_{tSTC}}\right) - 1} \tag{6.25}$$

$$I_{02} = \frac{I_{ph_{STC}} - \left(V_{oc_{STC}}/R_{sh}\right)}{\exp\left(\frac{V_{oc_{STC}}}{\alpha_2 V_{tSTC}}\right) - 1} \tag{6.26}$$

$$I_{03} = \frac{I_{ph_{STC}} - \left(V_{oc_{STC}}/R_{sh}\right)}{\exp\left(\frac{V_{oc_{STC}}}{\alpha_3 V_{tSTC}}\right) - 1} \tag{6.27}$$

where K_v is the voltage temperature coefficient (V/°C). There are nine unknown parameters: I_{ph}, I_{01}, I_{02}, I_{03}, α_1, α_2, α_3, R_s, and R_{sh}.

Table 6.2 Summary of static models of PV modules.

Model	Unknown parameters	Pros and cons
Ideal 1-D model	Photogenerated current I_{ph} and reverse saturation current I_0 and α	Easy to simulate. The great disparity between the results and the practical system. Cannot simulate the partial shading conditions with series-connected PV cells
1-D series model	Photogenerated current I_{ph}, series resistance R_s, and reverse saturation current I_0 and α	Easy to simulate with acceptable results for large PV arrays
1-D shunt model	Photogenerated current I_{ph}, shunt resistance R_{sh}, and reverse saturation current I_0 and α	Results are better for large-scale PV plants
Practical 1-D model	Photogenerated current I_{ph}, series resistance R_s, shunt resistance R_{sh}, and reverse saturation current I_0 and α	Good compromise between accuracy and performance
Ideal 2-D model	Photogenerated current I_{ph}, reverse saturation current I_{01} and I_{02}, and the current I_{D1}, I_{D2}, and α for D_1 and D_2	Simulation complexity is more than the 1-D ideal model. It is not able to simulate partial shading in a PV string.
Practical 2-D model	Photogenerated current I_{ph}, series resistance R_s, shunt resistance R_{sh}, reverse saturation current I_{01} and I_{02}, and the current I_{D1}, I_{D2}, and α for D_1 and D_2	Better accuracy at low irradiance levels. Accurate but slow in simulation
Practical 3-D model	Photogenerated current I_{ph}, series resistance R_s, shunt resistance R_{sh}, reverse saturation current I_{01}, I_{02}, and I_{03}, and the current I_{D1}, I_{D2}, and α for D_1 and D_2	Accurate but slow in simulation. Mostly used for lab-scale analysis of PV cells/modules.

6.3.8 Summary of PV static models

The static models reviewed are summarized in Table 6.2

6.4 Dynamic models of a PV module

Dynamic modeling of solar cells involves circuit elements to model the nonlinear dynamic behavior of a PV module. Typically, this requires the inclusion of junction capacitance at appropriate points of a PV model. Dynamic modeling is also adopted to model the impact of defects like hot spots in a solar module. Researchers have developed various kinds of dynamic models that, to the best of authors' knowledge, are limited to single- and two-diode models. The models that are explained in this chapter are as follows:

- Simplified 1-D dynamic model
- Detailed 1-D dynamic model
- Simplified 2-D dynamic model

6.4.1 Simplified 1-D model

The most simplified approach is to use a 1-D model for the inclusion of dynamic parameters as shown in Fig. 6.11. An energy storage element is used in addition to a 1-D practical model. The authors in Suskis and Galkin (2013) have used a capacitor. The presented model is simplified in the sense that by using only the 1-D model the reverse bias characteristics of a PV module are not included. However, the inclusion of junction capacitance C_j (in black box) provides an opportunity to model the dynamics associated with the PV module.

6.4.2 Detailed 1-D model

This electrical model, shown in Fig. 6.12, incorporates the dynamics of a PV module in the forward as well as in the reverse bias regions (Kim et al., 2013). To create distinction, the diode used in the 1-D model is split into forward-biased diode D_f and reverse-biased diode D_r. Furthermore, this model incorporates reverse breakdown, series inductance, and a lumped capacitance C_p. The capacitance C_p is included to model three capacitances, i.e., junction capacitance C_j, diffusion C_d, and breakdown capacitance C_{bd} (Mai et al., 2017). Moreover, the lumped

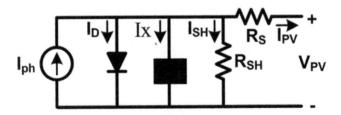

Figure 6.11 Simplified 1D dynamic model.

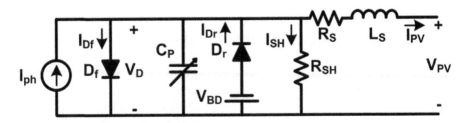

Figure 6.12 Detailed 1-D model. Detailed 1-D model.

capacitance C_p is not a fixed capacitance; rather a variable capacitance is used to showcase the impact of the environmental conditions on C_j, C_d, and C_{bd}. The governing equations for this model are as follows:

$$I_{ph}(G, T) = \left[I_{sc} \left(\frac{R_s + R_{sh}}{R_{sh}} \right) + K_i (T - T_{STC}) \right] \frac{G}{G_{STC}} \quad (6.28)$$

The diode D_f models the forward-biased operation of a PV module. The current through it is based on the Shockley diode equation presented in Eq. (6.2), while the thermal voltage is given in Eq. (6.3). The reverse saturation current I_0 is given as Eq. (6.29). The other diode in this model is D_r that represents the reverse bias characteristics of a PV module. It is mathematically expressed as Eq. (6.30).

$$I_0 = \frac{I_{sc} + K_i (T - T_{STC})}{\exp \left(\frac{V_{oc} + K_V (T - T_{STC})}{\alpha V_t} \right) - 1} \quad (6.29)$$

$$I_{Dr} = I_0 \exp \left(\frac{K_r V_{bd}}{\alpha V_t} \right) \left[\exp \frac{-K_r V_D}{\alpha V_t} - 1 \right] \quad (6.30)$$

The capacitance C_p models the nonlinear values of junction, diffusion, and breakdown capacitance as expressed in Eq. (6.31).

$$C_p = C_J + C_d + C_{bd} \quad (6.31)$$

The output voltage and current can be calculated using Kirchoff's law and by using the expressions for capacitor current and voltage.

6.4.3 Simplified 2-D Dynamic model of a PV module

A simplified 2-D dynamic model of a PV module may consist of a complete 2-D static model in addition to the capacitance as presented in (Grgic et al., 2018). The model presented is also shown in Fig. 6.13

This model incorporates the junction and diffusion capacitances of both the diodes as a separate element. This results in four capacitors connected in parallel

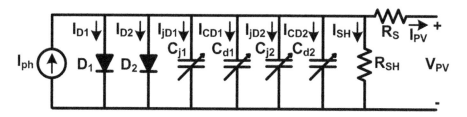

Figure 6.13 PV module 2-D dynamic. Simplified 2-D dynamic model of a PV module.

with the two diodes and the photocurrent source. This model is developed to simulate the impact of environmental conditions on the dynamic behavior of the diode. The output equation using this model is as follows:

$$I_{pv} = I_{ph} - I_{01}\left(e^{\frac{V_d}{\alpha_1 V_t}-1}\right) - I_{02}\left(e^{\frac{V_d}{\alpha_1 V_t}-1}\right) - \frac{V_d}{R_{sh}} - \frac{d}{dt}\left(q_{j1} + q_{d1} + q_{j2} + q_{d2}\right)$$

$$(6.32)$$

In Eq. (6.32), the junction and diffusion capacitance currents are obtained using Eqs. (6.33) and (6.34), respectively

$$i_{cj} = \frac{dq_j}{dt} = \frac{dV_d C_j}{dt} = \left(C_j + V_d\frac{dC_j}{dV_d}\right)\frac{dV_d}{dt} = C_{jeff}\frac{dV_d}{dt} \qquad (6.33)$$

$$i_{cd} = \frac{dq_d}{dt} = \frac{dV_d C_d}{dt} = \left(C_d + V_d\frac{dC_d}{dV_d}\right)\frac{dV_d}{dt} = C_{deff}\frac{dV_d}{dt} \qquad (6.34)$$

where C_{jeff} represents the effective junction capacitance and C_{deff} is the effective diffusion capacitance.

6.5   Discussion and conclusion

Various kinds of solar models have been developed and proposed by the researchers. A common development mechanism that can be adopted in almost all kinds of PV models is to evaluate the impact of a PV model on different scales of PV plants. Another approach is to model the impact of surface irregularities on output power. It is pertinent to mention that the PV models generate equations that are not trivial to solve, and hence numerical methods and computer-aided techniques are required to solve a given set of equations. This holds true for an ideal PV model as well. It is also noted that the use of the ideal PV model is not suitable to study the series connection of two or more PV modules. The unknown parameters of the PV models can be computed using metaheuristic and stochastic algorithms as explained in the next chapter.

References

Adeel, M., Hassan, A. K., Sher, H. A., & Murtaza, A. F. (2021). A grade point average assessment of analytical and numerical methods for parameter extraction of a practical PV device. *Renewable and Sustainable Energy Reviews*, *142*110826. Available from https://doi.org/10.1016/j.rser.2021.110826, https://www.sciencedirect.com/science/article/pii/S1364032121001210.

Ahmad, R., Murtaza, A. F., & Ahmed Sher, H. (2019). Power tracking techniques for efficient operation of photovoltaic array in solar applications—A review. *Renewable and Sustainable Energy Reviews, 101*, 82−102. Available from https://doi.org/10.1016/j.rser.2018.10.015, https://www.sciencedirect.com/science/article/pii/S1364032118307196.

Ahmad, R., Murtaza, A. F., Ahmed Sher, H., Tabrez Shami, U., & Olalekan, S. (2017). An analytical approach to study partial shading effects on PV array supported by literature. *Renewable and Sustainable Energy Reviews, 74*, 721−732. Available from https://doi.org/10.1016/j.rser.2017.02.078, https://www.sciencedirect.com/science/article/pii/S1364032117303088.

Babu, B. C., & Gurjar, S. (2014). A novel simplified two-diode model of photovoltaic (PV) module. *IEEE Journal of Photovoltaics, 4*(4), 1156−1161. Available from https://doi.org/10.1109/JPHOTOV.2014.2316371.

Bashahu, M., & Nkundabakura, P. (2007). Review and tests of methods for the determination of the solar cell junction ideality factors. *Solar Energy, 81*(7), 856−863. Available from https://doi.org/10.1016/j.solener.2006.11.002, https://www.sciencedirect.com/science/article/pii/S0038092X06002842.

Caracciolo, F., Dallago, E., Finarelli, D. G., Liberale, A., & Merhej, P. (2012). Single-variable optimization method for evaluating solar cell and solar module parameters. *IEEE Journal of Photovoltaics, 2*(2), 173−180. Available from https://doi.org/10.1109/JPHOTOV.2011.2182181.

Chin, V. J., Salam, Z., & Ishaque, K. (2015). Cell modelling and model parameters estimation techniques for photovoltaic simulator application: A review. *Applied Energy, 154*, 500−519. Available from https://doi.org/10.1016/j.apenergy.2015.05.035, https://www.sciencedirect.com/science/article/pii/S0306261915006455.

Eugene, D. C., & Richard, A. S. (2014). *Understanding the global energy crisis*. Purdue University Press.

Gray, J. L. (2003). The physics of the solar cell. *Handbook of photovoltaic science and engineering*, 61−112. Available from https://doi.org/10.1002/0470014008.ch3.

Grgic, I., Betti, T., Marasovic, I., Vukadinovic, D., Basic, M., Novel Dynamic model of a photovoltaic module. 3rd International Conference on Smart and Sustainable Technologies (SpliTech). (2018),

Jeffery, L. G., & Luque, H. (2011). The physics of the solar cell. *Handbook of photovoltaic science and engineering*.

Jordehi, A. R. (2016). Parameter estimation of solar photovoltaic (PV) cells: A review. *Renewable and Sustainable Energy Reviews, 61*, 354−371. Available from https://doi.org/10.1016/j.rser.2016.03.049, https://www.sciencedirect.com/science/article/pii/S1364032116300016.

Kim, K. A., Xu, C., Jin, L., & Krein, P. T. (2013). A dynamic photovoltaic model incorporating capacitive and reverse-bias characteristics. *IEEE Journal of Photovoltaics, 3*(4), 1334−1341. Available from https://doi.org/10.1109/JPHOTOV.2013.2276483.

Mahmoud, Y., & Xiao, W. (2018). Evaluation of shunt model for simulating photovoltaic modules. *IEEE Journal of Photovoltaics, 8*(6), 1818−1823. Available from https://doi.org/10.1109/JPHOTOV.2018.2869493.

Mai, X. H., Kwak, S.-K., Jung, J.-H., & Kim, K. A. (2017). Comprehensive electric-thermal photovoltaic modeling for power-hardware-in-the-loop simulation (PHILS) applications. *IEEE Transactions on Industrial Electronics, 64*(8), 6255−6264. Available from https://doi.org/10.1109/TIE.2017.2682039.

Masters, G. M. (2013). Renewable and efficient electric power systems. John Wiley & Sons.

Nadeem, A., Sher, H. A., & Murtaza, A. F. (2020). Online fractional open-circuit voltage maximum output power algorithm for photovoltaic modules. *IET Renewable Power Generation, 14*(2), 188−198. Available from https://doi.org/10.1049/iet-rpg.2019.0171.

Nadeem, A., Sher, H. A., Murtaza, A. F., & Ahmed, N. (2021). Online current-sensorless estimator for PV open circuit voltage and short circuit current. *Solar Energy, 213*, 198−210. Available from https://doi.org/10.1016/j.solener.2020.11.004, https://www.sciencedirect.com/science/article/pii/S0038092X20311452.

Salam, Z., Ishaque, K., Taheri, H., (2010) An improved two-diode photovoltaic (PV) model for PV system. 2010 Joint International Conference on Power Electronics, Drives and Energy Systems & 2010 Power India. https://doi.org/10.1109/PEDES.2010.5712374 1−5.

Sher, H.A., Addoweesh, K.E., Al-Haddad, K., (2015). Performance enhancement of a flyback photovoltaic inverter using hybrid maximum power point tracking. IECON 2015—41st Annual Conference of the IEEE Industrial Electronics Society. https://doi.org/10.1109/IECON.2015.7392947 005369−005373.

Sher, H. A., Murtaza, A. F., Addoweesh, K. E., & Chiaberge, M. (2015). Pakistan's progress in solar PV based energy generation. *Renewable and Sustainable Energy Reviews, 47*, 213−217. Available from https://doi.org/10.1016/j.rser.2015.03.017.

Suskis, P., & Galkin, I., (2013). Enhanced photovoltaic panel model for MATLAB-simulink environment considering solar cell junction capacitance. IECON 2013—39th Annual Conference of the IEEE Industrial Electronics Society. https://doi.org/10.1109/IECON.2013.6699374 1553-572X 1613−1618.

Villalva, M. G., Gazoli, J. R., & Filho, E. R. (2009). Comprehensive approach to modeling and simulation of photovoltaic arrays. *IEEE Transactions on Power Electronics, 24*(5), 1198−1208. Available from https://doi.org/10.1109/TPEL.2009.2013862.

A comparative study of metaheuristic optimization algorithms to estimate PV cell equivalent circuit parameters

7

Fawzi Mohammed Munir Al-Naima[1,2] and
Hussam Khalil Ibrahim Rushdi[1]
[1]Department of Computer Engineering, Al-Nahrain University, Baghdad, Iraq,
[2]Department of Electrical Power Engineering, Al-Kut University College, Wasit, Iraq

7.1   Introduction

The efficiency of solar panels has been greatly improved recently, which has led to a growth in the use of solar photovoltaic (SPV) energy to generate electricity. Nowadays, especially in rural regions, SPV electricity production is considered an important alternative to traditional sources of power supplies. Solar energy's direct and indirect benefits provide the diversification of energy sources, the fast electrification of rural areas in developing nations, the minimum greenhouse gas emissions, the reclaiming of degraded land, and the growth in regional and global alternative energy (Oghogho et al., 2014; Shaikh, 2017). Accurate parameter estimation and modeling of SPV current-voltage (I-V) and power-voltage (P-V) characteristics are necessary to improve the system's SPV performance and increase cost-effectiveness.

The fundamental goal of accurate parameter estimation and modeling of I-V characteristics of SPV cells/modules is to enable the designer to improve the system's cost-effectiveness and performance. Even though the SPV modules are nonlinear and rather complicated, they have been very common electricity-generating components because they are simple to set up and use in recent years. The SPV module must be correctly modeled to obtain the I-V characteristic, which specifies how the electrical equivalent circuit of the module behaves under various operating circumstances.

Estimating the SPV module's parameters is necessary for modeling and simulation purposes to determine the module behavior. The SPV module includes several parallels and series-connected SPV cells. SPV cell modeling can be done with a single diode, double diode, or triple diode. While the double-diode and triple-diode circuits offer a better representation of the loss inside the depletion region caused by carrier recombination, the single-diode model still offers balanced model accuracy and simplicity (Dizqah et al., 2014).

Furthermore, several parameter estimation strategies based on metaheuristic methods have been recently introduced in the literature. To name a few, we selected

Performance Enhancement and Control of Photovoltaic Systems. DOI: https://doi.org/10.1016/B978-0-443-13392-3.00007-4

in our study the following eleven such algorithms: Gray Wolf Optimizer (GWO) (Mirjalili et al., 2014), Ant Lion Optimizer (ALO) (Mirjalili, 2015b), Moth Flame Optimization (MFO) (Mirjalili, 2015a), Whale Optimization Algorithm (WOA) (Mirjalili & Lewis, 2016), Dragonfly Algorithm (DA) (Mirjalili, 2016), Hybrid Particle Swarm Optimization and Gray Wolf Optimizer Algorithm (HPSOGWO) (Singh & Singh, 2017), Harris Hawks Optimization (HHO) (Heidari et al., 2019), Marine Predators Algorithm (MPA) (Faramarzi et al., 2020), African Vultures Optimization Algorithm (AVOA) (Abdollahzadeh et al., 2021), COOT algorithm (Naruei & Keynia, 2021), and Ali Baba and the Forty Thieves algorithm (AFT) (Braik et al., 2022). It is now important to evaluate each technique in terms of its advantages, disadvantages, suitability for a given case involving parameter estimation, and level of complexity. Therefore, a thorough study of parameter estimate approaches is suggested in this chapter in considering the significance of these techniques in improving SPV cell performance. It is possible to sum up the main contributions made in this review chapter as follows:

This chapter explores optimization techniques in further detail, examining a group of eleven recently published metaheuristic algorithms and their modifications. The study combines the many objective functions with the root mean square error (RMSE) getting special attention. Additionally, the ideal algorithm for each case study is offered. Finally, a comparison has been made between these eleven algorithms. The acquired results demonstrate that the AFT method may deliver optimum values with minimized RMSE for the various SPV types under consideration. The study concludes that the AFT is a trustworthy optimization approach that can be suggested with assurance for parameter estimation of SPV circuit models.

7.2 Methodology

7.2.1 Problem formulation for the single-diode model

The single-diode model (SDM) equivalent circuit of the PV solar cell is shown in Fig. 7.1. It is possible to use the following mathematical formulae to determine the SDM's output current:

$$I_L = I_{ph} - I_d - I_{sh} \tag{7.1}$$

$$I_L = I_{ph} - I_{sd} \times \left[\exp\left(\frac{q(V_L + R_s I_L)}{(n \times k \times T)} \right) - 1 \right] - \frac{(V_L + R_s I_L)}{R_{sh}} \tag{7.2}$$

where I_L denotes the output current from the solar cell SDM, I_{ph} represents the photogenerated current, I_{sh} is the leakage current in the PN junction, I_{sd} indicates the diode saturation current of the SDM, R_{sh} is the shunt resistance, R_s is the series resistance, n is the diode ideality factor, q is the charge of the electron, K denotes the Boltzmann constant, and T is the cell temperature in degrees, Kelvin.

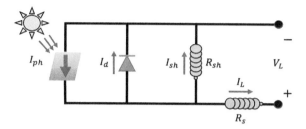

Figure 7.1 Equivalent circuit for SPV of the single-diode model (SDM).

The five unknown parameters needed to calculate the SDM using the previous mathematical formula are I_{ph}, I_{sd}, n, R_s, and R_{sh}.

7.2.2 Objective function

An essential objective function is the RMSE of the SPV characteristics between the estimated and the measured data. As a result, the decision variables of the vector (X) are estimated during each optimization iteration. The following is a formula that may be used to compute the RMSE:

$$RMSE(X) = \sqrt{\frac{1}{N}\sum_{i=1}^{N} f(V_L, L_L, X)^2} \qquad (7.3)$$

$$f(V_L, I_L, X) = I_{ph} - I_{sd}\left[\exp\left(\frac{q(V_L + R_s I_L)}{n \times k \times T}\right) - 1\right] - \frac{(V_L + R_s I_L)}{R_{sh}} - I_L \qquad (7.4)$$

where N represents the number of data, V_L is the measured voltage, I_L is the estimated current, and X denotes the decision vector of variables that are determined as follows: $X = (I_{ph}, I_{sd}, n, R_s, R_{sh})$.

7.3 Estimation of SPV cell circuit parameters

The estimation of the SPV cell parameters is a complex, multidimensional problem that can have several local optimum solutions and depends on the data obtainable, i.e., it can be resolved in either of two different ways: using the data given by the manufacturers or using experimental measurement data (I-V and P-V characteristic curves).

The most popular method of determining SPV characteristics is analytical if the manufacturer data are available. It determines the I-V characteristics of SPV cells or modules, considering essential points of the characteristic curve and straightforward mathematical expressions. These critical points are the short circuit point (SC), the maximum power point (MPP), and the open circuit point (OC).

On the other hand, the numerical method is the most popular when experimental data are provided. It often takes into consideration all of the characteristic curve's points, making it possible to determine SPV parameters that are more precise. Calculating the PV parameters can be performed using metaheuristic or deterministic techniques. Thus, analytical, metaheuristic, and deterministic approaches can be used to classify the estimation of the SPV parameters.

7.3.1 Analytical methods

Analytical approaches are best described by their clarity, low operational costs, and ease of use (Xiong et al., 2018). They determine the SPV parameters by solving the mathematical formulas considering particular characteristic points of the I-V curves. However, the accuracy of the analytical approaches significantly depends on the chosen characteristic points and the requirement of making some approximations or simplifications (Nunes et al., 2018). Additionally, the validity of the acquired parameters is limited because this approach usually provides the manufacturer's data to be obtained under the STC, i.e., different types of data than the I-V curves require to be modeled and simulated. An iterative approach was suggested by Wang et al. (2017) to determine the SDM parameters directly from the manufacturer's datasheets without using implicit formulas.

7.3.2 Numerical methods

The deterministic and metaheuristic types of numerical approaches have already been described. Deterministic techniques are highly effective in a local search, but they also demand convexity, continuity, and differentiation (Merchaoui et al., 2018) and frequently quickly converge to local minima. The estimation of SPV module parameters is a nonlinear, multimodal problem, making its accuracy unpredictable. The initial position significantly impacts its performance; hence, if the initial position is carried out outside of the ideal solution, its performance will be lower (Chen, Wu, et al., 2016). Deterministic approaches include the Newton-Raphson method (NRM) (Easwarakhanthan et al., 1986), Levenberg-Marquardt algorithm (LM) (Tossa et al., 2014), pattern search (PS) (AlHajri et al., 2012), and Nelder-Mead simplex (NMS) (Nelder & Mead, 1965).

In recent years, metaheuristic techniques have been widely applied to handle various engineering problems, including estimating SPV parameters, to avoid the drawbacks of deterministic methods. Many metaheuristic algorithms are based on natural techniques where a population collaborates to develop useful solutions by minimizing the objective function. They are appropriate for global optimization issues (Nunes et al., 2019; Wu et al., 2018). The main benefits of these approaches are their conceptual clarity, their ability to handle multidimensional problems with more than one solution, and the fact that they do not impose limits on the problem formulation (Chen et al., 2018; Chen, Yu, et al., 2016). However, the computational cost of this kind of technique may be less effective depending on the mechanism built into each metaheuristic.

Eleven effective metaheuristic algorithms have been selected from the literature for specific optimization problems with varying control parameters. The metaheuristics are mainly effective in solving multidimensional issues and performing global searches. They might, however, prematurely converge to locally optimal solutions depending on a population and demand a huge amount of time due to the stochastic search technique (Oliva et al., 2017). Additionally, the system's effectiveness is highly dependent on the correct adjustment of the control parameters and the harmony of the intensification and diversification mechanisms. The diversification process involves developing new solutions in unexplored search space regions, while the intensification mechanism focuses on creating new solutions in already researched areas.

7.3.3 Hybrid methods

Recently, hybrid approaches combining different methodologies have been discussed. This can be done by combining analytical and numerical methods or various metaheuristic approaches. A common practice is to combine two or more metaheuristics with different search engines to create a hybrid method that balances diversification and enhancement mechanisms. On the contrary, other metaheuristics have a better diversification mechanism, favoring global demand. However, the practice of hybridization requires the adjustment of many control parameters, which has to be done experimentally or through trial and error, which can reduce the efficiency and accuracy of the algorithm (Guo et al., 2018). Another disadvantage of hybrid methods is their need for a higher computational cost, which is why, in the problem of estimating the PV parameters, some authors use analytical expressions to calculate some parameters, allowing to reduce the computational cost considerably but with the consequence of compromising the precision of the results. An example of such a hybrid method is the Hybrid Particle Swarm Optimization and Gray Wolf Optimizer Algorithm (HPSOGWO), which combines the Particle Swarm Optimization (PSO) (Kennedy & Eberhart, 1995) with the Gray Wolf Optimizer Algorithm (GWO).

7.3.4 The selected metaheuristic optimization algorithms

7.3.4.1 Ant Lion Optimizer (ALO)

The Ant Lion Optimizer is a swarm intelligence optimization method created by Mirjalili that mimics the natural foraging behavior of ant lions (Mirjalili, 2015b). Ant lions naturally travel along a path made of sand and build pits to enclose ants. Ant lions capture ants, construct their holes, and wait for other prey whenever ants randomly move inside them (another ant). Ant lions are ants that search for food and are highly adapted to upgrade and save the best solution. They are used as problem-solving solutions.

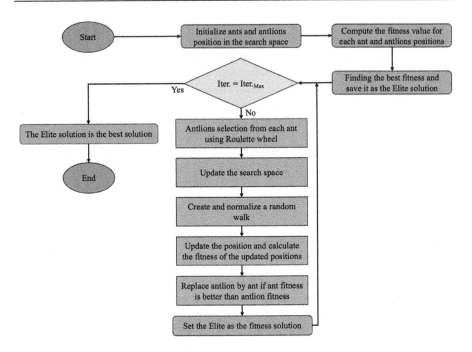

Figure 7.2 Ant Lion Optimizer algorithm.

The algorithm applies the following criteria when it is being optimized: (1) Ants use various random travels to move about the search area; these travels are appropriate for the ants throughout all dimensions and thus are affected by the location of traps. (2) The size of the gap in the trap that the ant lion builds depends on how to fit it; the more significant that gap, the greater the chances that ants will be captured. (3) Then the ants move toward ant lions, and random walking is dynamically reduced. (4) After hunting, the ant lion moves to a new site close to where the prey is captured and rebuilds its snare to accommodate the changes. Fig. 7.2 depicts the ALO algorithm's work process.

7.3.4.2 Moth Flame Optimization (MFO)

The Moth Flame Optimization (MFO) algorithm is an evolutionary-based algorithm created by Mirjalili (2015a). This method's concept is based on simulating how moths fly in spirals around the moon's light. The MFO method is a global optimization approach to calculate the fitness variance without requiring mathematical processes.

This method represents potential solutions with every moth specified by its location in the search area. The MFO algorithm's artificial lights indicate the locations visited most successfully during an iteration. Depending on where they are, moths fly toward the flames to obtain the objective function value. The flowchart of the MFO algorithm is shown in Fig. 7.3.

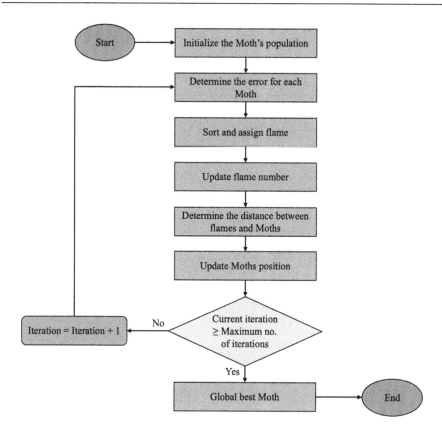

Figure 7.3 Moth Flame Optimization algorithm.

7.3.4.3 Whale Optimization Algorithm (WOA)

Mirjalili and Lewis presented this technique (Mirjalili & Lewis, 2016), which is a population-based optimization method that emulates the common behavior of humpback whales. In addition to being large and interesting in their skills to find food, humpback whales use a bubble-net foraging technique to hunt their prey, which includes krill and tiny fish. This hunting habit is the inspiration for WOA's three stages of operation. It begins by searching for the prey and surrounds the target before attacking it. The way humpback whales move around their prey is either one of two movements: shrinking movement or spiral movement. Fig. 7.4 shows the flowchart of the WOA algorithm.

7.3.4.4 Dragonfly Algorithm (DA)

The Dragonfly Algorithm (DA), a recently developed algorithm, is focused on the swarming attitudes and behaviors of dragonflies, which include both dynamic and static swarming during migration (Mirjalili, 2016). In the first step, dragonflies create tiny groups and hunt inside condensed regions, and in the second step, swarms

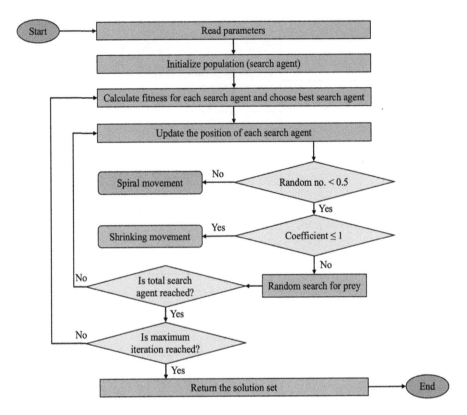

Figure 7.4 Whale Optimization Algorithm.

fly far distances. Exploration and exploitation are based on the principles of these features. The corrective behavior optimization of dragonflies inside a swarm involves dispersal to prevent collisions, alignment for speed matching, coherence for flying toward the neighborhood's center, attraction to food, and distractions from the predator. Fig. 7.5 shows a flowchart of the DA algorithm procedure.

7.3.4.5 Hybrid Particle Swarm Optimization and Gray Wolf Optimizer (PSOGWO)

7.3.4.5.1 Particle Swarm Optimization (PSO)

Developed by Kennedy and Eberhart in 1995, PSO is a stochastic optimization technique that takes inspiration from nature (Kennedy & Eberhart, 1995). Informed through the expected behavior of swarms and birds flying, it is a population-based, computationally efficient technique. According to the algorithm's approach, a swarm of particles explores the search area and updates both their optimal solutions and the best solution found by the swarms to find the most optimal solution.

The swarm is dynamically initiated for the N-dimensional search area as particles with location and velocity. The particle's velocity reflects the change rate of

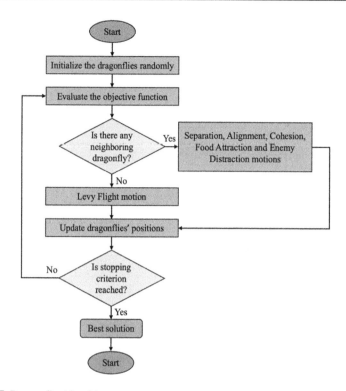

Figure 7.5 Dragonfly Algorithm.

the particle's location concerning its current location, and the particle's location shows the feasible solution. The particles adjust their positions concerning the location of the best particle. The flowchart of the PSO algorithm is shown in Fig. 7.6.

7.3.4.5.2   Gray Wolf Optimizer (GWO)

The active lifestyles of gray wolves inspired the Gray Wolf Optimization (GWO) method's basic principle as hunters and survivors (Mirjalili et al., 2014). The hierarchical social structure and the hunting technique were considered to create the mathematical model for the proposed method. The hunting order of gray wolves is accurately represented to determine the optimum explanation for an optimization process. Gray wolves are split into four categories, alpha, beta, delta, and omega, according to the trophic levels of the food chain. Gray wolves of the alpha kind occupy the highest spot in the food chain.

The alpha group's gray wolf gives the beta group instructions, and the latter group finishes the job. Located last and used as a target is the omega group GWO. It is also important to take into account the wolf's hunting behavior, which can be characterized as follows: (1) First, the gray wolf focuses on discovering its prey; (2) next, it explores the prey; and (3) finally, it attacks the prey. Fig. 7.7 depicts the GWO algorithm flowchart.

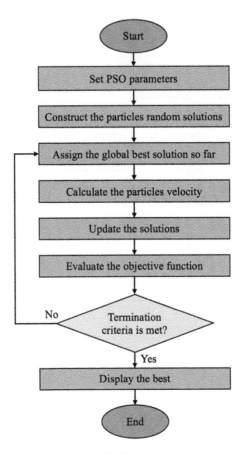

Figure 7.6 Particle Swarm Optimization algorithm.

7.3.4.5.3 Hybrid Particle Swarm Optimization and Gray Wolf Optimizer algorithm (HPSOGWO)

When dealing with nonlinear problems that are nonconvex or nonsmooth, many optimization techniques struggle with premature convergence or a slow convergence rate. While some methods more thoroughly explore the search area and suffer slow convergence, others do not successfully identify the best solution. Hence, different algorithms must be combined to create a hybrid optimization algorithm to balance exploitation and exploration correctly.

The HPSOGWO algorithm created by Şenel et al. (2019) has been used to work with different optimization issues. The GWO technique is used because the PSO frequently becomes stuck in a local solution. The Particle Swarm method distributes particles randomly, with little chance of preventing them from becoming stuck in a local solution. These random locations may lead the search to leave the overall best solution. So rather than sending the particles into random locations, the GWO might use its powerful exploration ability to move them to locations that the GWO

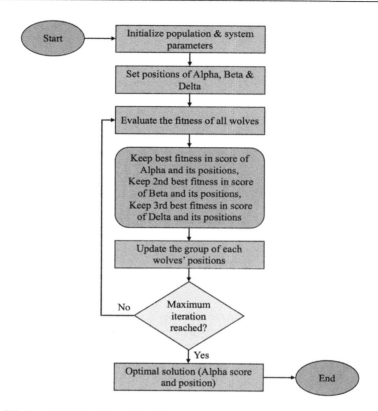

Figure 7.7 Gray Wolf Optimizer algorithm.

significantly enhances. As such, it would minimize the chance that the particles will become stuck in a local solution. The basic procedure of the HPSOGWO algorithm is shown in the flowchart of Fig. 7.8.

7.3.4.6 Harris Hawks Optimization (HHO)

The Harris Hawks Optimization (HHO) technique is a metaheuristic technique motivated by the chasing techniques and natural activity of Harris hawks (Heidari et al., 2019). By working together, numerous hawks hunt the rabbit from various angles, leading them to travel in the direction of the rabbit they found simultaneously. Depending on the continuously changing circumstances of the rabbits and their escape techniques, Harris hawks use various hunting strategies.

In the HHO algorithm, Harris hawks are considered the solution space, and the rabbit is considered the best suitable solution space (or optimal solution) at each step. In the beginning, hawks are spread randomly throughout the search area. The hawks then select their new location inside the region that the other crew members go to by using one of two ways to determine the rabbit's location. Simple rules, fewer parameters, and powerful local optimization abilities are all features of the HHO. The

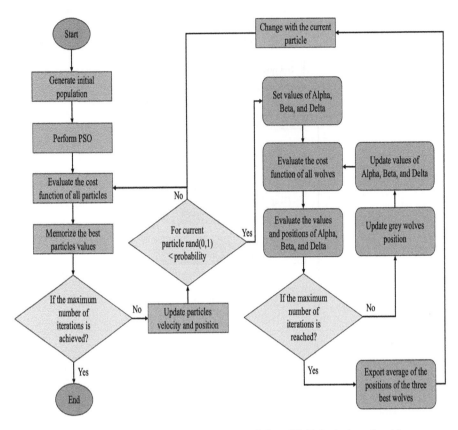

Figure 7.8 Hybrid Particle Swarm Optimization and Gray Wolf Optimizer algorithm.

population-based optimization process of HHO consists of three phases: exploration, transition, and exploitation. The flowchart in Fig. 7.9 depicts the HHO procedure.

7.3.4.7 Marine Predators Algorithm (MPA)

The marine predators algorithm (MPA) is an optimization technique inspired by nature. It refers to the principles that control the best hunting strategies and encounter probabilities between predators and prey in marine environments (Faramarzi et al., 2020). Marine predators, including sharks, fishes, and marlines, use the Lévy technique to search for food in prey-scarce areas. However, in prey-abundant regions, they switch to Brownian motion (Humphries et al., 2010). The optimal encounter rate strategy between predators and prey is affected by the type of movement and speed ratio (Bartumeus et al., 2002).

The following points describe the fundamental principles behind the interactions, memories, and hunting behavior of marine predators:

- For environments with low numbers of prey, marine predators apply the Lévy technique, while in environments with high numbers of prey, they apply Brownian motion.

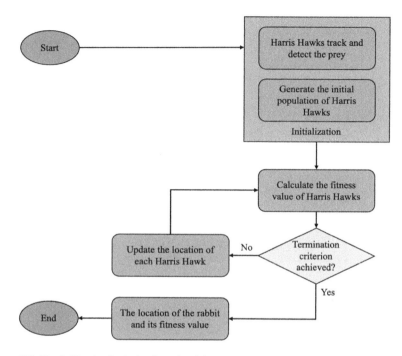

Figure 7.9 Harris Hawks Optimization algorithm.

- During their entire lifetime as they travel through various environments, they demonstrate the exact amounts of Lévy and Brownian motion.
- For the predator, Lévy is still the optimal technique when the speed ratio is low ($s = 0.1$); the prey moves into either Brownian motion or Lévy motion.
- The optimal technique for a predator pursuing prey that moves according to a Lévy distribution becomes Brownian when the predator's speed ratio is constant ($s = 1$), while other scenarios depend on the system's size.
- The best strategy for predators in high-speed scenarios is to remain still, while prey exhibits Lévy or Brownian motion.
- They use their superior memory to remember their friends and the best hunting spots.

Depending on these points of interest, the MPA algorithm can be organized into three phases that take into consideration various speed ratios while simulating the lifetime of the prey and predator: (1) in high-speed ratio or when prey is running faster than any predator; (2) throughout unit speed ratio or when both the prey and the predator are trying to move at nearly the same speed; and (3) in low-speed ratio, whenever the predator becomes moving much faster than the prey. The flowchart of the MPA algorithm is shown in Fig. 7.10.

7.3.4.8 African Vultures Optimization Algorithm (AVOA)

Abdollahzadeh et al. (2021) developed a unique population-based optimization method known as AVOA, which imitates the living and hunting habits of the

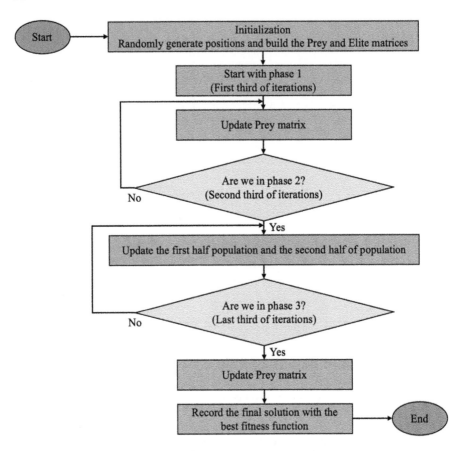

Figure 7.10 Marine Predators Algorithm.

African vulture. African vultures tend to attack small animals but could attack a sick or weak animal and eat a human corpse. One exciting aspect of these birds is their unique bald heads, which are essential for controlling body temperature and defending against bacteria and disease.

Vultures frequently use circular flight when flying in their natural environment as they keep moving across great distances in search of better food sources. Once the vultures find food, they engage in fights to determine the order of eating. Weaker vultures wait until the more powerful ones are filled before they eat. Fig. 7.11 shows the flowchart of the AVOA.

7.3.4.9 COOT algorithm

Previous studies have figured out different algorithms for finding the optimum values of unknown parameters (Darmansyah & Robandi, 2017). In this subsection, we have chosen the COOT optimization algorithm to determine the value of

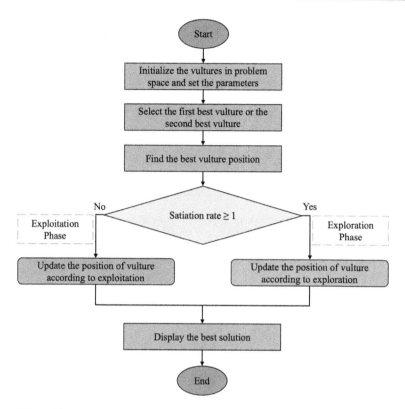

Figure 7.11 African Vultures Optimization Algorithm.

unknown parameters. This algorithm is based on the movement behaviors of groups of coots on the water surface. Coots are small water birds that exhibit various group behaviors on the water surface to reach their destination, mainly by moving toward food or a specific location. These birds create four movements on water: chain, leader, random, and positional adjustment to the leader (Naruei & Keynia, 2021).

Here is a simple explanation of how the algorithm operates:

- Initialization: The prepared population is created by generating a random population optimal solution inside the settlement area.

After the generation of the first overall, the objective issue is calculated by comparing the goal function for each coot-specified location. A random selection is made for the number of leaders (N_L) and the number of coots (N_{Coot}). The main goal of the algorithm is to identify the coot and leader who is generally optimal for each stage.

- Position update: The coot's position on the water is modified during the iterations based on the aforementioned four movements. The COOT algorithm can be represented as shown in Fig. 7.12.

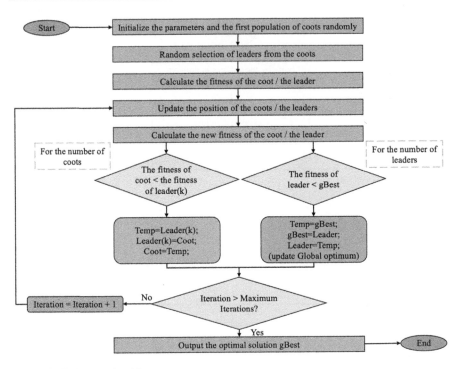

Figure 7.12 COOT algorithm.

7.3.4.10 Ali Baba and the Forty Thieves algorithm (AFT)

This method's main objective is to provide a different optimization method that uses the story of Ali Baba and the forty thieves as a unified view of human behavior (Braik et al., 2022). The following steps are constructed out of this story to satisfy the fundamental propositions of this method:

- The forty thieves work together as a team to locate Ali Baba's home, but the validity of the information they receive from someone, even one of them, is uncertain.
- The forty thieves will walk a certain distance from a starting point to locate Ali Baba's home.
- Marjaneh used various methods to keep Ali Baba safe from the thieves' arrival multiple times.

It is possible to optimize a fitness function by linking Marjaneh's decisions and the actions of the thieves. Fig. 7.13 depicts the flowchart of the AFT algorithm.

7.4   Simulation results and discussion

This chapter applies and evaluates several algorithms to determine optimal values for SPV model parameters. Three types of solar cells are considered: real data of a 55-mm-diameter commercial RTC France solar cell (1000 W/m², 33°C), real data of a Photowatt PWP-201 solar cell (1000 W/m², 45°C), and simulated data of an

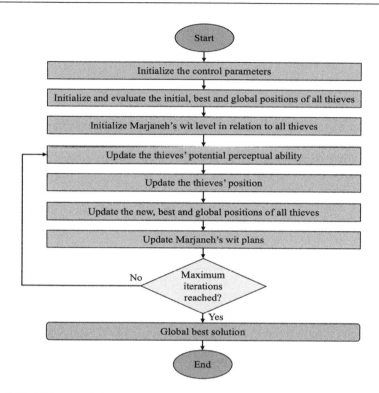

Figure 7.13 Ali Baba and the Forty Thieves algorithm.

Table 7.1 Input data for the SPV cell.

Simulated parameters	RTC France silicon		Photowatt PWP-201		HTS-144M6H480	
	Lower	**Upper**	**Lower**	**Upper**	**Lower**	**Upper**
I_{ph} (A)	0	1	0	2	0	20
I_{sd} (μA)	0	1	0	50	0	1
R_{sh} (Ω)	0	100	0	2000	0	1000
R_s (Ω)	0	0.5	0	2	0	2
n	1	2	1	50	0	2
Population	50		50		50	
Iterations	1000		1000		2000	

HTS-144M6H480 solar cell (1000 W/m², 25°C). Table 7.1 displays the parameter lower and upper boundaries for the SPV module that were taken into consideration.

The study compared eleven algorithms (GWO, ALO, MFO, WOA, DA, HPSOGWO, HHO, MPA, AVOA, COOT, and AFT) by running each algorithm 30 times on each SPV module. The accuracy, robustness, and convergence rate were evaluated by examining the top RMSE values and convergence curves.

7.4.1 Results of SDM for RTC France solar cell

Five parameters (I_{ph}, I_{sd}, R_{sh}, R_s, and n) must be determined for an SDM. The parameter values calculated by the algorithms and RMSE for comparison are shown in Table 7.2. This table shows that the AFT algorithm has the lowest RMSE of (0.0010) and the shortest execution time of (2.271) seconds among all tested algorithms. In this case, the RMSE and the time values are taken as a comparison index for different developed methods by the researchers.

Fig. 7.14 illustrates the algorithms' convergence speeds and demonstrates that the AFT can obtain minimal RMSE quicker than all other tested algorithms.

Table 7.2 Parameter estimation of the SDM for the RTC France solar cell.

Algorithm	I_{ph} (A)	I_{sd} (μA)	R_{sh} (Ω)	R_s (Ω)	n	RMSE	Time (s)
GWO	0.7698	0.7105	17.9126	0.0306	1.5677	0.0071	2.634
ALO	0.7603	0.7265	97.9895	0.0330	1.5675	0.0019	9.339
MFO	0.7608	0.9641	100.0000	0.0316	1.6002	0.0024	2.696
WOA	0.7597	0.4942	95.3344	0.0349	1.5250	0.0014	3.166
DA	0.7652	1.0000	76.5135	0.0315	1.6039	0.0039	28.523
HPSOGWO	0.7686	1.0000	6.6750	0.0000	1.6217	0.0382	3.206
HHO	0.7633	0.2758	31.6740	0.0364	1.4660	0.0020	6.209
MPA	0.7567	0.0592	90.9801	0.0434	1.3270	0.0048	5.175
AVOA	0.7599	0.2109	58.2107	0.0383	1.4392	0.0015	2.732
COOT	0.7604	0.4866	73.5639	0.0348	1.5235	0.0013	2.845
AFT	0.7608	0.3230	53.7185	0.0364	1.4812	0.0010	2.271

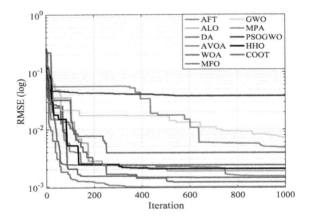

Figure 7.14 Convergence curves of eleven algorithms for RTC solar cell.

Fig. 7.15 shows the characteristic curves of current-voltage (I-V) and power-voltage (P-V) for the SDM using the best-optimized parameters acquired by the AFT algorithm. It is worth noting that the AFT's calculated data closely match the simulated data collected from the datasheet across the entire voltage range.

The difference between the calculated optimum current values for the AFT algorithm and the measured data from the manufacturer data sheet is shown in Fig. 7.16. At $v = 0.5833$ V, the maximum error for the tested range is 0.0025 A, which is practically acceptable.

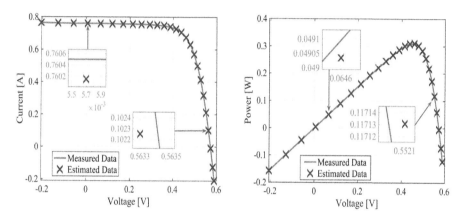

Figure 7.15 Characteristic curves for RTC solar cells based on parameters estimated from the best AFT algorithm.

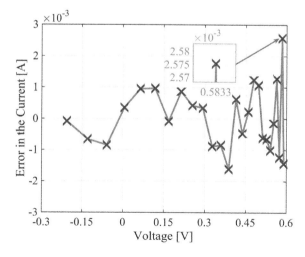

Figure 7.16 The error between the measured and estimated currents of the AFT algorithm for RTC solar cell.

The RMSE values of 30 runs for the RTC France solar cell for the SDM are shown in Fig. 7.17. Since the RMSE results for AFT are consistently lower than other techniques, this figure displays the algorithm's robustness.

7.4.2   Results of SDM for Photowatt PWP-201 solar cell

For this module, the optimum parameters for the control variables for the best run of the AFT algorithm and the comparative methods are given in Table 7.3. Their convergence rates are shown in Fig. 7.18 for comparison. It can be seen from this figure that the AFT algorithm has the lowest values for the RMSE and the time.

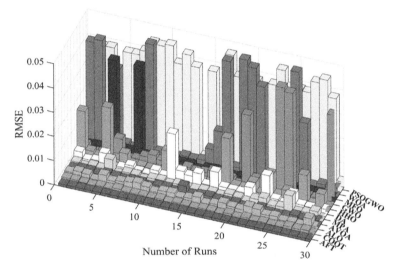

Figure 7.17 RMSE values of thirty independent runs for RTC solar cell.

Table 7.3 Parameter estimation of the SDM for the Photowatt PWP-201 solar cell.

Algorithm	I_{ph} (A)	I_d (μA)	R_{sh} (Ω)	R_s (Ω)	n	RMSE	Time (s)
GWO	1.0648	50.0000	106.9916	0.0026	1.7416	0.0484	2.701
ALO	1.0857	32.2788	247.0706	1.2387	1.6426	0.0442	9.019
MFO	1.0362	16.4963	1000.0000	0.9791	1.5418	0.0068	2.819
WOA	1.1121	0.3559	48.6889	0.1059	1.1751	0.0889	3.520
DA	1.0804	6.9414	69.5387	0.0068	1.4620	0.0682	34.864
HPSOGWO	1.0553	35.9430	300.4324	0.7723	1.6631	0.0156	3.576
HHO	1.1288	21.5185	76.8854	0.6352	1.5988	0.0421	6.331
MPA	1.0205	0.0777	883.3135	1.5583	1.0375	0.0140	5.401
AVOA	1.0342	0.0419	435.6718	1.6143	1.0000	0.0145	3.792
COOT	1.0354	14.9439	1000.0000	0.9904	1.5282	0.0064	2.948
AFT	1.0314	5.0381	998.1008	1.1540	1.3921	0.0028	2.490

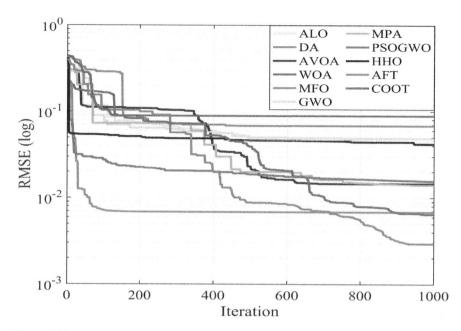

Figure 7.18 Convergence curves of eleven algorithms for the Photowatt PWP-201 solar cell.

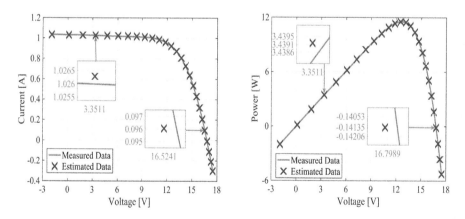

Figure 7.19 Characteristic curves for the Photowatt PWP-201 solar cell based on parameters estimated from the best AFT algorithm.

Fig. 7.19 illustrates the simulated data of the AFT as the best algorithm's I-V and P-V characteristic curves.

Fig. 7.20 displays that the maximum error for the tested voltage range is fairly acceptable at 0.0053 A and $v = 11.8018$ V.

Fig. 7.21 shows the RMSE values for the 30 runs for the Photowatt PWP-201 solar cell of SDM, illustrating the effectiveness of the AFT algorithm in obtaining the lowest RMSE values compared to other algorithms.

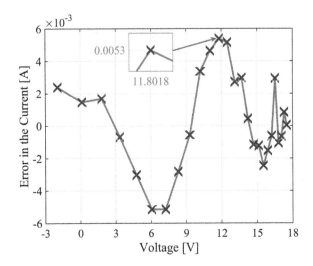

Figure 7.20 The error between the measured and estimated currents of the AFT algorithm for the Photowatt PWP-201 solar cell.

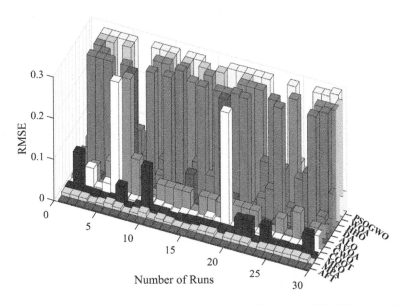

Figure 7.21 RMSE values of thirty independent runs for Photowatt PWP-201 solar cell.

7.4.3   Results of SDM for HTS-144M6H480 solar cell

Table 7.4 displays the optimal RMSE and the parameters estimated from each algorithm and explains the comparison between the results for the SDM. It can be seen

Table 7.4 Parameter estimation of the SDM for the HTS-144M6H480 solar cell.

Algorithm	I_{ph} (A)	I_d (μA)	R_{sh} (Ω)	R_s (Ω)	n	RMSE	Time (s)
GWO	12.0172	0.0941	408.2221	0.1418	0.7416	0.1163	6.299
ALO	12.0153	0.6133	523.7137	0.0951	0.8237	0.1446	29.205
MFO	11.9998	1.0000	1000.0000	0.0829	0.8481	0.1488	6.806
WOA	12.0062	0.4849	613.6048	0.1012	0.8123	0.1393	7.882
DA	12.0616	0.6108	250.5290	0.0913	0.8237	0.1556	107.836
HPSOGWO	12.0040	0.2750	582.5106	0.1166	0.7864	0.1302	8.163
HHO	12.2420	0.7571	80.14543	0.0641	0.8348	0.2202	15.485
MPA	11.9783	0.0575	999.9998	0.1553	0.7226	0.1022	12.620
AVOA	11.9903	0.3132	999.9679	0.1151	0.7922	0.1292	8.234
COOT	11.9727	0.0214	938.2286	0.1748	0.6874	0.0876	6.313
AFT	11.9597	0.0001	897.7353	0.2440	0.5658	0.0318	6.240

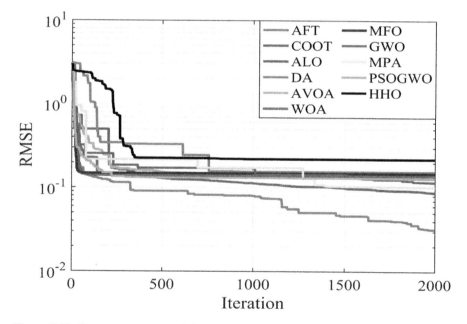

Figure 7.22 Convergence curves of eleven algorithms for HTS-144M6H480 solar cell.

that the AFT algorithm produced the minimum RMSE and time values in comparison with other tested algorithms.

The methods' convergence rates are shown in Fig. 7.22, demonstrating that the AFT technique can achieve the least RMSE quickly. Additionally, Fig. 7.23 displays the I-V and P-V curves for the SDM based on the estimated data from the best algorithm (AFT) at the best RMSE, confirming the high degree of agreement between the simulated and measured data.

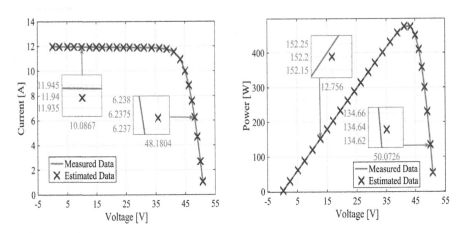

Figure 7.23 Characteristic curves for HTS-144M6H480 solar cell based on parameters estimated from the best AFT algorithm.

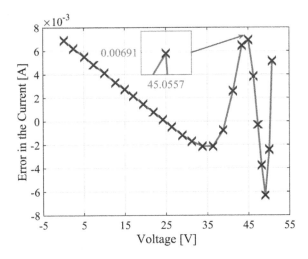

Figure 7.24 The error between the measured and estimated currents of the AFT algorithm for RTC solar cell.

Fig. 7.24 shows the maximum error over the tested voltage range is 0.0059 A at $v = 45.0557$ V, which is very acceptable.

Fig. 7.25 depicts the HTS-144M6H480 solar cell of SDM's RMSE values against the number of runs. It can be seen that the AFT method can achieve a minimum RMSE value in every run.

The AFT model is highly effective for equivalent circuit modeling of all three solar panels, as demonstrated in Figs. 7.16, 7.20, and 7.24.

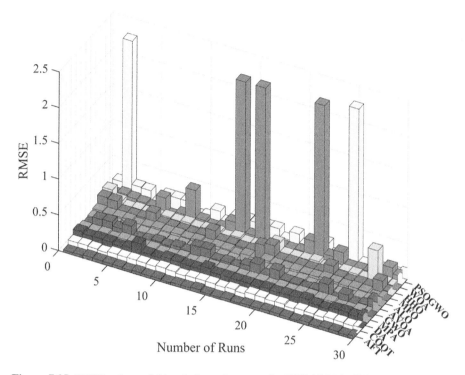

Figure 7.25 RMSE values of thirty independent runs for HTS-144M6H480 solar cell.

7.5   Conclusion

This chapter discusses various SPV models and their parameter estimation using metaheuristic optimization techniques from the literature. The article summarizes successful optimization methods and the objective advantages and disadvantages of different SPV models. The SPV model's starting parameters were determined using a single-diode model. The outcomes of each algorithm were summarized for easier decision-making and technique selection. The discussion of results demonstrates that the AFT is effective in estimating SPV module parameters and accurately modeling the optimization of I-V characteristics. The validity of the AFT technique in solving parameter estimation problems under various operating conditions is confirmed by data analysis using RMSE and time. This chapter assists researchers in selecting a suitable SPV model using proper measurement techniques. A future extension of this paper will compare the results of adopting the double-diode model (DDM) and multi-diode model (MDM) for the eleven optimization strategies mentioned earlier.

References

Abdollahzadeh, B., Gharehchopogh, F. S., & Mirjalili, S. (2021). African vultures optimization algorithm: A new nature-inspired metaheuristic algorithm for global optimization problems. *Computers and Industrial Engineering, 158*. Available from https://doi.org/10.1016/j.cie.2021.107408, https://www.journals.elsevier.com/computers-and-industrial-engineering.

AlHajri, M. F., El-Naggar, K. M., AlRashidi, M. R., & Al-Othman, A. K. (2012). Optimal extraction of solar cell parameters using pattern search. *Renewable Energy, 44*, 238−245. Available from https://doi.org/10.1016/j.renene.2012.01.082.

Bartumeus, F., Catalan, J., Fulco, U. L., Lyra, M. L., & Viswanathan, G. M. (2002). Optimizing the encounter rate in biological interactions: Lévy versus Brownian strategies. *Physical Review Letters, 88*(9), 4. Available from https://doi.org/10.1103/PhysRevLett.88.097901.

Braik, M., Ryalat, M. H., & Al-Zoubi, H. (2022). A novel meta-heuristic algorithm for solving numerical optimization problems: Ali Baba and the forty thieves. *Neural Computing and Applications, 34*(1), 409−455. Available from https://doi.org/10.1007/s00521-021-06392-x, http://link.springer.com/journal/521.

Chen, X., Xu, B., Mei, C., Ding, Y., & Li, K. (2018). Teaching−learning−based artificial bee colony for solar photovoltaic parameter estimation. *Applied Energy, 212*, 1578−1588. Available from https://doi.org/10.1016/j.apenergy.2017.12.115, http://www.elsevier.com/inca/publications/store/4/0/5/8/9/1/index.htt.

Chen, X., Yu, K., Du, W., Zhao, W., & Liu, G. (2016). Parameters identification of solar cell models using generalized oppositional teaching learning based optimization. *Energy, 99*, 170−180. Available from https://doi.org/10.1016/j.energy.2016.01.052, http://www.elsevier.com/inca/publications/store/4/8/3/.

Chen, Z., Wu, L., Lin, P., Wu, Y., & Cheng, S. (2016). Parameters identification of photovoltaic models using hybrid adaptive Nelder-Mead simplex algorithm based on eagle strategy. *Applied Energy, 182*, 47−57. Available from https://doi.org/10.1016/j.apenergy.2016.08.083, http://www.elsevier.com/inca/publications/store/4/0/5/8/9/1/index.htt.

Darmansyah, & Robandi, I. (2017), Photovoltaic parameter estimation using Grey Wolf Optimization. In *2017 3rd International conference on control, automation and robotics, ICCAR 2017* (pp. 593−597). Institute of Electrical and Electronics Engineers Inc. Indonesia. Available from: https://doi.org/10.1109/ICCAR.2017.7942766, 9781509060870.

Dizqah, A. M., Maheri, A., & Busawon, K. (2014). An accurate method for the PV model identification based on a genetic algorithm and the interior-point method. *Renewable Energy, 72*, 212−222. Available from https://doi.org/10.1016/j.renene.2014.07.014, http://www.journals.elsevier.com/renewable-and-sustainable-energy-reviews/.

Easwarakhanthan, T., Bottin, J., Bouhouch, I., & Boutrit, C. (1986). Nonlinear minimization algorithm for determining the solar cell parameters with microcomputers. *International Journal of Solar Energy, 4*(1), 1−12. Available from https://doi.org/10.1080/01425918608909835.

Faramarzi, A., Heidarinejad, M., Mirjalili, S., & Gandomi, A. H. (2020). Marine predators algorithm: A nature-inspired metaheuristic. *Expert Systems with Applications, 152*. Available from https://doi.org/10.1016/j.eswa.2020.113377, https://www.journals.elsevier.com/expert-systems-with-applications.

Guo, W.-y, Zhang, X., Liu, K.-x, & Zhang, J.-j (2018). Image enhancement based on improved antlion optimization algorithm. *DEStech Transactions on Computer Science and Engineering. (CNAI)*. Available from https://doi.org/10.12783/dtcse/cnai2018/24191.

Heidari, A. A., Mirjalili, S., Faris, H., Aljarah, I., Mafarja, M., & Chen, H. (2019). Harris hawks optimization: Algorithm and applications. *Future Generation Computer Systems*, *97*, 849−872. Available from https://doi.org/10.1016/j.future.2019.02.028.

Humphries, N. E., Queiroz, N., Dyer, J. R. M., Pade, N. G., Musyl, M. K., Schaefer, K. M., Fuller, D. W., Brunnschweiler, J. M., Doyle, T. K., Houghton, J. D. R., Hays, G. C., Jones, C. S., Noble, L. R., Wearmouth, V. J., Southall, E. J., & Sims, D. W. (2010). Environmental context explains Lévy and Brownian movement patterns of marine predators. *Nature*, *465*(7301), 1066−1069. Available from https://doi.org/10.1038/nature09116.

Kennedy, J., & Eberhart, R. (1995). *Particle swarm optimization*, . *IEEE international conference on neural networks—Conference proceedings* (Vol. 4, pp. 1942−1948). IEEE, undefined.

Merchaoui, M., Sakly, A., & Mimouni, M. F. (2018). Particle swarm optimisation with adaptive mutation strategy for photovoltaic solar cell/module parameter extraction. *Energy Conversion and Management*, *175*, 151−163. Available from https://doi.org/10.1016/j.enconman.2018.08.081, https://www.journals.elsevier.com/energy-conversion-and-management.

Mirjalili, S. (2015a). Moth-flame optimization algorithm: A novel nature-inspired heuristic paradigm. *Knowledge-Based Systems*, *89*, 228−249. Available from https://doi.org/10.1016/j.knosys.2015.07.006, https://www.journals.elsevier.com/knowledge-based-systems.

Mirjalili, S. (2015b). The ant lion optimizer. *Advances in Engineering Software*, *83*, 80−98. Available from https://doi.org/10.1016/j.advengsoft.2015.01.010, http://www.journals.elsevier.com/advances-in-engineering-software/.

Mirjalili, S. (2016). Dragonfly algorithm: A new meta-heuristic optimization technique for solving single-objective, discrete, and multi-objective problems. *Neural Computing and Applications*, *27*(4), 1053−1073. Available from https://doi.org/10.1007/s00521-015-1920-1, http://link.springer.com/journal/521.

Mirjalili, S., & Lewis, A. (2016). The Whale Optimization Algorithm. *Advances in Engineering Software*, *95*, 51−67. Available from https://doi.org/10.1016/j.advengsoft.2016.01.008, http://www.journals.elsevier.com/advances-in-engineering-software/.

Mirjalili, S., Mirjalili, S. M., & Lewis, A. (2014). Grey Wolf Optimizer. *Advances in Engineering Software*, *69*, 46−61. Available from https://doi.org/10.1016/j.advengsoft.2013.12.007, http://www.journals.elsevier.com/advances-in-engineering-software/.

Naruei, I., & Keynia, F. (2021). A new optimization method based on COOT bird natural life model. *Expert Systems with Applications*, *183*115352. Available from https://doi.org/10.1016/j.eswa.2021.115352.

Nelder, J. A., & Mead, R. (1965). A simplex method for function minimization. *The Computer Journal*, *7*(4), 308–313. Available from https://doi.org/10.1093/comjnl/7.4.308.

Nunes, H. G. G., Pombo, J. A. N., Bento, P. M. R., Mariano, S. J. P. S., & Calado, M. R. A. (2019). Collaborative swarm intelligence to estimate PV parameters. *Energy Conversion and Management*, *185*, 866−890. Available from https://doi.org/10.1016/j.enconman.2019.02.003, https://www.journals.elsevier.com/energy-conversion-and-management.

Nunes, H. G. G., Pombo, J. A. N., Mariano, S. J. P. S., Calado, M. R. A., & Felippe de Souza, J. A. M. (2018). A new high performance method for determining the parameters of PV cells and modules based on guaranteed convergence particle swarm optimization. *Applied Energy*, *211*, 774−791. Available from https://doi.org/10.1016/j.apenergy.2017.11.078, http://www.elsevier.com/inca/publications/store/4/0/5/8/9/1/index.htt.

Oghogho, I., Sulaimon, O., a, A. B., Egbune, D., & Abanihi, K. (2014). Solar energy potential and its development for sustainable energy generation in Nigeria: A road map to achieving this feat. *International Journal of Engineering and Management Sciences*.

Oliva, D., Abd El Aziz, M., & Ella Hassanien, A. (2017). Parameter estimation of photovoltaic cells using an improved chaotic whale optimization algorithm. *Applied Energy, 200*, 141−154. Available from https://doi.org/10.1016/j.apenergy.2017.05.029, http://www. elsevier.com/inca/publications/store/4/0/5/8/9/1/index.htt.

Şenel, F. A., Gökçe, F., Yüksel, A. S., & Yiğit, T. (2019). A novel hybrid PSO−GWO algorithm for optimization problems. *Engineering with Computers, 35*(4), 1359−1373. Available from https://doi.org/10.1007/s00366-018-0668-5, https://link.springer.com/ journal/366.

Shaikh, M. R. S. (2017). A review paper on electricity generation from solar energy. *International Journal for Research in Applied Science and Engineering Technology, V* (IX), 1884−1889. Available from https://doi.org/10.22214/ijraset.2017.9272.

Singh, N., & Singh, S. B. (2017). Hybrid algorithm of particle swarm optimization and grey wolf optimizer for improving convergence performance. *Journal of Applied Mathematics, 2017*, 1−15. Available from https://doi.org/10.1155/2017/2030489.

Tossa, A. K., Soro, Y. M., Azoumah, Y., & Yamegueu, D. (2014). A new approach to estimate the performance and energy productivity of photovoltaic modules in real operating conditions. *Solar Energy, 110*, 543−560. Available from https://doi.org/10.1016/j.solener.2014.09.043, http://www.elsevier.com/inca/publications/store/3/2/9/index.htt.

Wang, G., Zhao, K., Shi, J., Chen, W., Zhang, H., Yang, X., & Zhao, Y. (2017). An iterative approach for modeling photovoltaic modules without implicit equations. *Applied Energy, 202*, 189−198. Available from https://doi.org/10.1016/j.apenergy.2017.05.149, http://www.elsevier.com/inca/publications/store/4/0/5/8/9/1/index.htt.

Wu, L., Chen, Z., Long, C., Cheng, S., Lin, P., Chen, Y., & Chen, H. (2018). Parameter extraction of photovoltaic models from measured I-V characteristics curves using a hybrid trust-region reflective algorithm. *Applied Energy, 232*, 36−53. Available from https://doi.org/10.1016/j.apenergy.2018.09.161, http://www.elsevier.com/inca/publications/store/4/0/5/8/9/1/index.htt.

Xiong, G., Zhang, J., Yuan, X., Shi, D., & He, Y. (2018). Application of symbiotic organisms search algorithm for parameter extraction of solar cell models. *Applied Sciences, 8*(11), 2155. Available from https://doi.org/10.3390/app8112155.

Modeling and comparative study of half-cut cell and standard cell photovoltaic modules under partial shading conditions

8

C.H. Siow[1], Rodney H.G. Tan[1] and T. Sudhakar Babu[2]
[1]Department of Electrical & Electronic Engineering, UCSI University, Kuala Lumpur, Malaysia, [2]Department of Electrical and Electronics Engineering, Chaitanya Bharati Institute of Technology, Hyderabad, Telangana, India

8.1 Introduction

As the world's population, industrialization, urbanization, and technological advancement continue to grow at a rapid pace, the total energy demand will continuously rise. To ensure a sufficient supply of energy, scientists and researchers have been working tirelessly for decades to seek for ways to convert energy into forms that could be used for the general public's benefit. The world's energy consumption currently still mostly relies on nonrenewable sources of energy, primarily fossil fuels. The quantity of energy that can be collected from the planet is finite, and this indicates that the globe will be faced with an energy deficit in the foreseeable future. When looking for sustainable energy sources, among all the renewable energy sources that may be utilized, sunlight from the sun stands out as the ultimate existence, as it is a renewable source of energy that will not disappear shortly and it is free from creating pollution to the environment (Sharma et al., 2015).

Most of the energy on the planet comes from the sun in one form or another. Humans take advantage of the energy generated by the sun in a variety of ways. French physicist Edmund Becquerel discovered that certain materials when exposed to sunlight generate a very small quantity of electric current in the year 1839. When sunlight is captured, the process of PVs happens. Photovoltaics is a direct way of converting solar energy, while solar cells are utilized to convert solar energy directly into electricity with no pollution and no moving parts, which makes them exceptionally durable, trustworthy, and long-lasting (Vieira et al., 2020). The term "Photovoltaics" is derived from the direct conversion of light, which also implies from photon to electricity. It refers to materials that exhibit the photoelectric effect, absorb photons of light, and release electrons as a result of their absorption. It is possible to generate an electric current from the capture of these free electrons, which can be converted into electricity. In 1954, Bell Labs developed the world's first commercially effective silicon solar PV module. With years of advancements in PV module cell research, most

Performance Enhancement and Control of Photovoltaic Systems. DOI: https://doi.org/10.1016/B978-0-443-13392-3.00008-6

monocrystalline silicon PV modules are now capable of achieving the maximum efficiency of 26.7% in the lab (Green et al., 2019).

Improvement in the overall performance of solar PV modules has been proposed and tested over the years. From implementing the bypass diode to dealing with partial shading problems, different materials used for solar cell fabrication, various cell series-parallel configurations, and encapsulation methods were some of the efforts done by researchers from all around the world. In recent years, the half-cut cell PV module has achieved the commercial stage and is being introduced into the solar PV market. Half-cut cell technology was first introduced by REC, Australia, to the market. These impressive technologies brought them the 2015 Intersolar Award in Photovoltaics. Then, other market-leading solar PV manufacturers such as JA Solar, Jinko Solar, and Q cells also followed to introduce their half-cut cell modules to the market. The Half-Cut cell module brings advantages by halving the cell current, resulting in a reduction in resistive losses as the current is reduced by half, thereby enhancing the overall performance of the module. The half-cut cells have many benefits such as higher power output and are less prone to partial shade conditions due to shadow casting and soiling, leading to reduced overall power loss (Chiodetti & Dupuis, 2019).

8.1.1 Solar irradiance

Knowing the amount of sunshine available at a particular location at any given time is crucial information for solar system design. Solar irradiance is the power per area provided at the time, while solar insolation is the energy per area delivered over a specific period. To obtain solar irradiance data, measurements of direct, diffuse, and global irradiance are made at regular intervals throughout the day at various locations around the world. Direct irradiance is the quantity of solar energy that reaches the planet's surface without being diluted or absorbed by other factors. Diffuse irradiance is described as the sunlight that has been scattered by molecules and particles in the atmosphere, such as clouds, yet has nevertheless managed to reach the planet's surface; the condition is shown in Fig. 8.1. The sum of diffuse and direct solar irradiance is referred to as global solar irradiance, and W/m^2 is the unit of measurement. Under the standard test condition (STC), the solar PV cells were assumed to be exposed to 1000 W/m^2 solar irradiance. Under any shading conditions, only diffuse irradiance is assumed (Boland et al., 2008).

8.1.2 Partial shading

During clear sky conditions with no shading, a solar PV module generates current based on solar irradiance power received from the sun. When it comes to partial shading conditions, a small area of shadow cast on the solar PV module can cause the entire output power to decrease drastically. The output current drops proportionally to the shaded area. Even with just a single cell of the module shaded, the output current generation is limited to the output of the shaded cell produced. Fig. 8.2 shows a solar PV module shaded by a nearby structure; the PV module will suffer a

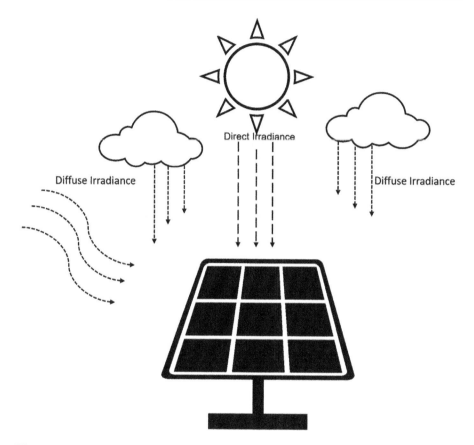

Figure 8.1 Direct and diffuse irradiance.

major power loss under this condition, leading to a reduction in energy production by the solar PV system. Depending on how the shade is cast across the PV module surface, the amount of power loss varies accordingly.

The first investigation of partial shading on the PV system is found in the research work presented by Lashway (1988). The amount of shade area cast on the solar PV module has a direct relation to the reduction in energy production. The decrease in solar PV module output current is proportional to the amount of shade area cast over the module.

Photovoltaic modules are made of solar cells that are connected in series. A single piece of solar cell has a voltage of around 0.5 volts; cells are connected in series in a module to increase the voltage suitable for charge controller or inverter operation. Since the solar cell is connected in series, the highest current generated from the module will be limited by the shaded cell current. When the multiple modules are connected in series to achieve a system voltage of 600–1000 V for string inverter operation, the shading problem becomes prevalent, since shading of one

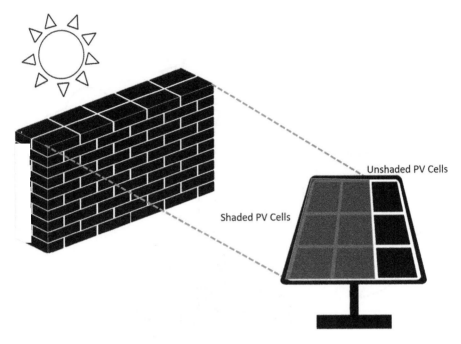

Figure 8.2 Partial-shaded PV module.

cell would have an impact on the entire string. The shaded cells may change to reverse-biased, resulting in the shaded cells acting as loads and dissipating power as heat generated by the unshaded cells in the module, causing hot spots that will eventually damage the module.

8.1.3 Bypass diode

The shading on a PV module caused by shadow or uneven soiling will result in the reduction of the current generation. A bypass diode is added in a PV module so that the higher current generated by the unshaded cells can bypass the shaded cells. Bypass diodes allow each module to be divided into sections or groups, decreasing the current generation loss caused by shaded cells. A bypass diode is connected in parallel with the solar cell group to provide a lower resistance pathway for higher current-generating cell current to bypass the lower current-generating shaded cell in the solar PV module. The implementation of a bypass diode makes sure the series-connected solar cells continue to supply the higher current at the expense of a lower voltage as shown in Fig. 8.3. A bypass diode configures in reverse bias between the solar cell groups. Ideally, each solar cell in the module would need a bypass diode, but this will increase the cost of the solar PV module and it is not economically or commercially viable. With all the considerations, a commercial standard PV module available in the market will have up to three bypass diodes.

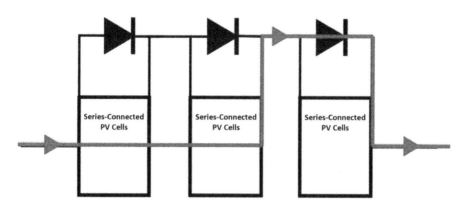

Figure 8.3 Current bypassing shaded cell group through a bypass diode.

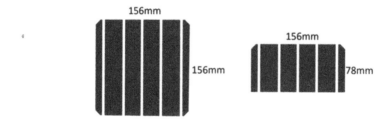

Figure 8.4 The physical layout of the standard cell and half-cut cell.

8.1.4 Half-cut cell

The solar PV module constructed using half-cut cells is less vulnerable to the shaded cell. The power equation $P = I^2R$ shows the power losses are the product of the resistance R and the square of current I^2. Cutting the solar cell into two halves will eventually split the cell current in half. Solar cells originally with the dimensions of 156 mm × 156 mm become 156 mm × 78 mm for the half-cut cell module (Sarniak, 2020). Fig. 8.4 shows the physical layout of the normal cell and the half-cut cell.

The half-cut cells double the total number of cells on the module. The basic concept is that adding cells in series will accumulate voltage and adding a second string of cells in parallel accumulates current (Joshi et al., 2019). To overcome the encountered condition, manufacturers have figured out a way that connects two strings of half-cut cells in parallel; this architecture helps the voltage at the emitting side of the cell panel to be identical to the standard cell but with half of the internal current. The half-cut cell module and standard-cell module configuration with three bypass diodes are shown in Fig. 8.5.

This problem can be improved by the special architecture configuration of the half-cut cell PV module. Half-cut PV modules can perform unaffected for the top

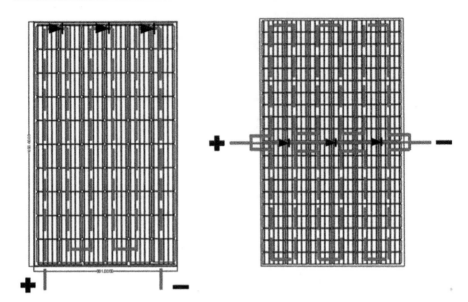

Figure 8.5 Solar cell and bypass diode configuration for standard and half-cut cells.

half of the panel if the bottom half is shaded or vice versa. When two strings are connected in parallel, isolation can be done for the lower current shaded cell to just a portion. But not all kinds of partial shading patterns can be improved by the half-cut design module. There are several research works done to investigate the partial shading effect for solar PV power systems.

8.1.5   Review of partial shading research

Different types of research work have been conducted by different researchers from all around the world. Partial shading or partial shadowing has been identified as the main cause of reducing the energy yield of solar PV systems. At K. U. Leuven, a 5-kWp PV system was erected. Three distinct subsystems make up the system: a central inverter, a string inverter, and several AC modules. Parts of the solar array are shaded by plants and other nearby impediments throughout the year. The dimensions of shadowing barriers were measured, and the expected shadowing losses were calculated using several methods (Woyte et al., 2003). They had investigated the equivalent circuit model and the circuit's characteristics under the partial shading effect. The researcher also clarified and proposed a net MPPT method for the sake of drastic power generation improvement, while the partial shading effect of a string-connected solar system evaluation was conducted by Andre Mermoud (2010). The PV array's electrical behavior under a partial shading effect was studied in some cases using the PVsyst tools.

The power generation of PV systems is greatly influenced by partial or total shadowing of their cells, which is dependent on the PV array arrangement, shading characteristics, and the existence of a bypass diode (Dolara et al., 2013). The researcher deals with the investigation of the impact of PS on both poly and monocrystalline PV module operation. In the experiment conducted for both shaded and unshaded, the I-V and P-V curves are measured and recorded. Moballegh and Jiang (2014) conducted experiments and modeling to evaluate the partial shading condition of the PV array. Ten shading profiles were used to validate the prediction technique in three different array configurations. The predicted power peak was within 5% of the real measurement for almost all the cases.

Das et al. (2017) focus on modeling and studying I-V and P-V curves of a PV array under different irradiance levels. The researchers had developed an array module in a MATLAB® environment and evaluated different array configurations, including SP, TCT, BL, and TCT. With the research work conducted, they validated that the TCT configuration is best suited under partial shading conditions. Salem and Awadallah (2016) presented a methodology for assessing and detecting the partial shading condition of PV arrays based on artificial neural networks (ANNs).

Sarniak (2020) modeled the functioning of the half-cut cell PV module under partial shading in the MATLAB package. The researcher modeled the half-cut cell with 120 cells and 6 bypass diodes in the Simulink, and verification tests were carried out for the PV module JAM60S03-320/PR using the I-V 400 m. Four different partial shadings whose conditions included shading of cells, shading of 8 cells, shading of 12 cells, and shading of 24 cells for a half-cut cell PV module were verified. Shaima et al. (2021) investigated by covering the monocrystalline module horizontally and testing it under different levels of irradiance. Numan et al. (2020) evaluated the single module working under the partial shading effect on either horizontal or vertical shading conditions, while the diagonal shading condition is not discussed.

Praveen and Suresh (2020) mention that partial shading causes significant power losses in PV arrays. The researchers had mentioned very detailed partial shading conditions under different irradiance rates, including corner shading, center shading, right side end shading, bottom-end shading, L-shape shading, frame shading, random shading, and diagonal shading. The research works cover most of the shading that will be faced in the real environment. The results shows that with different types of array configurations resulting in different outputs. The study shows that partial shading cast on a single PV panel will affect the overall output performance of an entire solar PV system consists of 540 panels.

Table 8.1 summarizes the literature about partial shading of the half-cut cell that had contributed to knowledge and the detailed study of the partial shading effect that is continuing. These studies demonstrate that assessing PV systems under partial shade conditions is not a novel concept, as evidenced by the large quantity of research that has been done on a similar subject. While most of the work is done using standard cells, it can be observed from the situation that not all of them investigate various partial shading patterns, which include vertical, horizontal, and

Table 8.1 Summary of research on standard and half-cut cell modules under various shade conditions.

References	Cell type	Horizontal shade	Vertical shade	Diagonal shade
Dolara et al. (2013), Numan et al. (2020), and Bonthagorla and Mikkili (2020)	Standard	Yes	Yes	Yes
Moballegh and Jiang (2014), Shaima et al. (2021)	Standard	Yes	Yes	No
Das et al. (2017), Mermoud and Lejeune (2019), and Kajihara and Harakawa (2005)	Standard	Yes	No	No
Salem and Awadallah (2016)	Standard	No	Yes	No
Chiodetti and Dupuis (2019)	Half-Cut	Yes	Yes	No
Sarniak (2020)	Half-Cut	No	Yes	No

diagonal shading patterns. For this research, studies for both standard cell and half-cut cell performance work under partial shade in all horizontal, vertical, and diagonal shading conditions will be modeled, graphs and results obtained will be documented, and the findings will be explained in detail.

8.2 Methodology

A standard solar PV module with 60 cells and a half-cut cell solar PV module with 120 series-connected solar cells were chosen for this chapter to model the power-voltage and current-voltage characteristics that show how different types of PV modules work under the partial shading effect. To model the PV module, factors such as characteristic parameters, solar irradiance, operating temperature, and shading profile need to be considered. Based on the semiconductor's theory, the current-voltage characteristic of PV cells was illustrated by the single-diode equivalent PV cell circuit shown in Fig. 8.6.

Eq. (8.1) shows the equation for the solar PV cell equivalent circuit. I_{ph} is the solar-generated current where I_s is the diode saturation current and n is the quality factor (diode emission coefficients), while R_s is the series resistance and R_p is the shunt resistance.

$$I = I_{ph} - I_s\left(e^{\frac{V+IR_s}{nv_t}}\right) - \frac{(V+IR_s)}{R_p} \tag{8.1}$$

The PV module distributed by manufacturers will often come with standard parameters listed in the datasheet such as the maximum power voltage and current, short circuit current, open-circuit voltage, and temperature coefficients.

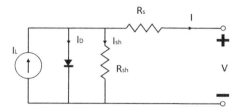

Figure 8.6 Solar cell equivalent circuit.

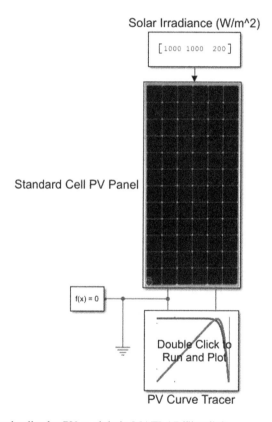

Figure 8.7 Standard cell solar PV module in MATLAB/Simulink.

8.2.1 Modeling of standard photovoltaic module

The solar PV module can be developed in the MATLAB Simulink environment. First, a constant block with the form of a 1×3 matrix was used to represent the solar irradiance in W/m^2 as an input of the solar cell PV panel shown in Fig. 8.7.

Conditions such as partial shading and soiling will cause the solar irradiance received by the panel to decrease. The solar irradiance input will be assigned to each solar cell group 1, 2, and 3 accordingly by the selector. The architecture of the circuit

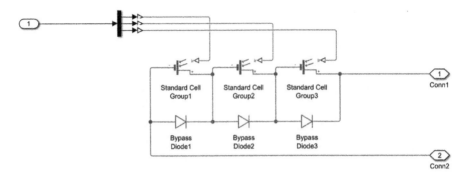

Figure 8.8 Simulink Subsystem of standard-cell solar PV module.

Table 8.2 PV panel parameters from the datasheet.

Module parameters	JAM60S01-300/PR Standard	JAM60S03-300/PR Half-Cut
Rated maximum power (Pmax)	300 W	300 W
Open circuit voltage (Voc)	39.85 V	39.05 V
Short circuit current (Isc)	9.75 A	9.90 A
Maximum power voltage (Vmp)	32.26 V	32.23 V
Maximum power current (Imp)	9.30 A	9.31 A
Module efficiency	18.3%	18.0%

construction is shown in Fig. 8.8. There are 20 solar cells connected in series to form a solar cell group; then each solar cell group is connected in series to form the complete panel circuit. Each group of solar cells is connected to a bypass diode in parallel.

There are five parameters to take into account for the module constructed, which are the short circuit current Isc, open circuit voltage Voc, ideality factor, series resistance, and cell temperature. Parameters such as Isc and Voc can be directly taken from the datasheet listed in Table 8.2.

The ideality factor and series resistance can be calculated using the PV array component in MATLAB/Simulink with the parameter given earlier. Set the parallel strings and series connected modules per string number to one, and leave the temperature coefficient default as the component setting shown in Fig. 8.9.

The open circuit voltage and series resistance were divided by the total number of cells, 60, as the solar cells are connected in series. The PV curve tracer will plot the current-voltage characteristic and power voltage characteristic curve.

8.2.2 Modeling of a half-cut cell photovoltaic module

For the modeling of the half-cut cell solar PV panel, the input of the constant block can be filled in the form of a 2×3 matrix, e.g., [1000 250 200; 200 200 200], as shown in Fig. 8.10.

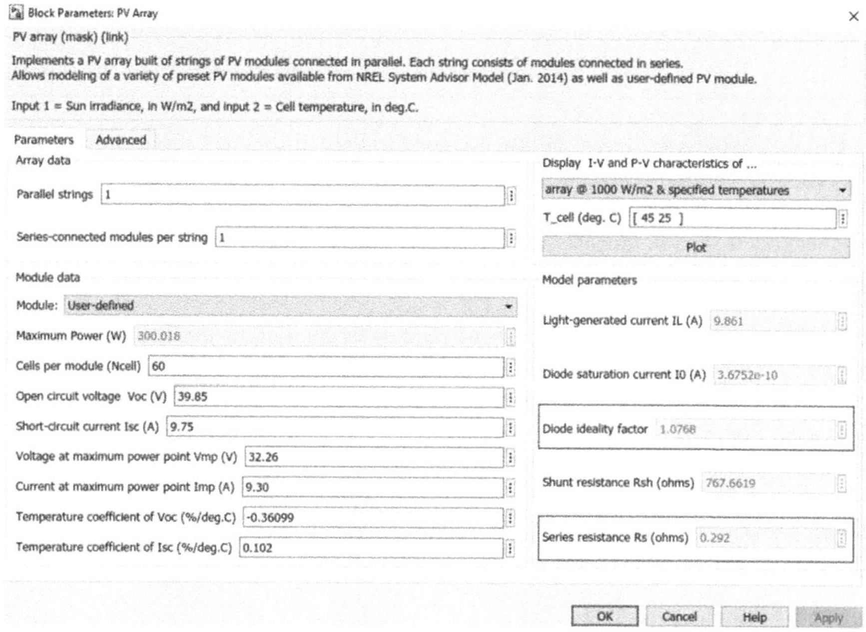

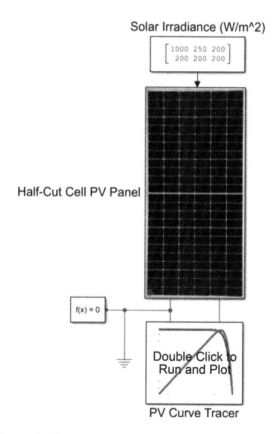

Block Parameters: PV Array ×

PV array (mask) (link)

Implements a PV array built of strings of PV modules connected in parallel. Each string consists of modules connected in series.
Allows modeling of a variety of preset PV modules available from NREL System Advisor Model (Jan. 2014) as well as user-defined PV module.

Input 1 = Sun irradiance, in W/m2, and input 2 = Cell temperature, in deg.C.

Parameters | Advanced

Array data

Parallel strings 1

Series-connected modules per string 1

Module data

Module: User-defined

Maximum Power (W) 300.018

Cells per module (Ncell) 60

Open circuit voltage Voc (V) 39.85

Short-circuit current Isc (A) 9.75

Voltage at maximum power point Vmp (V) 32.26

Current at maximum power point Imp (A) 9.30

Temperature coefficient of Voc (%/deg.C) -0.36099

Temperature coefficient of Isc (%/deg.C) 0.102

Display I-V and P-V characteristics of ...

array @ 1000 W/m2 & specified temperatures

T_cell (deg. C) [45 25]

Plot

Model parameters

Light-generated current IL (A) 9.861

Diode saturation current I0 (A) 3.6752e-10

Diode ideality factor 1.0768

Shunt resistance Rsh (ohms) 767.6619

Series resistance Rs (ohms) 0.292

OK | Cancel | Help | Apply

Figure 8.9 Block parameter of a PV array for a standard cell solar PV module.

Solar Irradiance (W/m^2)

$$\begin{bmatrix} 1000 & 250 & 200 \\ 200 & 200 & 200 \end{bmatrix}$$

Half-Cut Cell PV Panel

f(x) = 0

Double Click to Run and Plot

PV Curve Tracer

Figure 8.10 Half-cut cell solar PV module model in MATLAB/Simulink.

The solar irradiance input will be assigned to the half-cut cell group from 1 to 6, respectively. The half-cut cell solar PV panel used consists of a total of 120 solar cells. There are a total of six groups of solar cells, each having 20 cells connected together. Each section consists of three groups of solar cells connected in series. A total of two sections are connected in parallel with the bypass diodes connected within the circuit as shown in Fig. 8.11.

The parameter input of the half-cut module is similar to the modeling of the standard module with the Voc and Isc taken from Table 8.2. The PV array component in MATLAB/Simulink is used to calculate the ideality factor and the series resistance. Fig. 8.12 shows the block parameter of a half-cut cell solar PV module.

The short circuit current for the half-cut solar cell has to be divided by two due to the half size of the standard solar cell. The total half-cut cell number per group and the open circuit voltage for each half-cut cell can be determined in Eqs. (8.2) and (8.3), where N_s is the total number of standard cells in the module.

$$\text{Cells per Group} = \frac{N_s}{3} \tag{8.2}$$

$$\text{Voc per Cell} = \frac{Voc}{N_s} \tag{8.3}$$

The series resistance of each half-cut solar cell can be determined in Eq. (8.4), where Rs is the total series resistance of the entire panel obtained from the PV array model shown in Fig. 8.12.

$$\text{Rs per Cell} = \frac{2Rs}{N_s} \tag{8.4}$$

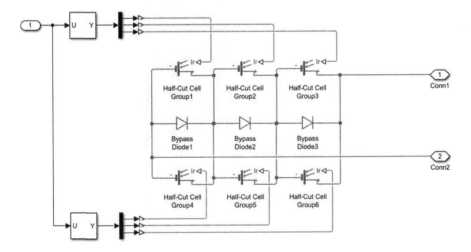

Figure 8.11 Simulink subsystem of a half-cut cell solar PV module.

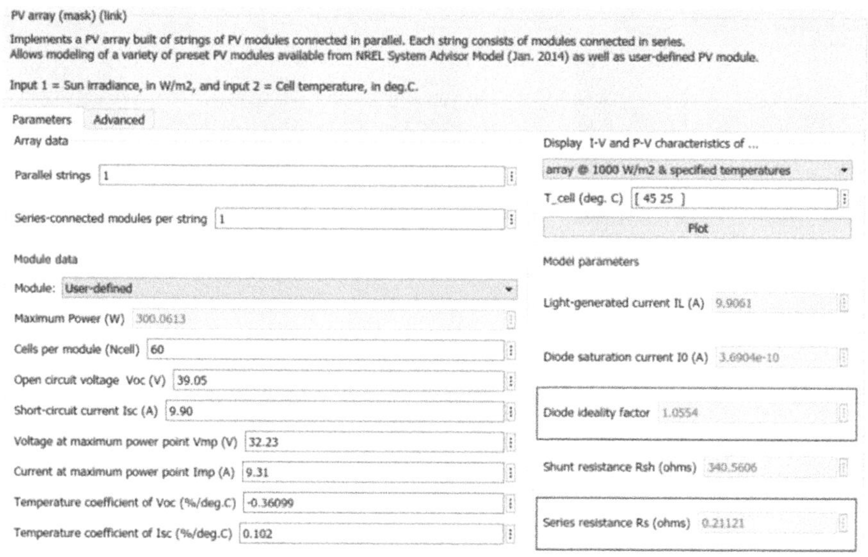

PV array (mask) (link)

Implements a PV array built of strings of PV modules connected in parallel. Each string consists of modules connected in series.
Allows modeling of a variety of preset PV modules available from NREL System Advisor Model (Jan. 2014) as well as user-defined PV module.

Input 1 = Sun irradiance, in W/m2, and input 2 = Cell temperature, in deg.C.

Parameters	Advanced

Array data

Display I-V and P-V characteristics of ...

Parallel strings 1

array @ 1000 W/m2 & specified temperatures ▾

T_cell (deg. C) [45 25]

Series-connected modules per string 1

Plot

Module data

Model parameters

Module: User-defined ▾

Light-generated current IL (A) 9.9061

Maximum Power (W) 300.0613

Cells per module (Ncell) 60

Diode saturation current I0 (A) 3.6904e-10

Open circuit voltage Voc (V) 39.05

Short-circuit current Isc (A) 9.90

Diode ideality factor 1.0554

Voltage at maximum power point Vmp (V) 32.23

Current at maximum power point Imp (A) 9.31

Shunt resistance Rsh (ohms) 340.5606

Temperature coefficient of Voc (%/deg.C) -0.36099

Temperature coefficient of Isc (%/deg.C) 0.102

Series resistance Rs (ohms) 0.21121

Figure 8.12 Block parameter of a PV array for a half-cut cell solar PV module.

8.3 Results and discussion

The model simulated takes place under the standard testing condition, which takes 1000 W/m^2 of solar irradiance for the solar cell without shade. For the shaded solar cells, the global irradiance received for the cells that come from the diffuse irradiance is taken at around 20% of the standard testing condition. A systematic shading pattern is proposed to test the standard and half-cut module as shown in Fig. 8.13. The systematic shading conditions are classified into three types: vertical shading, horizontal shading, and diagonal shading. The different shading patterns of the standard module are represented by A, B, and C in Fig. 8.13. A represents unshaded, 33%, 66%, and 100% shaded for the horizontal shading condition. B represents unshaded, 25%, 50%, 75%, and 100% shaded for the vertical shading condition. C represents unshaded, 25%, 50%, 75%, and 100% shaded for the diagonal shading condition of a standard-cell module.

One of the most comprehensive ways to test the PV array is the measurement of an I-V characteristic curve. The I-V characteristic curve shows the behavior of the PV system under all load conditions from open circuit to short circuit, while the P-V curve can be used to monitor the power loss of the PV module effectively. Fig. 8.14 shows the I-V curve and P-V curve for a standard-cell PV module.

The different shading patterns of the half-cut cell are represented by A, B, and C. A represents unshaded, 33%, 66%, and 100% shaded for horizontal shading conditions. B represents 25%, 50%, 75%, and 100% shaded for vertical shading conditions. C represents unshaded, 25%, 50%, 75%, and 100% shaded for diagonal shading conditions of a half-cut cell module as shown in Fig. 8.15.

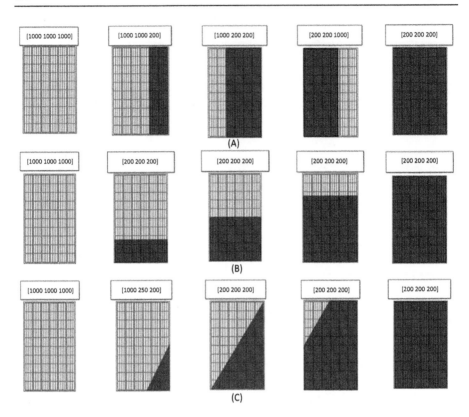

Figure 8.13 Shading conditions on a standard-cell module. (A) horizontal shade, (B) vertical shade, (C) diagonal shade.

Different solar irradiance inputs are used to simulate the module under each partial shading condition. For the unshaded cells, an irradiance of 1000 W/m^2 is applied, whereas the fully shaded cells are subjected to an irradiance of 200 W/m^2. For the cell that is partially shaded, a 250 W/m^2 is applied. The solar irradiance input for each condition is shown in Figs. 8.13 and 8.15. The result I-V curve and P-V curve generated following the shaded condition of the half-cut cell PV module are shown in Fig. 8.16.

With the result shown, the short circuit current decreases when the cell is shaded, while the maximum power output decreases with the increase of shading on the module. When the irradiance on the solar PV module is nonuniform, multiple steps will be observed. This effect comes with the activation of the bypass diode. When the bypass diode is activated, the shaded cells are bypassed, and the unshaded cells will cause multiple peaks in the P-V curve and multiple steps for the I-V curve. The percentage difference in power losses between the standard cell and the half-cut cell is shown in Table 8.3.

The result shows that the overall performance of the half-cut cell PV module is better than the standard cell PV module. The most obvious difference occurs when

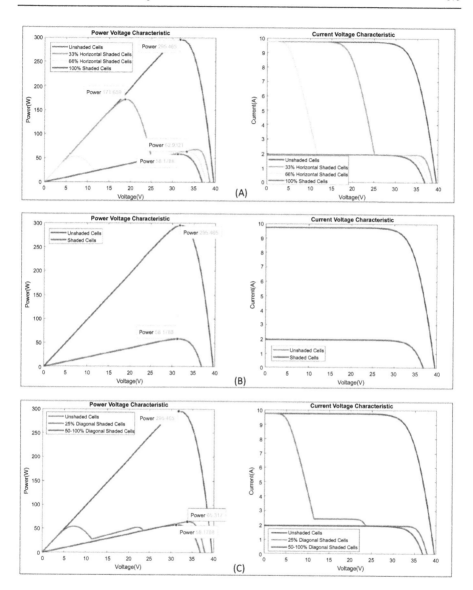

Figure 8.14 I-V characteristics for standard-cell partial shading. (A) horizontal shade, (B) vertical shade, (C) diagonal shade.

facing the 50% vertical shading and 25% diagonal shading compared to both PV modules. There is a power drop of 80% when the standard-cell PV module faces 25% and 50% vertical shading, while a half-cut cell drops by 40.39% when dealing with 25% and 50% vertical shading. For the 25% diagonal shading of the standard cell module, the loss of power is 77.9%, while there is only a power drop of

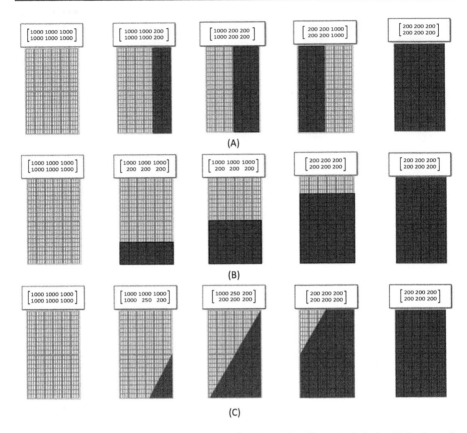

Figure 8.15 Shading conditions of a half-cut cell PV module. (A) vertical shade, (B) horizontal shade, (C) diagonal shade.

36.47% for the same condition for the half-cut cell. For the horizontal shading, both the standard and half-cut cells act similarly due to the configuration of the solar cell and the bypass diode of the module.

Half-cut cell PV modules perform better in the vertical shading region by more than 39.61% compared to the standard cell PV modules in similar shading conditions, while 41.43% higher power in the partial shading pattern diagonally in the half-cut cell than in the standard cell. The architecture of the half-cut cell module can make sure the half-cut PV panel works better under partial shading conditions; the main difference between the half-cut cell and standard cell PV is the doubled section of the panel, which can be bypassed when the solar cells encounter shading or cell failure conditions. The half-cut cell modules offered by the manufacturer had reached a cost that is competitive with standard cell modules. But there are cons that increase the risk of bad contact during manufacturing as the soldering connection of solar cells is doubled. A half-cut cell module also requires more manufacturing processes where they need to undergo the process of cutting the

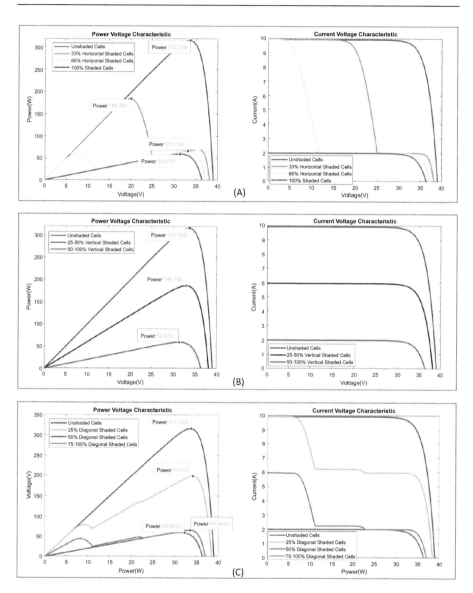

Figure 8.16 I-V characteristics for half-cut cell partial shading. (A) horizontal shade, (B) vertical shade, (C) diagonal shade.

solar cell in half. However, it has better performance against partial shading, and a damaged cell allows a half-cut cell module to quickly gain a marketplace in the PV module industry.

Both the standard-cell and half-cut cell PV modules can be tested experimentally in real-life conditions. The irregular shape of shading lays on both types of

Table 8.3 Power loss comparison for standard-cell and half-cut cell PV modules.

Shading condition	Standard-cell power loss	Half-cut cell power loss	Power loss difference
Full shade (100%)	80%	80.63%	0.63%
Horizontal shade (33%)	41.83%	41.96%	0.13%
Horizontal shade (66%)	78.8%	79.07%	0.27%
Vertical shade (25%)	80%	40.39%	39.61%
Vertical shade (50%)	80%	40.39%	39.61%
Vertical shade (75%)	80%	80.63%	0.63%
Diagonal shade (25%)	77.9%	36.47%	41.43%
Diagonal shade (50%)	80%	78.71%	1.29%
Diagonal shade (75%)	80%	80.63%	0.63%

modules will give different power and current generation. A half-cut cell module is suitable for rooftop installation in urban cities, where high-rise buildings, transmission towers, electrical poles, trees, and other obstacles create irregular shadows on the PV module. A half-cut cell module can effectively increase energy production yield compared to standard-cell modules under certain circumstances of partial shading.

The standard and half-cut cell PV panel Simulink models presented in this chapter are made available by the authors for the readers to download from the following MathWorks official MATLAB central file exchange link: https:// www.mathworks.com/matlabcentral/fileexchange/75088-standard-cell-pv-panel.

The half-cut cell PV panel Simulink model presented in this chapter is made accessible by the authors for readers to download from the following MathWorks official MATLAB central file exchange link: https://www.mathworks.com/matlab-central/fileexchange/115530-half-cut-cell-pv-panel.

8.4 Conclusion

In summary, the overall performance of the half-cut cell PV module is better than the standard-cell PV module. The half-cut PV module can increase the yield of energy production under vertical and diagonal partial shading. The half-cut cell module advantage can be better exploited in urban city rooftops or building integrated PV installations, where high-rise buildings, transmission towers, electrical poles, trees, and other obstacles create irregular shadows cast on the PV module.

References

Andre Mermoud, T. L. (2010). *Partial shadings on PV arrays: By-pass diode benefits analysis.* EU PVSEC.

Boland, J., Ridley, B., & Brown, B. (2008). Models of diffuse solar radiation. *Renewable Energy, 33*(4), 575–584. Available from https://doi.org/10.1016/j.renene.2007.04.012.

Bonthagorla, P. K., & Mikkili, S. (2020). Performance analysis of PV array configurations (SP, BL, HC and TT) to enhance maximum power under non-uniform shading conditions. *Engineering Reports, 2*(8). Available from https://doi.org/10.1002/eng2.12214.

Chiodetti, M., & Dupuis, J. (2019). *Half-cell module behaviour and its impact on the yield of a PV plant.* EU PVSEC.

Das., Mohapatra, A., & Nayak, B. (2017). Modeling and characteristic study of solar photovoltaic system under partial shading condition. *Materials Today: Proceedings, 4*(14), 12586–12591. Available from https://doi.org/10.1016/j.matpr.2017.10.066.

Dolara, A., Lazaroiu, G. C., Leva, S., & Manzolini, G. (2013). Experimental investigation of partial shading scenarios on PV (photovoltaic) modules. *Energy, 55*, 466–475. Available from https://doi.org/10.1016/j.energy.2013.04.009.

Green, M. A., Dunlop, E. D., Levi, D. H., Hohl-Ebinger, J., Yoshita, M., & Ho-Baillie, A. W. Y. (2019). Solar cell efficiency tables (version 54). *Progress in Photovoltaics: Research and Applications, 27*(7), 565–575. Available from https://doi.org/10.1002/pip.3171.

Joshi, A., Khan, A., & Afra, S. P. (2019). Comparison of half cut solar cells with standard solar cells. In *Advances in science and engineering technology international conferences (ASET).* Available from https://doi.org/10.1109/ICASET.2019.8714488.

Kajihara, A., & Harakawa, T. (2005). Model of photovoltaic cell circuits under partial shading. In *Proceedings of the IEEE international conference on industrial technology* (Vol. 2005, pp. 866–870). Available from https://doi.org/10.1109/ICIT.2005.1600757.

Lashway, C. (1988). Photovoltaic system testing techniques and results. *IEEE Transactions on Energy Conversion, 3*(3), 503–506. Available from https://doi.org/10.1109/60.8058.

Mermoud, A., & Lejeune, T. (2019). Partial shadings on PV arrays: By-pass diode benefits analysis. In *25th European photovoltaic solar energy conference.*

Moballegh, S., & Jiang, J. (2014). Modcling, prediction, and experimental validations of power peaks of PV arrays under partial shading conditions. *IEEE Transactions on Sustainable Energy, 5*(1), 293–300. Available from https://doi.org/10.1109/TSTE.2013.2282077.

Numan, A., Dawood, Z., & Hussein, H. (2020). Theoretical and experimental analysis of photovoltaic module characteristics under different partial shading conditions. *International Journal of Power Electronics and Drive Systems, 11*(3), 1508. Available from https://doi.org/10.11591/ijpeds.v11.i3.pp1508-1518.

Praveen, K., & Suresh, M. (2020). Performance analysis of PV array configurations (SP, BL, HC and TT) to enhance maximum power under non-uniform shading conditions. *Engineering Reports, 2*(8), 1–22. Available from https://doi.org/10.1002/eng2.12214.

Salem, F., & Awadallah, M. A. (2016). Detection and assessment of partial shading in photo-voltaic arrays. *Journal of Electrical Systems and Information Technology*, *3*(1), 23−32. Available from https://doi.org/10.1016/j.jesit.2015.10.003.

Sarniak, M. (2020). Modeling the functioning of the half-cells photovoltaic module under partial shading in the Matlab package. *Applied Sciences*, *10*(7). Available from https://doi.org/10.3390/app10072575.

Shaima, K., Abdulridha., Saad, A., Tuma, O. A., & Abdulrazzaq. (2021). Study of the partial shading effect on the performance of silicon PV panels string. *Journal of Applied Sciences and Nanotechnology*, *1*, 32−42.

Sharma, S., Jain, K. K., & Sharma, A. (2015). Solar cells: In research and applications— A review. *Materials Sciences and Applications*, *06*(12), 1145−1155. Available from https://doi.org/10.4236/msa.2015.612113.

Vieira, R., de Araújo, F., Dhimish, M., & Guerra, M. (2020). A comprehensive review on bypass diode application on photovoltaic modules. *Energies*, *13*(10), 2472. Available from https://doi.org/10.3390/en13102472.

Woyte, A., Nijs, J., & Belmans, R. (2003). Partial shadowing of photovoltaic arrays with different system configurations: Literature review and field test results. *Solar Energy*, *74*(3), 217−233. Available from https://doi.org/10.1016/s0038-092x(03)00155-5.

Experimental investigation of the interaction between PV soiling loss and atmospheric and climatic conditions: a case study of the semiarid climate of Morocco

9

Mounir Abraim[1,2], Massaab El Ydrissi[1,2], Omaima El Alani[1,2], Hicham Ghennioui[1], Abdellatif Ghennioui[1,2], Mohamed Boujoudar[1,2] and Alae Azouzoute[3]

[1]Laboratory of Signals, Systems, and Components, Sidi Mohamed Ben Abdellah University, Fez, Morocco, [2]Green Energy Park research platform (IRESEN/UM6P), Ben Guerir, Morocco, [3]National Renewable Energy Laboratory, PV Materials Reliability and Durability Group, Denver, CO, United States

9.1 Introduction

Soiling is the process of accumulation of dust particles and other contaminants on the surface of solar collectors (Ilse et al., 2018). This phenomenon is mostly pronounced in the desert, arid, and semiarid regions, where most large-scale solar projects are deployed (Abraim, Salihi, et al., 2022; Aïssa et al., 2022). The accumulated soiling prevents solar irradiance from reaching the internal PV cells due to absorption and scattering of the incoming light (Abraim, El Gallassi, et al., 2022; Azouzoute et al., 2020; Bellmann et al., 2020). This leads to a significant loss of PV glass transmission and, consequently, a loss of energy that can exceed 1%/day (Sayyah et al., 2014). The effect of soiling on PV plant performance has been widely addressed in the literature (Al-Addous et al., 2019; Dehghan et al., 2022; Gostein et al., 2014; Sayyah et al., 2014; Younis & Alhorr, 2021). Most of these studies agreed that soiling depends on the local weather conditions (Conceição et al., 2022; Figgis & Helal, 2022; Hasan et al., 2022; Hussain et al., 2022). Understanding the soiling rate accumulation as a function of weather conditions can provide guidance for the selection of potential sites for large-scale solar projects. Moreover, it can serve to develop more accurate PV soiling prediction models. Only a few relevant studies address the question of the relation between PV soiling and weather conditions. For instance, Micheli and Muller (2017) investigated the possibility of predicting PV soiling losses using only meteorological parameters in 20 sites across the United States. The particulate matter (PM) was found to be the best soiling predictor coefficient of determination ($R^2 = 0.82$). The high R^2

Performance Enhancement and Control of Photovoltaic Systems. DOI: https://doi.org/10.1016/B978-0-443-13392-3.00009-8

was obtained since the annual average of the daily mean value was used. This eliminates large variations and random fluctuations and thus considerably strengthens the correlation. Soiling was found to be a very complex phenomenon and not easy to model (Hussain et al., 2022; Laarabi et al., 2022). A promising approach to predict soiling loss as a function of environmental variables is the use of artificial neural networks (ANNs). ANNs can capture complex interactions between environmental variables and daily changes in soiling rate accumulation (Javed et al., 2017; Zitouni et al., 2021). However, this type of model is not transportable and a model developed for one site may not necessarily be relevant to another site. Kimber et al. (2006) propose a simplified, site-independent method for estimating the time series of daily and annual soiling losses by combining site soiling rate with rainfall patterns. This method assumes that soiling accumulates at a fixed rate and that cleaning only takes place when daily total rainfall exceeds a given threshold. However, this assumption is not realistic, since soiling is characterized by a high daily and seasonal variability (Javed et al., 2020). Furthermore, this method requires giving the site daily average soiling rate as input. This information is not available for all regions and requires to have a history of ground-soiling data. It should be noted that the effectiveness of rain cleaning cannot be defined based only on total daily rainfall, as other factors are also involved (Hanrieder et al., 2021). Another site-independent soiling estimation model was developed by the Humboldt State University (HSU) (Coello & Boyle, 2019). The HSU soiling model calculates accumulated particulate mass given ambient airborne particulate matter concentrations ($PM_{2.5}$ and PM_{10}) and rain data. The total accumulated particulate mass is then used to estimate the soiling loss using settling velocities. This model also has some limitations such as the need to provide the deposition or settling velocity of particulates, which significantly affects the accuracy of the model. It also assumes perfect cleaning when the rainfall exceeds a given threshold, which is not entirely correct, as explained earlier.

The purpose of this study is to provide experimental results on the influence of environmental factors on soiling rate accumulation in the semiarid climate of Morocco and also to help identify and recapitulate the important interactions between PV soiling and the main environmental factors. To this end, a DustIQ PV optical soiling sensor was used to measure transmission loss due to soiling, and a high-precision weather station was also installed at the same location to collect weather data. The use of the DustIQ is beneficial to this study as it measures transmission loss due to soiling with an internal optical system, making it independent of sunlight and sky conditions (Korevaar et al., 2017; Wolfertstetter et al., 2021). This improves the stability of the data and minimizes the uncertainty of the measurement caused by fluctuating sky conditions.

This paper is structured as follows: Section 9.2 describes the site of study, the experimental setup, the data source, and the data preprocessing procedure. Section 9.3 provides the results and discussion of the performed analysis, including the main interactions and relationships between PV soiling and weather factors in the semiarid climate of Morocco. Furthermore, a multiple linear regression (MLR) model was fitted to combine all the parameters as a linear function of PV soiling. In Section 9.4, a conclusion of the results is given.

Figure 9.1 Green Energy Park research facility in Ben Guerir.

9.2 Experimental setup and dataset

Data collection for this experimental study was conducted at the Green Energy Park research facility located in Ben Guerir, Morocco (Fig. 9.1). According to the Köppen-Geiger climate classification, Ben Guerir corresponds to the Bsh (hot semi-arid) climate (El Alani et al., 2022; El Ydrissi et al., 2020). This site is not only characterized by high solar potential but also by high dust accumulation on PV modules, making it a good choice for soiling studies (Abraim, Salihi, et al., 2022; Abraim, El Gallassi, et al., 2022; Azouzoute et al., 2021).

A real-time PV optical soiling sensor and a high-precision weather station were exposed from February 27, 2018, to June 6, 2019, at the same time and under the same conditions, to collect PV soiling and weather measurements. Dust aerosol concentration is one of the important factors contributing to the accumulation of soiling on PV solar panels (Conceição et al., 2018; Fountoukis et al., 2018). Aerosol optical depth (AOD) is a good indicator of the particulate aerosol mass concentration in the atmosphere (Capdevila et al., 2016; Conceição et al., 2018; Prasad et al., 2022). Therefore, this parameter was used in this study. The approach and parameters used in this study are shown in Fig. 9.2 and further described in the following subsections.

9.2.1 Weather parameters

Measurements of the site's weather conditions were carried out thanks to a high-precision meteorological station installed at the Green Energy Park in Ben Guerir, Morocco (see Fig. 9.3). This station is equipped with several accurate and calibrated sensors. The detailed technical specifications of each sensor are presented in

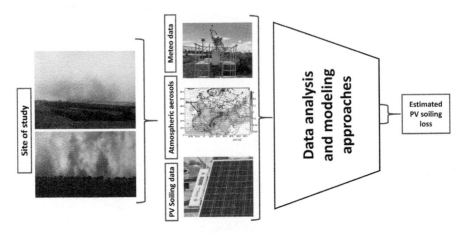

Figure 9.2 The approach and parameters used in this study.

Figure 9.3 High-precision weather station installed at the Green Energy Park.

Table 9.1. The main weather parameters used in this study are air temperature (T_a), relative humidity (RH), wind speed (WS), and rain.

The dust aerosol optical depth at 550 nm (AOD) was obtained from the CAMS-AOD Copernicus database and provided by the European Centre for Medium-Range Weather Forecasts (ECMWF) using MODIS. The weather parameters were measured with a one-minute time step. The daily mean was then calculated for T_a,

Table 9.1 Meteorological station sensor specifications.

Parameter	Sensor	Measurement range	Accuracy
Temperature	Campbell Scientific CS215	−40°C to +70°C	± 0.3°C at 25°C ± 0.4°C over +5°C to +40°C ± 0.9°C over −40°C to +70°C
Relative humidity	Campbell Scientific CS215	0%−100%	± 2% over 10%−90% ± 4% over 0%−100%
Wind speed	NRG #40C Anemometer	1−96 m/s	<0.1 m/s over 5−25 m/s
Rain	Tipping Bucket Rain Gauge Young 52202	0.1 mm per tip	± 2% up to 25 mm/h ± 3% up to 50 mm/h

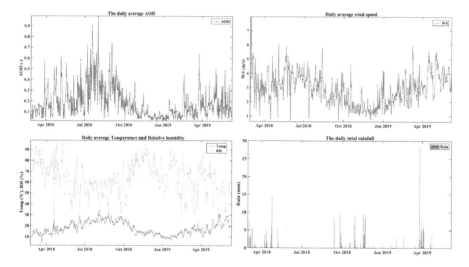

Figure 9.4 The daily values of weather parameters.

WS, RH, and AOD. In the case of rain, the daily sum was considered. The daily values of the weather parameters used in this study are illustrated in Fig. 9.4.

9.2.2 PV soiling measurements

The measurement of the soiling effect on the optical efficiency of PV solar collectors was performed using the DustIQ PV optical soiling sensor from Kipp&Zonen. This sensor has two separate optical sensors that quantify the accumulation of soiling through the scattered blue LED light received from the accumulated soiling particles on the top of its glass panel (see Fig. 9.5).

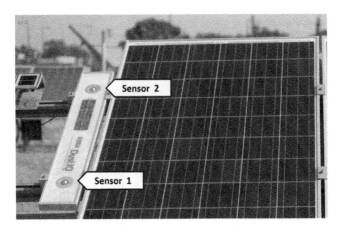

Figure 9.5 DustIQ installed at Green Energy Park.

The scattered LED light is measured by photodiodes, and the output signals are then converted by an internal processing unit into a transmission loss (*TL*) and a soiling ratio (*SR*), where

$$SR_t = 1 - TL_t \tag{9.1}$$

The *TL* represents the transmission loss of *PV* glass due to soiling accumulation, and *SR* is the remaining optical efficiency of a soiled PV glass. These parameters are the most widely used in the literature to indicate the effect of soiling on the performance of PV technology (Conceição et al., 2022; Gostein et al., 2014; Nepal et al., 2019).

Using the DustIQ for this specific study has many advantages over conventional methods of measuring PV soiling. The DustIQ performs soiling measurements without the need for solar irradiance, making it independent of solar angle of incidence (AOI) and sky conditions. This reduces the uncertainty caused by fluctuating sky and irradiance and leads to a more accurate analysis where soiling is only dependent on the influence of weather conditions.

During the measurement campaign, the soiling ratio was measured with a time step of one minute to comply with the guidelines of the International Electrotechnical Commission (IEC) International Standard 61724-1 for monitoring the performance of PV systems (International Electrotechnical Commission, 2017). When the DustIQ is mounted vertically, the two soiling sensors will produce different soiling ratio readings (Abraim, Salihi, et al., 2022), as the soiling tends to accumulate more in the lower parts due to gravity and morning dew that partially moves the soiling particles downward (Gostein et al., 2015). To calculate the daily *SR* and prevent data noise caused by dew formation, the average value of the data between

11:00 a.m. and 4:00 p.m. is calculated when it is guaranteed that no dew is present on the measuring surface of the instrument (Eq. 9.2).

$$SR = \frac{1}{N} \sum_{h=11\text{a.m}}^{h=4\text{p.m}} \frac{SR1 + SR2}{2} \qquad (9.2)$$

where $SR1$ and $SR2$ are the readings from sensor 1 and sensor 2, respectively. N is the number of selected daily data points.

The focus of this study is to determine the interaction and relationship between weather conditions and daily variation in soiling. Therefore, the soiling rate ΔSR is calculated to present the daily variation of SR. This parameter is more meaningful since the soiling ratio of the next day SR_{d+1} will be compared with that of the present day SR_d (Eq. 9.3).

$$\Delta SR_d = SR_{d+1} - SR_d \qquad (9.3)$$

Negative ΔSR values signify performance reduction due to soiling accumulation, while positive values indicate performance recovery thanks to cleaning. The daily measured SR and the calculated ΔSR are shown in Fig. 9.6.

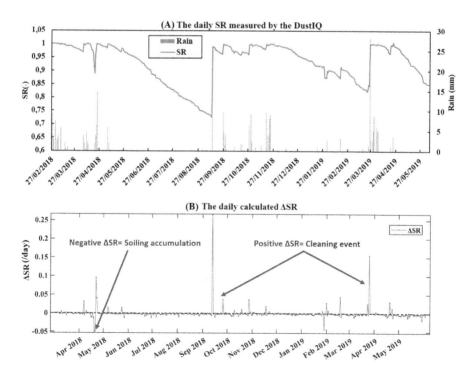

Figure 9.6 (A) The daily measured soiling ratio and (B) the calculated daily soiling rate.

9.3   Results and discussion

9.3.1   The interaction between PV soiling and weather conditions

Atmospheric aerosol is one of the significant parameters that define the rate of soiling accumulation. Nevertheless, the other weather factors (i.e., WS, RH, T_a, and rain) are also involved in the soiling accumulation process. Fig. 9.7A displays the scatter plot of ΔSR against AOD, and Fig. 9.7B shows the scatter plots of ΔSR against each weather parameter, where the plots are labeled with the value of their Pearson correlation coefficient. The Pearson's correlation test was applied to measure the linear correlation between the different weather parameters and the ΔSR, and it is defined in Eq. (9.4).

$$CC = \frac{\sum \left(\Delta SR - \Delta \overline{SR}\right)(y - \bar{y})}{\sqrt{\sum \left(\Delta SR - \Delta \overline{SR}\right)^2 \sum (y - \bar{y})^2}} \tag{9.4}$$

where CC represents the Pearson correlation coefficient, ΔSR is the soiling rate, $\overline{\Delta SR}$ is the mean of the values of ΔSR, y is the targeted weather parameter, and $\bar{y}$ is the mean of the values of the targeted weather parameter. It is worth noting that a CC value close to 1 or -1 means a strong positive or negative correlation, respectively, while a CC value around 0 means a very weak correlation.

The main goal of this first analysis is to understand the interaction between PV soiling and aerosols and to test the possibility of deriving an indication of soiling rate accumulation from the AOD data alone. Fig. 9.7A shows that, although there is no direct and linear correlation between ΔSR and AOD, cleaning events only occur when AOD is less than 0.5 (relatively low aerosol concentration), and no cleaning events can be observed when AOD values are high. Furthermore, the scatter plots in Fig. 9.7B and the results of Pearson's correlation test show a weak correlation

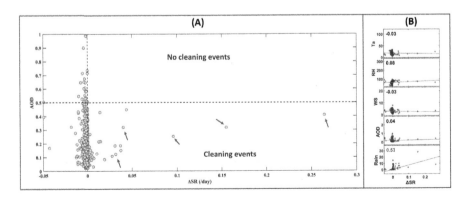

Figure 9.7 (A) The scatterplot of the daily ΔSR against AOD and (B) the result of Pearson's correlation tests between ΔSR and each weather parameter.

between each environmental parameter and the soiling rate ΔSR, indicating that it is difficult to explain the daily variation in soiling from weather conditions. Rain was found to have a significant correlation with ΔSR compared to the other parameters, which can be explained by the fact that rain is the most influential natural cleaning agent.

The very weak correlations between each independent weather parameter and ΔSR are surprising, as the related literature always associates the soiling phenomenon with site weather conditions (Chanchangi et al., 2021; Figgis, & Scabbia, & Aissa, 2022). These unexpected findings raise the possibility for further analysis to study the PV soiling behavior as a function of more than one weather parameter, as the soiling rate accumulation could be the effect of the interactions and combinations of several parameters. Fig. 9.8 represents scatter plots of ΔSR against AOD and every weather parameter, where the color intensities show the value of the weather parameters. As shown, PV soiling is a very complex phenomenon that cannot be easily predicted directly from the environmental variables. However, many insights and hypotheses can still be obtained from the collected data.

It can be observed from Fig. 9.8A that rain is the most significant weather parameter contributing to PV module cleaning. However, the cleaning efficiency of rain is not fully correlated with the total amount of rain, as other factors are also involved (e.g., rain intensity, soiling type, etc.). Moreover, a rain event that occurs when the solar panels are already free of soiling will not result in a significant cleaning effect and will not be visible in the data (e.g., 8 mm of rain has a more significant cleaning effect than 28 mm). When light rain (0.1 mm < rain < 2 mm) occurs during a high aerosol load in the atmosphere (see Fig. 9.9), ΔSR can reach -5%/day; this phenomenon is called red rain (dusty and turbid rain). This phenomenon makes the dust particles stickier and more difficult to be cleaned by the upcoming rain events, and it takes more time and effort to be cleaned manually (see Fig. 9.9).

In Fig. 9.8B, a clear link between ΔSR and temperature can be identified, as cleaning events are mainly related to low daily mean temperature ($T_a < 20°C$). This temperature is generally related to rainy days, while soiling accumulation days are mainly characterized by relatively high temperatures. This can be explained by the dry climate and the lack of precipitation (summer days). Days with an intermediate temperature may also exhibit soiling accumulation, possibly due to the influence of humidity. As shown in Fig. 9.8C, days with a high relative humidity between 50% and 75% may somehow favor soiling accumulation, since the humidity is not low enough to minimize adhesion forces and allow the detachment of soiling particles through wind, nor high enough to clean the surface (dew formation and gravity). In contrast, days of higher relative humidity (RH > 75%) are in most cases related to cleaning events, thanks to rain or dew formation and gravity. Wind speed is also a factor that can define the soiling accumulation rate. As shown in Fig. 9.8D, it is hard to extract a clear relationship between this parameter and ΔSR. However, the moments when the wind speed is relatively high are most likely related to the soiling accumulation due to dust transport. The effect of wind speed on soiling rate accumulation could be different, as other meteorological factors are also involved.

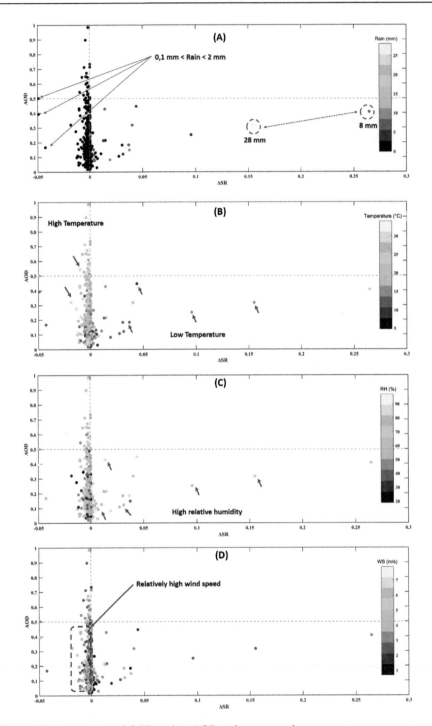

Figure 9.8 Scatter plots of △SR against AOD and every weather parameter.

Figure 9.9 (A) High dust-loaded atmosphere at the Green Energy Park and (B) the effect of red rain on the PV modules.

To recap, the use of tri-variate scatterplots allows for the deduction of some assumptions that may help in understanding the overall relationship between weather conditions and the rate of soiling accumulation. Otherwise, it is difficult to determine a firm conclusion about the direct correlation between each weather parameter and the rate of soiling accumulation. This emphasizes the complexity of PV soiling, since the rate of soiling accumulation depends on the interaction and combination of several weather parameters.

9.3.2 Multiple linear regression model

The previous analysis reveals that soiling rate accumulation is not related to each environmental parameter independently, but rather is the result of the combination and interaction between several variables. Therefore, to model the soiling phenomenon as a function of multiple weather parameters at the same time, a multiple linear regression (MLR) model was applied in this study. In this model, the ΔSR was simply expressed as a linear function of different environmental variables. The equation for this model is presented in Eq. (9.5).

$$\Delta SR = \beta_0 + \beta_1 \times T_a + \beta_2 \times RH + \beta_3 \times WS + \beta_4 \times AOD + \beta_5 \times Rain \qquad (9.5)$$

where β_0, β_1, β_2, ..., β_5 are the model parameters. The least-square linear regression method implemented in the Scikit-learn Python framework was used to fit the model and to obtain the parameters. This method minimizes the residual sum of squares between the observed target in the dataset and the target predicted by the linear approximation. The input dataset is similar to that used in the previous analysis. Table 9.2 shows the obtained parameters after fitting the model.

To assess the ability of the MLR model, the most common statistical metrics were used, mainly the root mean square error (RMSE), the mean absolute error (MAE), and the coefficient of determination R^2. As can be seen from Fig. 9.10, the developed model was able to track the trend of daily soiling variation ($R^2 = 52\%$).

Table 9.2 The resulted MLR model parameters.

Parameter	Value
β_0	1.2001×10^{-5}
β_1	3.9009×10^{-5}
β_2	-8.1683×10^{-5}
β_3	-7.4982×10^{-4}
β_4	0.0032
β_5	0.0043

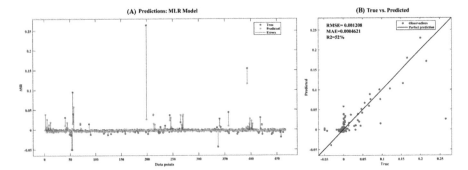

Figure 9.10 (A) MLR model predictions of ΔSR in comparison with the actual ΔSR and (B) regression plot of the MLR model.

The RMSE and MAE were found to be 0.001208 and 0.0004621, respectively. However, the error is still relatively high (see Fig. 9.10A). Therefore, and as expected, the PV soiling phenomenon is very complex to be modeled using a simple model such as MLR.

As mentioned earlier, the soiling phenomenon is very complex and cannot be easily predicted with a few parameters. Therefore, more powerful models (e.g., artificial neural networks) should be applied. There is also a need to provide more data to cover a wider range of possible effects and environmental scenarios. It is also worth noting that PV soiling is not only driven by environmental conditions, as there are other factors that influence soiling-related PV performance losses, for example, the physicochemical properties of the dust, the probability of dew formation, and the exposure time since the last cleaning or rainfall, which can promote the cementation of dust particles. Gathering information on these factors could be challenging, time-consuming, and costly, but it will certainly lead to a better understanding of the soiling accumulation process and improve the predictive performance of soiling models. Nevertheless, for more reliable data, it is highly recommended to install a soiling sensor in operational solar plants and in prefeasibility studies.

9.4 Conclusion

This study explores the interactions between PV soiling loss and site weather conditions. The results obtained indicate that PV soiling is a very complex phenomenon that cannot be easily modeled using simple meteorological data and simple models. However, the analysis of the collected weather and soiling data provides some insights into the general interaction between soiling rate accumulation and weather parameters. It was found that each weather parameter could explain a small portion of the daily variation in soiling rate. For example, temperature can indicate the season, rain can explain performance recovery through cleaning, and wind speed and relative humidity can contribute to either soiling accumulation or cleaning depending on their value. Soiling rate accumulation depends on the synergistic and interdependent effects of all-weather parameters. Using the MLR model to include all factors in a single equation improves the correlation. Nevertheless, the error obtained is still significant, as other factors also contribute to the soiling accumulation process, such as the physicochemical properties of the dust, the probability of dew formation, and the exposure time since the last cleaning or rainfall. Therefore, more sophisticated models such as ANNs and more data and parameters are needed to better understand the soiling phenomenon and improve the predictive performance of soiling models. Further research is needed to physically describe the soiling accumulation process in different climates and to identify the relevant parameters for PV soiling in the case of each climate type, as well as the possibility of developing a site-independent soiling model. Moreover, a soiling monitoring instrument is recommended for efficient planning of the cleaning strategy of PV solar plants.

Acknowledgments

This work was funded by the Green Energy Park. The authors are very appreciative of their support and the opportunity to work on their research platform.

References

Abraim, M., El Gallassi, H., El Alani, O., Ghennioui, H., Ghennioui, A., Hanrieder, N., & Wilbert, S. (2022). Comparative study of soiling effect on CSP and PV technologies under semi-arid climate in Morocco. *Solar Energy Advances*, 2, 100021. Available from https://doi.org/10.1016/j.seja.2022.100021.

Abraim, M., Salihi, M., El Alani, O., Hanrieder, N., Ghennioui, H., Ghennioui, A., El Ydrissi, M., & Azouzoute, A. (2022). Techno-economic assessment of soiling losses in CSP and PV solar power plants: A case study for the semi-arid climate of Morocco. *Energy Conversion and Management*, 270. Available from https://doi.org/10.1016/j.enconman.2022.116285, https://www.journals.elsevier.com/energy-conversion-and-management.

Al-Addous, M., Dalala, Z., Alawneh, F., & Class, C. B. (2019). Modeling and quantifying dust accumulation impact on PV module performance. *Solar Energy*, *194*, 86−102. Available from https://doi.org/10.1016/j.solener.2019.09.086, http://www.elsevier.com/inca/publications/store/3/2/9/index.htt.

El Alani, O., Hajjaj, C., Ghennioui, H., Ghennioui, A., Blanc, P., Saint-Drenan, Y.-M., & El Monady, M. (2022). Performance assessment of SARIMA, MLP and LSTM models for short-term solar irradiance prediction under different climates in Morocco. *International Journal of Ambient Energy*, 1−17. Available from https://doi.org/10.1080/01430750.2022.2127889.

Azouzoute, A., Merrouni, A. A., Garoum, M., & Bennouna, E. G. (2020). *Energy reports. Soiling loss of solar glass and mirror samples in the region with arid climate* (6, pp. 693−698). Elsevier Ltd. Morocco. Available from http://www.journals.elsevier.com/energy-reports/, 10.1016/j.egyr.2019.09.051, 23524847.

Azouzoute, A., Zitouni, H., El Ydrissi, M., Hajjaj, C., Garoum, M., Bennouna, E. G., & Ghennioui, A. (2021). Developing a cleaning strategy for hybrid solar plants PV/CSP: Case study for semi-arid climate. *Energy*, *228*, 120565. Available from https://doi.org/10.1016/j.energy.2021.120565.

Aïssa, B., Scabbia, G., Figgis, B. W., Garcia Lopez, J., & Bermudez Benito, V. (2022). PV-soiling field-assessment of Mars™ optical sensor operating in the harsh desert environment of the state of Qatar. *Solar Energy*, *239*, 139−146. Available from https://doi.org/10.1016/j.solener.2022.04.064, http://www.elsevier.com/inca/publications/store/3/2/9/index.htt.

Bellmann, P., Wolfertstetter, F., Conceição, R., & Silva, H. G. (2020). Comparative modeling of optical soiling losses for CSP and PV energy systems. *Solar Energy*, *197*, 229−237. Available from https://doi.org/10.1016/j.solener.2019.12.045, http://www.elsevier.com/inca/publications/store/3/2/9/index.htt.

Capdevila, H., Naidoo, V., & Graeber, M. (2016). *Soiling forecast and measurements for large PV power generation projects in dessert environments. Conference record of the IEEE photovoltaic specialists conference* (pp. 2071−2075). Germany: Institute of Electrical and Electronics Engineers Inc. Available from http://doi.org/10.1109/PVSC.2016.7749994, 9781509027248.

Chanchangi, Y. N., Ghosh, A., Baig, H., Sundaram, S., & Mallick, T. K. (2021). Soiling on PV performance influenced by weather parameters in Northern Nigeria. *Renewable Energy*, *180*, 874−892. Available from https://doi.org/10.1016/j.renene.2021.08.090, http://www.journals.elsevier.com/renewable-and-sustainable-energy-reviews/.

Coello, M., & Boyle, L. (2019). Simple model for predicting time series soiling of photovoltaic panels. *IEEE Journal of Photovoltaics*, *9*(5), 1382−1387. Available from https://doi.org/10.1109/JPHOTOV.2019.2919628, https://ieeexplore.ieee.org/xpl/RecentIssue.jsp?punumber = 5503869.

Conceição, R., González-Aguilar, J., Merrouni, A. A., & Romero, M. (2022). Soiling effect in solar energy conversion systems: A review. *Renewable and Sustainable Energy Reviews*, *162*. Available from https://doi.org/10.1016/j.rser.2022.112434, https://www.journals.elsevier.com/renewable-and-sustainable-energy-reviews.

Conceição, R., Silva, H. G., Mirão, J., Gostein, M., Fialho, L., Narvarte, L., & Collares-Pereira, M. (2018). Saharan dust transport to Europe and its impact on photovoltaic performance: A case study of soiling in Portugal. *Solar Energy*, *160*, 94−102. Available from https://doi.org/10.1016/j.solener.2017.11.059, http://www.elsevier.com/inca/publications/store/3/2/9/index.htt.

Dehghan, M., Rashidi, S., & Waqas, A. (2022). Modeling of soiling losses in solar energy systems. *Sustainable Energy Technologies and Assessments*, *53*, 102435. Available from https://doi.org/10.1016/j.seta.2022.102435.

Figgis, B., & Helal, M. (2022). Condensation characterization for PV soiling. *IEEE Journal of Photovoltaics*, *12*(6), 1522−1526. Available from https://doi.org/10.1109/JPHOTOV.2022.3202087, https://ieeexplore.ieee.org/xpl/RecentIssue.jsp?punumber = 5503869.

Figgis, B., Scabbia, G., & Aissa, B. (2022). Condensation as a predictor of PV soiling. *Solar Energy*, *238*, 30−38. Available from https://doi.org/10.1016/j.solener.2022.04.025, http://www.elsevier.com/inca/publications/store/3/2/9/index.htt.

Fountoukis, C., Figgis, B., Ackermann, L., & Ayoub, M. A. (2018). Effects of atmospheric dust deposition on solar PV energy production in a desert environment. *Solar Energy*, *164*, 94−100. Available from https://doi.org/10.1016/j.solener.2018.02.010, http://www.elsevier.com/inca/publications/store/3/2/9/index.htt.

Gostein, M., Caron, J. R., & Littmann, B. (2014). *Measuring soiling losses at utility-scale PV power plants*. IEEE. Available from http://doi.org/10.1109/PVSC.2014.6925056.

Gostein, M., Duster, T., & Thuman, C. (2015). *Accurately measuring PV soiling losses with soiling station employing module power measurements. IEEE 42nd photovoltaic specialist conference, PVSC 2015*. United States: Institute of Electrical and Electronics Engineers Inc. 9781479979448. Available from https://doi.org/10.1109/PVSC.2015.7355993.

Hanrieder, N., Wilbert, S., Wolfertstetter, F., Polo, J., Alonso, C., & Zarzalejo, L. F. (2021). *Why natural cleaning of solar collectors cannot be described using simple rain sum thresholds. Proceedings − ISES solar world congress 2021* (pp. 959−969). Spain: International Solar Energy Society 9783982040875. Available from https://doi.org/10.18086/swc.2021.37.02.

Hasan, K., Yousuf, S. B., Tushar, M. S. H. K., Das, B. K., Das, P., & Islam, M. S. (2022). Effects of different environmental and operational factors on the PV performance: A comprehensive review. *Energy Science and Engineering*, *10*(2), 656−675. Available from https://doi.org/10.1002/ese3.1043, http://onlinelibrary.wiley.com/journal/10.1002/(ISSN)2050-0505.

Hussain, N., Shahzad, N., Yousaf, T., Waqas, A., Javed, A. H., Khan, M. A., & Shahzad, M. I. (2022). Study of soiling on PV module performance under different environmental parameters using an indoor soiling station. *Sustainable Energy Technologies and Assessments*, *52*, 102260. Available from https://doi.org/10.1016/j.seta.2022.102260.

Ilse, K. K., Figgis, B. W., Naumann, V., Hagendorf, C., & Bagdahn, J. (2018). Fundamentals of soiling processes on photovoltaic modules. *Renewable and Sustainable Energy Reviews*, *98*, 239−254. Available from https://doi.org/10.1016/j.rser.2018.09.015.

International Electrotechnical Commission. (2017). *Photovoltaic system performance*. Technical Committee 82. International Electrotechnical Commission.

Javed, W., Guo, B., & Figgis, B. (2017). Modeling of photovoltaic soiling loss as a function of environmental variables. *Solar Energy*, *157*, 397−407. Available from https://doi.org/10.1016/j.solener.2017.08.046, http://www.elsevier.com/inca/publications/store/3/2/9/index.htt.

Javed, W., Guo, B., Figgis, B., Martin Pomares, L., & Aïssa, B. (2020). Multi-year field assessment of seasonal variability of photovoltaic soiling and environmental factors in a desert environment. *Solar Energy*, *211*, 1392−1402. Available from https://doi.org/10.1016/j.solener.2020.10.076, http://www.elsevier.com/inca/publications/store/3/2/9/index.htt.

Kimber, A., Mitchell, L., Nogradi, S., & Wenger, H. (2006). *The effect of soiling on large grid-connected photovoltaic systems in California and the Southwest Region of the United States. Conference record of the 2006 IEEE 4th world conference on photovoltaic energy conversion,*

WCPEC-4 (2, pp. 2391−2395). United States: IEEE Computer Society 9781424400164. Available from https://doi.org/10.1109/WCPEC.2006.279690.

Korevaar, M., Mes, J., Nepal, P., Snijders, G., & Mechelen, V. X. (2017). *Novel soiling detection system for solar panels. 33rd European Photovoltaic Solar Energy Conference and Exhibition* (pp. 2349−2351). EU PVSEC. Available from http://doi.org/10.4229/EUPVSEC20172017-6BV.2.11.

Laarabi, B., Sankarkumar, S., Rajasekar, N., El Baqqal, Y., & Barhdadi, A. (2022). Modeling investigation of soiling effect on solar photovoltaic systems: New findings. *Sustainable Energy Technologies and Assessments*, *52*, 102126. Available from https://doi.org/10.1016/j.seta.2022.102126.

Micheli, L., & Muller, M. (2017). An investigation of the key parameters for predicting PV soiling losses. *Progress in Photovoltaics: Research and Applications*, *25*(4), 291−307. Available from https://doi.org/10.1002/pip.2860, http://onlinelibrary.wiley.com/journal/10.1002/(ISSN)1099-159X.

Nepal, P., Korevaar, M., Ziar, H., Isabella, O., & Zeman, M. (2019). Accurate soiling ratio determination with incident angle modifier for PV modules. *IEEE Journal of Photovoltaics*, *9*(1), 295−301. Available from https://doi.org/10.1109/JPHOTOV.2018.2882468, http://eds.ieee.org/jpv.html.

Prasad, A. A., Nishant, N., & Kay, M. (2022). Dust cycle and soiling issues affecting solar energy reductions in Australia using multiple datasets. *Applied Energy*.

Sayyah, A., Horenstein, M. N., & Mazumder, M. K. (2014). Energy yield loss caused by dust deposition on photovoltaic panels. *Solar Energy*, *107*, 576−604. Available from https://doi.org/10.1016/j.solener.2014.05.030, http://www.elsevier.com/inca/publications/store/3/2/9/index.htt.

Wolfertstetter, F., Esquelli, A., Wilbert, S., Hanrieder, N., Blum, N., Korevaar, M., Bergmans, T., Zarzalejo, L., Polo, J., Alami-Merrouni, A., & Ghennioui, A. (2021). Incidence angle and diffuse radiation adaptation of soiling ratio measurements of indirect optical soiling sensors. *Journal of Renewable and Sustainable Energy*, *13*(3). Available from https://doi.org/10.1063/5.0048001, http://scitation.aip.org/content/aip/journal/jrse.

El Ydrissi, M., Ghennioui, H., Bennouna, E. G., & Farid, A. (2020). Techno-economic study of the impact of mirror slope errors on the overall optical and thermal efficiencies- case study: Solar parabolic trough concentrator evaluation under semi-arid climate. *Renewable Energy*, *161*, 293−308. Available from https://doi.org/10.1016/j.renene.2020.07.015, http://www.journals.elsevier.com/renewable-and-sustainable-energy-reviews/.

Younis, A., & Alhorr, Y. (2021). Modeling of dust soiling effects on solar photovoltaic performance: A review. *Solar Energy*, *220*, 1074−1088. Available from https://doi.org/10.1016/j.solener.2021.04.011, http://www.elsevier.com/inca/publications/store/3/2/9/index.htt.

Zitouni, H., Azouzoute, A., Hajjaj, C., El Ydrissi, M., Regragui, M., Polo, J., Oufadel, A., Bouaichi, A., & Ghennioui, A. (2021). Experimental investigation and modeling of photovoltaic soiling loss as a function of environmental variables: A case study of semi-arid climate. *Solar Energy Materials and Solar Cells*, *221*, 110874. Available from https://doi.org/10.1016/j.solmat.2020.110874.

Wide-spectrum solar radiation utilization techniques and applications

10

Abdalrhman A. Kandil, Mohamed S. Salem, Mohamed M. Awad and Gamal I. Sultan
Mechanical Power Engineering Department, Faculty of Engineering, Mansoura University, Mansoura, Egypt

10.1 Splitting system

Solar radiation is a broad range of wavelengths, with nearly all of the sun's energy concentrated within it. However, conventional photovoltaic (PV) cells can only respond to a specific subset of these wavelengths, based on the absorption and collection characteristics of the materials within the cell. As a result, a substantial section of the spectrum is simply converted into heat (Shockley & Queisser, 1961).

To overcome this limitation, scientists have proposed spectrum-splitting systems that can maximize the utilization of solar radiation. These systems generally include concentrators, along with a splitting device, a thermal receiver, and a photonic receiver as shown in Fig. 10.1.

Solar concentrators focus sunlight onto a small area, which increases the intensity of the light and allows PV cells to convert more of it into electricity. Beam splitters divide the sunlight into different wavelengths, which can then be directed to different types of PV cells that are optimized for those wavelengths (Caltzidis et al., 2021). Spectrum-splitting systems can increase solar power efficiency by up to 50%. They are still in the early stages of development, but they have the potential to revolutionize the solar power industry. Despite the potential advantages of beam-splitting systems, they are not yet commonly used on a commercial scale for several reasons. Firstly, the manufacturing process for the materials used in the splitting process can be complex and challenging to produce. Secondly, determining the precise wavelengths suitable for splitting can also pose a challenge. Lastly, the high cost of these materials is another factor that has limited their commercial use.

To overcome these challenges and facilitate the development of beam-splitting systems, it is crucial to explore and present different techniques used in the splitting process. This information can provide researchers and industry professionals with a comprehensive guide for further research and development, potentially leading to improved cost-effectiveness and wider implementation of beam-splitting systems.

Performance Enhancement and Control of Photovoltaic Systems. DOI: https://doi.org/10.1016/B978-0-443-13392-3.00011-6

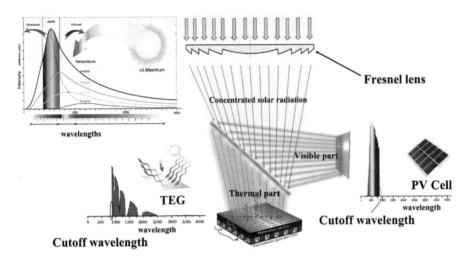

Figure 10.1 Example of applying a beam splitter with PV and thermal applications (Kandil, Awad, et al., 2023).

10.2 Beam-splitting techniques

Different techniques can be used in order to split the solar radiation, each with its own unique set of applications. Understanding these techniques is essential to construct an appropriate system that meets specific needs. In the next section, we discuss the different techniques used for splitting the solar spectrum and highlight some of their potential applications (Kandil, Salem, et al., 2023). By gaining a comprehensive understanding of these techniques, we can effectively design and implement a spectrum-splitting system that maximizes the utilization of solar radiation. The choice of which technique to use for splitting the solar spectrum depends on the specific application. By understanding the different available techniques, we can effectively design and implement a spectrum-splitting system that maximizes the utilization of solar radiation.

10.3 Prisms

Prisms are specialized optical devices that utilize the process of refraction to split the beam into its constituent components. The occurrence of this phenomenon can be attributed to the wavelength-dependent variation of the refractive index. As the white light enters the prism, it is refracted and split into its spectral components, with each wavelength bending at a slightly different angle (Stefancich et al., 2012). This results in a spectrum of colors, ranging from red to violet, being projected onto a screen or surface.

Dispersive prisms are commonly used in spectroscopy and other scientific applications where it is essential to separate and analyze different wavelengths of light. The ability of a dispersive prism to separate light into its constituent colors has significant implications for spectrum-splitting technology, as it can be used to divide solar radiation into specific wavelength ranges, allowing for greater efficiency in PV conversion (Thio & Park, 2019). Li et al. developed a novel prism-based model that incorporates antireflection coatings and microprisms to achieve a transmissivity of approximately 80%.

The system was tested in real-world scenarios and demonstrated its effectiveness in splitting beam radiation with minimal beam deflection. This design is particularly effective when used in conjunction with concentrated multijunction cells. Li, Michel, et al. (2019) proposed an innovative spectrum-splitting optical system that incorporates microprism arrays to enhance the performance of micro-CPV modules. The optical design achieved impressive optical transmission rates of 74.9% (experiment) and 80.9% (model) even without the use of antireflection coatings. By reducing beam deflection, this design enables efficient spectrum matching and facilitates the cost-effective and scalable production of micro-CPV modules.

10.4 Holographic concentrators

Holographic films are a type of specialized optical film that can be used in several applications, involving spectrum-splitting systems. They are manufactured through the recording of interference patterns formed between two beams of light. Holographic films have shown great potential in spectrum-splitting systems, where they can be used to split and focus solar radiation with high efficiency and minimal loss (Ju et al., 2017). With their unique recording process and optical properties, holographic films offer a promising pathway toward developing more efficient and versatile solar energy conversion solutions.

Holographic splitters offer several advantages over traditional methods. They are small, lightweight, and low cost, and they have high transmissivity. They are effective in concentrated photovoltaic/thermal (CPV/Th) systems.

A study by Meckler (1985) presented a novel CPV/Th system that incorporated a holographic splitter and a parabolic dish. In this configuration, the visible portion of the solar spectrum was efficiently directed to a solar cell integrated with the dish, while the infrared portion was effectively utilized in a thermal Rankine cycle. The experimental results showcased outstanding overall efficiency by synergistically combining the thermal Rankine cycle and the PV cell.

Marín-Sáez et al. (2019) developed a holographic system employing a silicon PV cell. The system design utilized a cylindrical holographic lens system with concentrators generated through an algorithmic approach. Experimental testing of the system was conducted under outdoor conditions, achieving a solar radiance concentration ratio of 3.48 suns. The corresponding output current density reached approximately 1460 A/m^2.

These findings highlight the potential of holographic splitters in developing high-performance spectrum-splitting systems that are efficient, lightweight, and cost-effective. By utilizing these systems, solar radiation can be effectively and efficiently converted into electricity and thermal energy, offering a promising pathway toward achieving a sustainable and clean energy future.

10.5   Dichroic beam splitters

A dichroic beam splitter is a commonly used device that splits the beam into two components: one that is transmitted and another that is reflected, each with its distinct wavelength. Various kinds of dichroic splitters are available, including cold and hot mirrors, which are two of the most important types (Mittelman et al., 2022). A cold mirror, for instance, is a splitter that reflects about 95% of the visible part while transmitting almost 85% of infrared radiation (Kaluba et al., 2020). Conversely, hot mirrors can reflect longer wavelengths while transmitting shorter ones. Researchers have conducted numerous studies via both types in spectrum-splitting systems to enhance their performance.

Several advantages come with using dichroic beam splitters in spectrum-splitting systems. These devices are cost-effective, lightweight, and highly efficient. They allow full utilization of the spectrum by splitting the beam radiation into its constituent wavelengths. This approach maximizes the use of solar energy by converting both visible and infrared light into electricity and thermal energy. Ultimately, dichroic beam splitters play a crucial role in the development of advanced spectrum-splitting systems, enabling more efficient and sustainable energy conversion from solar radiation.

Piarah and Djafar (2019) compared the characterization of both hot and cold mirrors. They tested both types under a 50-W halogen bulb. The study found that a cold mirror was more effective than a hot mirror, producing 100.53×10^{-3} W of energy compared to the hot mirror's 68.77×10^{-3} W. Based on these findings, it is recommended that cold mirrors be used as beam splitters in future research.

10.6   Nanofluid splitting system

One promising method to overcome these problems is using nanomaterials. Lately, nanofluids have been widely evaluated as beam splitters for many applications. Nanofluids offer the ability to suitably manage the wavelength. DeJarnette et al. (2015) examined the plasmonic nanofluid splitter for a CPV/Th system. To get better control of energy, the proposed system was constructed to be a two-pass fluid with heat recovery from solar. To produce electricity and hot water on a commercial and industrial scale, Looser et al. (2014) established a PV cell with a splitting system. Firstly, the fluid was used to cool and absorb infrared wavelengths to decrease the PV temperature.

As mentioned earlier, numerous fluids have been evaluated for possible degradation and long-term performance. Karami et al. (2015) examined the performance of using CuO nanofluid by changing the volume fraction and the flow rates of nanoparticles under the natural sun. The study aimed to determine if the system efficiency would improve by changing the aforementioned characteristics. Another similar experiment was implemented under the real sun by Gupta et al. (2015) using Al_2O_3/water as a nanofluid. It can be concluded that decreasing the flow rate would be more efficient for lower temperatures and higher efficiency, whereas increasing flow rates could be essential to reduce heat loss at higher temperatures.

10.7 Beam-splitting analysis

The following section describes the splitting process and the governing equation that can help researchers model the system for future work.

The splitting system mainly consists of a PV cell, a beam concentrator, and a beam splitter device. The concentrator focuses the sunlight onto the PV cell, while the dichroic mirror splits the light into two parts: One part is used to generate electricity, and the other part is used for thermal purposes.

The formulation of equations for each component is accomplished, taking into account various assumptions in the theoretical analysis. These assumptions include considering isotropic properties of PV materials that remain constant with temperature, disregarding thermal contact resistances, assuming a constant resistance for the PV cell, maintaining constant optical properties of each layer across different wavelengths, and neglecting heat loss from all sides of the PV cell due to its thin sheet thickness.

The study focuses on investigating the distribution of energy at the different parts of the system after splitting. The temperature distribution is estimated using the energy equation. In order to simulate the behavior of the solar cell following the splitter, a hybrid approach involving both electrical and thermal equations is employed.

The research is structured into three primary sections, encompassing thermal analysis, electrical analysis, and the integration of thermal and electrical models with the beam-splitting system.

The procedure for dividing the incident beam radiation at a specific wavelength entails evaluating the overall radiation (Q_{tot}) received by the concentrator (Fresnel lens), as illustrated in the following equation (Yin et al., 2018):

$$Q_{tot} = \int_{280}^{4000} A_{opt} G(\lambda) d\lambda \quad \text{(Shockley \& Queisser, 1961)}$$

where $G(\lambda)$ [W/m^2] is the irradiance as a function of wavelength, A_{opt} [m^2] is the area of the concentrator, and λ is the wavelength. However, determining the incident radiation over the PV part after splitting Q_{PV} [W] is crucial, and it can be estimated as

$$Q_{PV} = \int_{280}^{\lambda_c} A_{opt} \eta_{opt} R_{splitter} \alpha_{PV} G(\lambda) d\lambda \quad \text{(Kandil, Salem, et al., 2023)}$$

In this equation, η_{opt} represents the total efficiency of the concentrator (Gupta, 1982), α_{PV} is the absorptivity of the PV, $R_{splitter}$ is the reflectivity of the splitter, and λ_c is the cutoff wavelength of the solar splitter. It is important to note that the cutoff wavelength has different values depending on the type of beam splitter.

To calculate the incident radiation over the thermal part Q_{Th} [W].

$$Q_{Th} = \int_{\lambda_c}^{4000} A_{opt}\eta_{opt}\tau_{splitter}G(\lambda)d\lambda \quad \text{(Caltzidis et al., 2021)}$$

In this equation, $\tau_{splitter}$ represents the transmissivity of the beam splitter. Fig. 10.2 shows the ratio between the thermal and visible split parts under different cutoff wavelengths.

10.8   PV analysis via a beam splitter

10.8.1   Thermal PV model

A photovoltaic cell is made up of multiple layers of materials that work together to absorb sunlight and generate an electric current. The amount of electricity that a PV cell can generate depends on its size, efficiency, and the amount of solar radiation it receives. The energy that is not converted into electricity is converted into unwanted thermal energy. This equation, $[(1 - \eta_{SC}) \times S_{SC}]$, shows the relationship between the electrical and thermal models of a PV cell. The efficiency of a PV cell, η_{SC}, is calculated using the electrical model. Each layer of a PV cell has specific thermophysical and optical properties, which are listed in Tables 10.1 and 10.2.

For each PV layer (i), the heat conduction equation in X−Y coordinates is given by

$$\nabla.(U_i\nabla T_i) + S_i = q \quad \text{(Kandil, Salem, et al., 2023)}$$

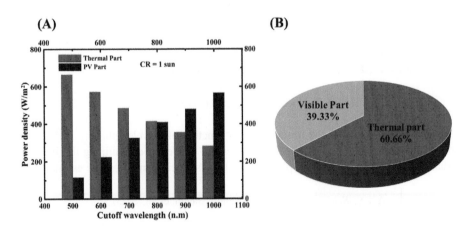

Figure 10.2 (A) A graph depicting the distribution of energy across various wavelengths, and (B) the energy split at a spec wavelength of 700 nm (Kandil, Awad, et al., 2023).

Table 10.1 Applications of using nanofluid as a beam splitter.

Application	Type of nanofluid splitting system	Results	References
CPV/thermal system with heat pipe	Ag/CoSO$_4$-propylene glycol nanofluid	Under five suns, the system efficiency achieved its maximum value of 73.20% The system efficiency was improved by about 10.4% compared to the conventional one	Han et al. (2020)
PV/thermal system	Ag@TiO$_2$ nanoparticles	While the PV conversion efficiency decreased, the thermal conversion efficiency witnessed an increase when the concentration of nanofluid was between 0 and 200 ppm.	Li, He, et al. (2019)
PV/thermal system	Ag/CoSO$_4$ + propylene glycol	The PV system achieved an efficiency of 30.2%, while the PV/T system demonstrated a thermal efficiency of 49.3%	Wang et al. (2022)
PV/thermal system	CoSO$_4$ + propylene glycol	The transmissivity decreased from 55.5% to 30.5% in the wavelength range of 200−975 nm. However, in the wider wavelength range of 200−2500 nm, the transmissivity exhibited a different trend; the transmissivity reduced from 68.8% to 7.7% as the optical path length increased.	Han et al. (2022)
Compare nanofluids in CPV/thermal system	ZnO + glycol	The correlation coefficient of the glycol-ZnO nanofluid as a beam splitter was found to be 0.218, which was 0.05 higher compared to the correlation coefficients of water/polypyrene and water/Ag/SiO$_2$ nanofluids, respectively.	Huaxu et al. (2020)
PV/thermal system	Ag@TiO$_2$ nanofluids + ethylene glycol/water	An overall efficiency of up to 83.7% under solar radiation 1000 W/m^2 was attained under a concentration of 200 ppm	He et al. (2019)

Table 10.2 Optical properties of PV materials (Kandil et al., 2022).

Material	Absorptivity	Emissivity	Reflectivity	Transmissivity
Glass cover	0.04	0.85	0.04	0.92
EVA layer	0.08	–	–	0.9
Tedlar	0.128	0.9	0.86	0.012
Silicon	0.92	–	–	0.2

Table 10.3 Thermophysical properties of PV materials (Kandil et al., 2022).

Layer	Density (kg/m^3)	Specific heat (J/kg K)	Thermal conductivity (W/m K)
Glass	3000	500	2
Tedlar	1200	1250	0.15
EVA	960	2090	0.311
ARC	2400	691	32
Silicon	2330	677	130

U_i [W/m K] denotes the thermal conductivity of the layer (i), T_i [K] is the layer's temperature, S_i [W/m^3] is the generated heat resulting from the split solar radiation [W/m^3] (Siddiqui & Arif, 2013), and q [W/m^2] is the rate of heat loss from the PV cell layers due to both conduction and radiation.

10.8.2 Electrical model

A PV electrical model is a mathematical representation of the electrical model for the PV module. This model can be used to estimate the output voltage, current, and power of the PV cell or module under different conditions of irradiance, temperature, and load. Table 10.3 presents some optical properties of silicon PV cell (Table 10.4).

The calculation of the short circuit current density of the PV cell under concentrated radiation is shown (Kandil, Awad, et al., 2023).

$$J_{SC} = CR \int_{280}^{\lambda_c} \eta_{opt} \tau_{PV} \alpha_{PV} G(\lambda) \frac{e\lambda}{h_p c} EQE(\lambda) d\lambda \quad \text{(Stefancich et al., 2012)}$$

where CR [suns] is the concentration ratio, τ_{PV} is the overall transmissivity of the cell, e is the element charge, h_p [J s] is the Planck constant, c [m/s] is the light speed, and EQE is the PV external quantum efficiency. The value of EQE is dependent on the type of PV.

To determine the open circuit voltage (V_{OC}) of the PV, we can use the following equation:

$$V_{OC} = \frac{h_p c}{\lambda_c} \frac{V_{OC,1}}{eE_g} + \frac{n_f k_B T_{PV}}{e} \ln(C) \quad \text{(Thio & Park, 2019)}$$

Table 10.4 Electrical characteristics of PV materials (Kandil et al., 2022).

Electrical properties	Value
Bandgap of PV, E_g (eV)	1.42
Boltzmann coefficient, k_B (J/K)	1.38065×10^{-23}
Current coefficient β_J (A/C/m^2)	0.73
Electron charge q (C)	1.6022×10^{-19}
Ideality factor of PV cell nf	1.33
Light speed c (m/s)	3.0×10^8
Open circuit voltage V_{OC} (V)	0.965
Planck constant h_p (J·s)	6.626×10^{-34}
Series resistance R_s (Ω)	12.16×10^{-3}
Efficiency temperature coefficient (K^{-1})	0.002

The equation involves various parameters such as the open circuit voltage at $CR = 1$ ($V_{OC.1}$), PV ideality factor (n_f), Boltzmann constant (k_B) [J/K], temperature of PV (T_{PV}) based on the thermal analysis, PV bandgap (E_g) [eV], and logarithm of concentration ratio ($\ln(C)$).

Regarding the electrical output power, it can be determined as shown.

$$P_{PV} = V_{OC}J_{SC}A_{PV}FF(1 - \beta_{ref}(T_{PV} - T_{Ref})) \quad \text{(Li, Michel, et al., 2019)}$$

It involves parameters such as PV area (A_{PV}), efficiency temperature coefficient of the PV cell (β_{ref}), fill factor (FF), and reference temperature (T_{Ref}). Both the fill factor and the normalized open circuit voltage can be obtained.

$$FF = \frac{v_{OC} - \ln(v_{OC} + 0.72)}{1 + v_{OC}} \quad \text{(Ju et al., 2017)}$$

$$v_{OC} = \frac{eV_{OC}}{k_B T_{PV}} \quad \text{(Meckler, 1985)}$$

Solved examples

Example 1

A beam-splitting system is utilized to split solar radiation of 800 W/m^2 with a concentration of 10 suns into two parts: One part with low wavelengths is directed to a PV module, and the other part is directed to a heat exchanger to heat the flow of water. The PV efficiency is set to be 0.2, at a cutoff wavelength of 700 nm, concerning the visible light, while the total PV efficiency is 0.13 concerning total radiation. The heat exchanger effectiveness is 0.7. All calculations are assumed to be per unit area. The following calculations are required:

1. Find the ratio between the visible and thermal parts at cutoff wavelengths of 500, 600, and 700 nm.
2. Determine the water outlet temperature at a cutoff wavelength of 700, given that the water flow rate is 0.05 kg/s, and the water inlet temperature is 298K.
3. Calculate the utilization factor of the entire system with respect to the total solar radiation.
4. Compare the PV alone with the PV/heat exchanger hybrid splitting system.

Solution

To calculate the ratios between visible and thermal parts, the following equations (Kandil, Awad, et al., 2023; Caltzidis et al., 2021) can be applied:

$$Ratio = \frac{Q_{PV}}{Q_{th}}$$

where

$$Q_{PV} = \int_{280}^{\lambda_c} A_{opt}\eta_{opt}R_{splitter}\alpha_{PV}G(\lambda)d\lambda$$

$$Q_{Th} = \int_{\lambda_c}^{4000} A_{opt}\eta_{opt}\tau_{splitter}G(\lambda)d\lambda$$

in the given equations, λ_c is the cutoff wavelength, while the area of both systems is assumed to be 1 m^2 based on the previous description. Furthermore, the other optical parameters have not been mentioned; they can be neglected.

Therefore, the aforementioned equation can be modified as

$$Ratio = \frac{\int_{280}^{\lambda_c} G(\lambda)d\lambda}{\int_{\lambda_c}^{4000} G(\lambda)d\lambda}$$

In order to solve the aforementioned integrations, we should get the values of solar radiation as a function of the wavelength, which can be found on the NASA website, and then any method of numerical integration can be applied. Since the required cutoff wavelengths are 500, 600, and 700 nm, these integrations can be obtained directly from Fig. 10.1.

Under a wavelength of 500 nm

$Q_{PV} = 100$ W/m^2 $Q_{th} = 700$

Under a wavelength of 600 nm

$Q_{PV} = 200$ W/m^2 $Q_{th} = 600$

Under a wavelength of 700 nm

$Q_{PV} = 300$ W/m^2 $Q_{th} = 500$ W/m^2

The ratio under 500 nm $= \frac{100}{700} = 0.142$

The ratio under 600 nm $= \frac{200}{600} = 0.33$

The ratio under 700 nm $= \frac{300}{500} = 0.6$

To calculate the water outlet temperature at a wavelength of 700 nm, the simple equation of heat storage can be applied:

$$\varepsilon * Q_{th} = \dot{m}C_p(T_{out} - T_{in})$$

$$T_{out} = T_{in} + \frac{C \times \varepsilon \times Q_{th}}{\dot{m}C_p}$$

where T_{in} is the water inlet temperature, T_{out} is the water outlet temperature, Q_{th} is the heat power from the thermal splitting part, C is the concentration ratio, $\dot{m}$ is the mass flow rate, and C_p is the water heat capacity that is equal to 4186 J/kg/K.

$$T_{out} = 298 + \frac{10 \times 0.7 \times 500}{0.05 \times 4186} = 314.7\text{K}$$

To calculate the utilization factor,

$$U.F = \frac{P_{th} + P_{PV}}{Q_{tot}}$$

For thermal power, it can be calculated from the heat transferred to the water, which is equal to

$$P_{th} = C \times \varepsilon \times Q_{th}$$

$$P_{th} = 10 \times 0.7 \times 500 = 3500 \text{ W}$$

For PV electrical power

$$P_{PV} = C \times \eta_{PV-splitting} \times Q_{PV}$$

$$P_{PV} = 10 \times 0.2 \times 300 = 600 \text{ W}$$

$$U.F = \frac{3500 + 600}{10 \times 800} = 0.5125$$

Regarding the PV alone

$$P_{PV} = C \times \eta_{PV} \times Q_{tot}$$

$$P_{PV} = 10 \times 0.13 \times 800 = 1040 \text{ W}$$

$$U.F = \frac{P_{PV}}{Q_{tot}}$$

$$U.F = \frac{1040}{10 \times 800} = 0.13$$

Comment:
The hybrid system has a higher utilization factor than the PV alone, which means that it is more efficient at converting solar radiation into electricity and heat. This is because the hybrid system can use a wide span of wavelengths of solar radiation, including the longer wavelengths that are not as efficient at producing

electricity. The hybrid system also can store the heat that it produces, which can be used later to generate electricity or to provide heating.

Example 2

A PV panel with a unity surface area (1 m^2) has an efficiency of 20% and receives 1000 W/m^2 of solar radiation. The panel is made up of a single layer of silicon with a thermal conductivity of 130 W/m K. The panel is surrounded by air with a heat transfer coefficient of 10 W/m^2 K. Assuming steady-state conditions, what is the temperature of the panel?

Solution:

The panel's electrical efficiency is 20%, so the amount of electrical power generated is

$$P = \eta_{sc} \times S = 0.2 \times 1000 \frac{W}{m^2} = 200 \ W/m^2$$

The remaining energy is converted into heat:

$$Q = (1 - \eta_{sc}) \times S = 0.8 \times 1000 \ W/m^2 = 800 \ W/m^2$$

Using the heat equation for a single layer, we can write as follows:

$$\nabla \cdot (U\nabla T) + Q = q$$

where U is the thermal conductivity of the silicon layer, T is the layer temperature, Q is the generated heat, q" is the heat flux from the panel to the surrounding air due to convection, and h is the heat transfer coefficient of air.

Assuming steady-state conditions, the heat flux into the panel is equal to the heat flux out of the panel:

$$q = h * (T - T_a)$$

where T_a is the ambient temperature.

Substituting these equations into the heat conduction equation and simplifying, we get

$$U\frac{\partial^2 T}{\partial x^2} + Q - h(T - T_a) = 0$$

Assuming 1-D heat transfer, we can integrate this equation to get

$$T(x) = \left(\frac{Q}{h} + T_a\right) + \left((T(0) - T_a) - \frac{Q}{h}\right) * \exp\left(\frac{-hX}{U}\right)X$$

where T (0) is the temperature at the surface of the panel (x = 0).

Assuming T (0) = Ta (i.e., the panel is in the equilibrium case with the surrounding air at the surface), we get

$$T(x) = T_a + \frac{Q}{h} * \left(1 - \exp\left(\frac{-hX}{U}\right)\right)$$

Plugging in the given values, we get
T(x) = 298 K + 800 W/m² / (10 W/m². K) * (1 − exp (− 10 W/m². K / 130 W/m. K * 1 m))

T(x) = 317 K

Therefore, the temperature of the panel is 317 K or 44°C.
Comment:
This problem demonstrates the application of the heat conduction equation and the heat transfer coefficient in calculating the temperature of a solar panel. It also highlights the importance of considering both the electrical efficiency and thermal properties of the panel when designing and optimizing solar energy systems. Additionally, it shows the significance of convective heat transfer in determining the temperature of the panel, which can impact its overall performance and durability.

Example 3
A photovoltaic cell with an area of 0.01 m² is exposed to solar radiation with a concentration ratio of 10 suns using a splitting system. The cutoff wavelength of the beam splitter is 874 nm. The cell has an optical efficiency of 0.9, a transmissivity of 0.95, an absorption coefficient of 0.9, and a quantum efficiency of 0.8 for wavelengths between 280 nm and 1100 nm. The cell temperature is 50°C, and the reference temperature is 25°C. Use the given electrical properties to calculate the following:

1. The PV short circuit current density.
2. The PV open circuit voltage.
3. The electrical output power of the cell.

For the short circuit current:
To determine the short circuit current, we can use the following equation (Stefancich et al., 2012):

$$J_{SC} = CR \int_{280}^{\lambda_c} \eta_{opt} \tau_{PV} \alpha_{PV} G(\lambda) \frac{e\lambda}{h_p c} EQE(\lambda) d\lambda$$

where $CR = 10$, $\tau_{PV} = 0.95$, $e = 1.6022 \times 10^{-19}$ C, $h_p = 6.626 \times 10^{-34}$ J s, $c = 3.0 \times 10^8$ m/s, $\lambda_c = 874$ nm, $E_g = 1.42$ eV, and $EQE(\lambda)$ is given as 0.8 for wavelengths between 280 nm and 1100 nm.

For the irradiance as a function of the wavelength, it can be estimated using the following formula:

$$G(\lambda) = 1.353 \times 10^{-3} \times \lambda^{-5} \times \left(\exp\left(\frac{1.438 \times 10^7}{\lambda \times T_{sun}} \right) - 1 \right)^{-1}$$

where $T_{sun} = 5762$ K. We can use numerical integration to evaluate the integral. Assuming a step size of 1 nm, we can calculate the short circuit current density as follows:

$$J_{SC} \approx 26.48 \ \text{A/m}^2$$

To calculate the open circuit voltage, we can use the following equation (Thio & Park, 2019):

$$V_{OC} = \frac{h_p c}{\lambda_c} \frac{V_{OC.1}}{eE_g} + \frac{n_f k_B T_{PV}}{e} \ln(C)$$

$$T_{PV} = 50°C + 273.15 = 323.15 \ \text{K}$$

Plugging in the values, we get

$$V_{OC} \approx 0.964 \ \text{V}$$

Therefore, the open circuit voltage is approximately 0.964 V.
To calculate the fill factor, we can use the following equation (Ju et al., 2017):

$$FF = \frac{v_{OC} - \ln(v_{OC} + 0.72)}{1 + v_{OC}}$$

where v_{OC} can be calculated as shown:

$$v_{OC} = \frac{eV_{OC}}{k_B T_{PV}}$$

Plugging in the values, we get

$$FF \approx 0.797$$

Therefore, the FF is approximately 0.797.
To evaluate the generated power of the cell, we can use the following equation (Li, Michel, et al., 2019):

$$P_{PV} = V_{OC} J_{SC} A_{PV} FF (1 - \beta_{ref}(T_{PV} - T_{Ref}))$$

Substituting the previously calculated values, we get

$$P_{PV} = 0.954 \times 28.07 \times 0.01 \times 0.797 \times (1 - 0.002 \times (50 - 25)) = 0.166 \text{ W}$$

Therefore, the generated power of the PV cell is 0.166 W.

Comment:

It is worth noting that in this calculation, the external quantum efficiency (EQE) of the PV cell was used as a constant of 0.8 for wavelengths between 280 nm and 1100 nm. However, in reality, EQE is a function of wavelength and can vary significantly across the spectral range of interest. Therefore, for more accurate results, a detailed spectral analysis of the EQE should be performed and incorporated into the calculation.

Problems

1. A beam splitter is used in a concentrated solar power system to split received solar radiation into visible and thermal components. The system consists of a PV cell with an area of 0.02 m², a Fresnel lens with a focal length of 0.5 m, and a beam splitter with a cutoff wavelength of 900 nm. The solar radiation is concentrated by a factor of 20 suns. The cell has an optical efficiency of 0.8, a transmissivity of 0.9, an absorption coefficient of 0.85, and a quantum efficiency of 0.75 for wavelengths between 300 nm and 1100 nm. What is the maximum concentration ratio that can be achieved with this system?

2. A beam splitter is used in a hybrid PV/PCM system to split incoming solar radiation into visible and thermal components. The system consists of a PV cell with an area of 0.015 m², a beam splitter with a cutoff wavelength of 850 nm, and a PCM with a melt temperature of 60°C. The solar radiation is concentrated by a factor of 15 suns. The cell has an optical efficiency of 0.85, a transmissivity of 0.92, an absorption coefficient of 0.88, and a quantum efficiency of 0.8 for wavelengths between 280 nm and 1100 nm. If the ambient temperature is 30°C, what is the maximum electrical output power that can be obtained from the PV cell?

3. A solar collector with an area of 1 m² is designed to use a beam splitter to split the received radiation into visible and thermal parts. The beam splitter has a cutoff wavelength of 700 nm. The collector has an optical efficiency of 0.85, a transmissivity of 0.9, an absorption coefficient of 0.95, and a thermal efficiency of 0.8. The collector is exposed to direct solar radiation with an intensity of 1000 W/m². Calculate the following:
   a. The amount of visible light incident on the collector.
   b. The thermal energy received by the collector.
   c. The total energy output of the collector.

4. A hybrid solar panel is designed to use a beam splitter. The panel consists of a PV cell with an area of 0.02 m² and a thermal absorber with an area of 0.1 m². The beam splitter has a cutoff wavelength of 850 nm. The panel has an optical efficiency of 0.9, a transmissivity of 0.95, an absorption coefficient of 0.8, and a thermal efficiency of 0.6. The panel is exposed to direct solar radiation with an intensity of 800 W/m². Calculate the following:
   a. The amount of visible light incident on the PV cell.
   b. The thermal energy is absorbed by the thermal absorber.
   c. The short circuit current density and open circuit voltage of the PV cell.
   d. The electrical output power of the panel.

5. A beam-splitting system with a concentration ratio of 20 suns is used to split solar radiation of 900 W/m^2 into visible and thermal parts. The visible part is focused into a PV cell with an efficiency of 0.18 at a cutoff wavelength of 650 nm, while the thermal part is directed to a heat exchanger with an effectiveness of 0.8. Determine the total electrical output power per unit area and the temperature rise of the water at a flow rate of 0.02 kg/s and an inlet temperature of 25°C.
6. A beam-splitting system is designed to split solar radiation of 1000 W/m^2 into visible and thermal parts with a concentration ratio of 15 suns. The visible part is directed to a PV cell with an efficiency of 0.25 at a cutoff wavelength of 600 nm, while the thermal part is directed to a heat exchanger with an effectiveness of 0.6. The PV cell and the heat exchanger are connected in series, and the load resistance is 50 Ω. Determine the maximum generated power and the corresponding voltage and current of the system.

$$Q_{PV} = \int_{280}^{\lambda_c} A_{opt}\eta_{opt}R_{splitter}\alpha_{PV}G(\lambda)d\lambda$$

$$Q_{Th} = \int_{\lambda_c}^{4000} A_{opt}\eta_{opt}\tau_{splitter}G(\lambda)d\lambda$$

References

Caltzidis, I., Kübler, H., Pfau, T., Löw, R., & Zentile, M. A. (2021). Atomic Faraday beam splitter for light generated from pump-degenerate four-wave mixing in a hollow-core photonic crystal fiber. *Physical Review A, 103*(4). Available from https://doi.org/10.1103/PhysRevA.103.043501, https://journals.aps.org/pra/abstract/10.1103/PhysRevA.103.043501.

DeJarnette, D., Brekke, N., Tunkara, E., Hari, P., Roberts, K., Otanicar, T. (2015). Design and feasibility of high temperature nanoparticle fluid filter in hybrid thermal/photovoltaic concentrating solar power. *Proceedings of SPIE - The International Society for Optical Engineering, 9559*. http://spie.org/x1848.xml, https://doi.org/10.1117/12.2186117. 1996756X SPIE, United States.

Gupta, P. K. (1982). Efficiency of fresnel lenses with respect to thermal losses. *Applied Energy, 12*(2), 87–98. Available from https://doi.org/10.1016/0306-2619(82)90020-4.

Gupta, H. K., Agrawal, G. D., & Mathur, J. (2015). Investigations for effect of Al$_2$O$_3$-H$_2$O nanofluid flow rate on the efficiency of direct absorption solar collector. *Case Studies in Thermal Engineering, 5*, 70–78. Available from https://doi.org/10.1016/j.csite.2015.01.002, http://www.journals.elsevier.com/case-studies-in-thermal-engineering/.

Han, X., Zhao, X., & Chen, X. (2020). Design and analysis of a concentrating PV/T system with nanofluid based spectral beam splitter and heat pipe cooling. *Renewable Energy, 162*, 55–70. Available from https://doi.org/10.1016/j.renene.2020.07.131, http://www.journals.elsevier.com/renewable-and-sustainable-energy-reviews/.

Han, X., Zhao, X., Huang, J., & Qu, J. (2022). Optical properties optimization of plasmonic nanofluid to enhance the performance of spectral splitting photovoltaic/thermal systems. *Renewable Energy, 188*, 573–587. Available from https://doi.org/10.1016/j.renene.2022.02.046, http://www.journals.elsevier.com/renewable-and-sustainable-energy-reviews/.

He, Y., Hu, Y., & Li, H. (2019). An Ag@TiO$_2$/ethylene glycol/water solution as a nanofluid-based beam splitter for photovoltaic/thermal applications in cold regions. *Energy Conversion and Management, 198.* Available from https://doi.org/10.1016/j.enconman.2019.111838.

Huaxu, L., Fuqiang, W., Dong, Z., Ziming, C., Chuanxin, Z., Bo, L., & Huijin, X. (2020). Experimental investigation of cost-effective ZnO nanofluid based spectral splitting CPV/T system. *Energy, 194.* Available from https://doi.org/10.1016/j.energy.2020.116913.

Ju, X., Xu, C., Han, X., Du, X., Wei, G., & Yang, Y. (2017). A review of the concentrated photovoltaic/thermal (CPVT) hybrid solar systems based on the spectral beam splitting technology. *Applied Energy, 187,* 534−563. Available from https://doi.org/10.1016/j. apenergy.2016.11.087, http://www.elsevier.com/inca/publications/store/4/0/5/8/9/1/index.htt.

Kaluba, V. S., Mohamad, K., & Ferrer, P. (2020). Experimental and simulated performance of hot mirror coatings in a parabolic trough receiver. *Applied Energy, 257.* Available from https://doi.org/10.1016/j.apenergy.2019.114020.

Kandil, A. A., Awad, M. M., Sultan, G. I., & Salem, M. S. (2022). Investigating the performance characteristics of low concentrated photovoltaic systems utilizing a beam splitting device under variable cutoff wavelengths. *Renewable Energy, 196,* 375−389. Available from https://doi.org/10.1016/j.renene.2022.06.129, http://www.journals.elsevier.com/renewable-and-sustainable-energy-reviews/.

Kandil, A. A., Awad, M. M., Sultan, G. I., & Salem, M. S. (2023). Performance of a photovoltaic/thermoelectric generator hybrid system with a beam splitter under maximum permissible operating conditions. *Energy Conversion and Management, 280.* Available from https://doi.org/10.1016/j.enconman.2023.116795.

Kandil, A. A., Salem, M. S., Awad, M. M., & Sultan, G. I. (2023). Adaptation of the solar spectrum to improve the use of sunlight: A critical review on techniques, applications, and current trends. *Advanced Sustainable Systems, 7*(6). Available from https://doi.org/10.1002/adsu.202300012, http://www.advsustainsys.com.

Karami, M., Akhavan-Bahabadi, M. A., Delfani, S., & Raisee, M. (2015). Experimental investigation of CuO nanofluid-based direct absorption solar collector for residential applications. *Renewable and Sustainable Energy Reviews, 52,* 793−801. Available from https://doi.org/10.1016/j.rser.2015.07.131.

Li, H., He, Y., Wang, C., Wang, X., & Hu, Y. (2019). Tunable thermal and electricity generation enabled by spectrally selective absorption nanoparticles for photovoltaic/thermal applications. *Applied Energy, 236,* 117−126. Available from https://doi.org/10.1016/j.apenergy.2018.11.085, http://www.elsevier.com/inca/publications/store/4/0/5/8/9/1/index.htt.

Li, D., Michel, J., Hu, J., & Gu, T. (2019). Compact spectrum splitter for laterally arrayed multi-junction concentrator photovoltaic modules. *Optics Letters, 44*(13), 3274−3277. Available from https://doi.org/10.1364/OL.44.003274, https://www.osapublishing.org/ol/abstract.cfm?uri = ol-44-13-3274.

Looser, R., Vivar, M., & Everett, V. (2014). Spectral characterisation and long-term performance analysis of various commercial heat transfer fluids (HTF) as direct-absorption filters for CPV-T beam-splitting applications. *Applied Energy, 113,* 1496−1511. Available from https://doi.org/10.1016/j.apenergy.2013.09.001.

Marín-Sáez, J., Chemisana, D., Atencia, J., & Collados, M. V. (2019). Outdoor performance evaluation of a holographic solar concentrator optimized for building integration. *Applied Energy, 250,* 1073−1084. Available from https://doi.org/10.1016/j.apenergy.2019.05.075, https://www.journals.elsevier.com/applied-energy.

Meckler, M. (1985). Fixed solar concentrator-collector-satelite receiver and co-generator. 490.

Mittelman, G., Kribus, A., Epstein, M., Lew, B., Baron, S., Flitsanov, Y., & Vitoshkin, H. (2022). Solar spectral beam splitting for photochemical conversion and polygeneration.

Energy Conversion and Management, *258*. Available from https://doi.org/10.1016/j.enconman.2022.115525.

Piarah, W. H., Djafar, Z., & Syafaruddin, M. (2019). The characterization of a spectrum splitter of Techspec AOI 50.0mm square hot and cold mirrors using a halogen light for a photovoltaic-thermoelectric generator hybrid. *Energies*, *12*(3). Available from https://doi.org/10.3390/en12030353, https://www.mdpi.com/1996-1073/12/3.

Shockley, W., & Queisser, H. J. (1961). Detailed balance limit of efficiency of p-n junction solar cells. *Journal of Applied Physics*, *32*(3), 510−519. Available from https://doi.org/10.1063/1.1736034.

Siddiqui, M. U., & Arif, A. F. M. (2013). Electrical, thermal and structural performance of a cooled PV module: Transient analysis using a multiphysics model. *Applied Energy*, *112*, 300−312. Available from https://doi.org/10.1016/j.apenergy.2013.06.030.

Stefancich, M., Zayan, A., Chiesa, M., Rampino, S., Roncati, D., Kimerling, L., & Michel, J. (2012). Single element spectral splitting solar concentrator for multiple cells CPV system. *Optics Express*, *20*(8), 9004−9018. Available from https://doi.org/10.1364/OE.20.009004, http://www.opticsinfobase.org/view_article.cfm?gotourl = http%3A%2F%2Fwww%2Eopticsinfobase%2Eorg%2FDirectPDFAccess%2FBF228CFC%2DAB15%2DBF6B%2DD1E9E8E762B453BA%5F231705%2Epdf%3Fda%3D1%26id%3D231705%26seq%3D0%26mobile%3Dno&org = .

Thio, S. K., & Park, S. Y. (2019). Dispersive optical systems for highly-concentrated solar spectrum splitting: Concept, design, and performance analyses. *Energies*, *12*(24). Available from https://doi.org/10.3390/en12244719, https://www.mdpi.com/1996-1073/12/24.

Wang, G., Ge, Z., & Lin, J. (2022). Design and performance analysis of a novel solar photovoltaic/thermal system using compact linear Fresnel reflector and nanofluids beam splitting device. *Case Studies in Thermal Engineering*, *35*. Available from https://doi.org/10.1016/j.csite.2022.102167.

Yin, E., Li, Q., & Xuan, Y. (2018). A novel optimal design method for concentration spectrum splitting photovoltaic−thermoelectric hybrid system. *Energy*, *163*, 519−532. Available from https://doi.org/10.1016/j.energy.2018.08.138, http://www.elsevier.com/inca/publications/store/4/8/3/.

Climate-specific bidding for solar photovoltaic-based power projects, considering the varied operation maintenance costs in large geographically divergent nations

11

Sanju John Thomas[1], Sheffy Thomas[2], Mohamed M. Awad[3] and Sudhansu S. Sahoo[1]
[1]School of Mechanical Sciences, Odisha University of Technology and Research, Bhubaneswar, Odisha, India, [2]Department of Electronics and Instrumentation, Federal Institute of Science and Technology (FISAT), Ernakulam, Kerala, India, [3]Mechanical Power Engineering Department, Faculty of Engineering, Mansoura University, Mansoura, Egypt

11.1 Introduction

Large geographically divergent nations have varied landscapes and seasons, while the policy and regulatory measures adapted for renewable energy integration are unilateral to meet timelines with international commitments (Chattopadhyay et al., 2015). India has got huge solar irradiation potential with average levels of 5.5 kWh/m²/day (Jamil et al., 2016), while geographically the land is divided into different regions based on climatic conditions (Chattopadhyay et al., 2015). The general categorizations of these climatic regions are hot and dry, warm and humid, moderate, composite, cold and cloudy, and cold and sunny (Chattopadhyay et al., 2016). Solar photovoltaic (PV) energy is widely accepted as a method to increase renewable energy share in the energy pool, considering technology readiness level and easy integration (Thomas et al., 2023). However, the solar PV power generation business is highly competitive and has reached rock bottom pricing of 2.44 INR/kWh in 2018 from 17.90 INR/kWh in 2010 (Thomas et al., 2022; Fundamentals and Innovations in Solar Energy, n.d.; New Research Directions in Solar Energy Technologies, Energy, Environment, and Sustainability, n.d.). The main parameters considered by the bidders in projects without subsidies/incentives are cheap land banks, logistical proximity, civil infrastructure, power evacuation facilities, and favorable policies for power intake (Thomas et al., 2022). The projects with land having good irradiation, level

Performance Enhancement and Control of Photovoltaic Systems. DOI: https://doi.org/10.1016/B978-0-443-13392-3.00012-8

terrain, water availability, and favorable weather conditions decrease the capital cost while blocking agricultural land (Thapar et al., 2018; Thomas et al., 2022). Ground-mounted solar PV plants are usually at a height of 1.5 m, while any additional requirement to adjust the angle of tilt, tides, water logging, and landscape irregularities will add to the capital cost (Panjawani et al., 2020). The main components of operational expenses are cleaning costs, labor, routine maintenance, and breakdowns. The life of solar PV panels is considered 25 years, for bidding calculations in solar PV projects; however, very few solar projects have reached the end stage to analyze life cycle economic calculations (Windarta et al., 2021). The annual degradation factor of solar PV panels is set by manufacturers as 0.5%−1.0%, while a quick degradation rate of 1.0%−3.0% happens in the first year due to field exposure (Reis et al., 2002). The degradation of solar panels is largely due to a reduction in maximum power (P_{max}), which for younger modules of less than 5 years fill factor (FF) due to microcracks and quality is the main reason. For older modules of age greater than 5 years, the reason besides FF is short circuit current (I_{sc}) due to symptomatic encapsulant discoloration (Chattopadhyay et al., 2015; Chattopadhyay et al., 2016). Interconnect breakage, encapsulant discoloration, delamination, snail trails, metallization, and backsheet degradation are the common defects of solar panels, but they vary in dimensions between climatic zones (Wohlgemuth et al., 2010). The defects are more common in hot and dry and wet and humid regions, while least in cold and cloudy regions (Chattopadhyay et al., 2015; Chattopadhyay et al., 2016). The accumulation of dust on the glass of solar panels affects the reliability of the performance of solar panels, accelerating defects. The mitigation mechanism of dust with proper cleaning mechanism provides a longer yield, which is directly linked to payback and internal rate of return (IRR) (Mani & Pillai, 2010; Rahman et al., 2015; Zaihidee et al., 2016). The average reduction in the efficiency of solar panels due to dust accumulation and associated impacts is between 11% and 33% in the world, while it is between 11% and 22% in India (Shenouda et al., 2022). In order to reduce operational cost, manual cleaning is adopted, which has lower dust-removing efficiency, while can damage the glass of solar panels (Al Shehri et al., 2016; Yuy et al., 2013). Water management, labor, and method of cleaning are factors that contribute to the operational expenses, while the effectiveness of cleaning directly contributes to the efficiency of the panels, longevity, avoidance of defects, and in turn payback (Hammad et al., 2018; Shah et al., 2020).

The Government of India's (GoI's) Jawaharlal Nehru National Solar Mission (JNNSM) provided accelerated solar PV integration (Bose & Sarkar, 2019). However, the policy has shifted from feed-in-tariff to e-reverse auctions where the price has substantially come down. The incentives and subsidies were cut since 42% of the plant cost is solar panels, which are generally imported. The rule to include local content requirement (LCR) in bidding has increased the number of patents in solar PV, but the project cost has increased (Probst et al., 2020). Thus, e-reverse auction is very competitive but will affect the quality of the components, leading to productivity and payback potential. Many such projects that have won at low prices stand canceled (Dubey et al., 2013; Probst et al., 2020). Thus in a cost-intensive solar PV installation scenario, the reverse auction to meet the lowest quoted price may bring long-term impacts, especially in a country with different climatic zones.

While field surveys on the performance of solar panels in different climatic zones in India are carried out (Chattopadhyay et al., 2015; Chattopadhyay et al., 2016; Dubey et al., 2013) and defects identified, there are no conclusive mitigation studies done to constrain defects, especially with respect to commercial implications. The current work hence has tried to accumulate climate zone-wise solar panel defects, corresponding mitigation strategies, and impact on operational cost. In order to identify the feasibility of a bidding price of INR 2.44/kWh, sensitivity analysis is done in varying capital cost, operational cost, annual incremental rate for operational cost and feed-in-tariff, rate of interest for debt and working capital, and annual degradation rate of solar panels. Based on sensitivity analysis, feasible IRR, levelized cost of electricity (LCOE), payback, and carbon credits with polycrystalline technology in a 70:30 debt-equity ratio are calculated.

11.2   Solar irradiation in India and different climatic zones

An estimated 5000 trillion kWh/year of solar equivalent is received by India, while the average daily global radiation is 5 kWh/m^2/day. The sunshine hours are estimated to be 2300 and 3200 hours per year in most parts of India (Khare et al., 2013). It is found that most of the regions have very good DNI of average 6.0−6.5 kWh/m^2/day, while parts of Rajasthan, Gujarat, central regions of Madhya Pradesh, Maharashtra, and the Deccan Plateau have more than average DNI (Fig. 11.1). The eastern parts of Ladakh have a very good DNI of above 7.0−7.5 kWh/m^2/day. The coastal regions have comparatively uniform DNI levels at 4.5−5 kWh/m^2/day, while the foothills of the Himalayas, parts of Uttar Pradesh, and northeastern states have DNI levels between 4.0 and 5.0 kWh/m^2/day.

India is classified into different climatic zones (Fig. 11.2; Chattopadhyay et al., 2015), which are hot and dry, warm and humid, composite, moderate, cold and sunny, and cold and cloudy. It is seen that most of the regions, which have DNI levels between 6.0 and 6.5 kWh/m^2/day, are classified as hot and dry, while the coastal regions and northeast are classified as warm and humid. The Ladakh region that has very good irradiation levels falls in the cold and cloudy zone, while the north-central region states predominantly fall under the composite category.

This categorization of zones was done by Chattopadhyay et al. (2015) based on Köppen studies on climate classification (Köppen, 1918). The National Solar Policy (Guidelines for Tariff Based Competitive Bidding Process for Procurement of Power from Grid Connected Solar PV Power Project) does not mention climate-specific bidding, while the solar panel standards are mentioned for crystalline silicon (c-Si) solar cell modules IEC 61215, the most commonly used technology

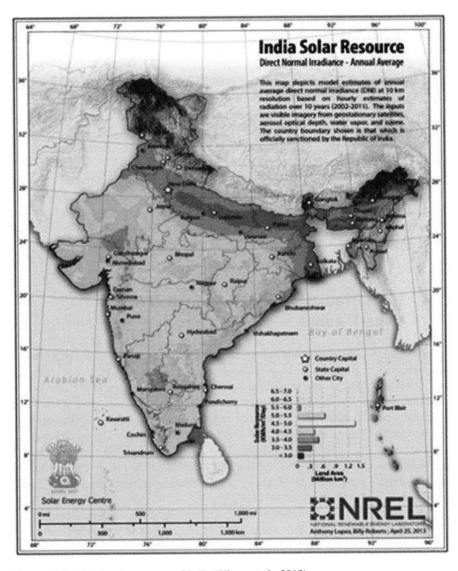

Figure 11.1 Solar irradiance map of India (Khare et al., 2013).

(Solar Energy Policies and Guidelines. Available at MNRE – SOLAR POLICIES, n.d.). There were few surveys and research done to identify the climate-specific operation efficiency and degradation levels of solar plants. Findings by Slamova et al. (2012) conclude that temperature, precipitation, humidity, altitude, thermal cycling, snow load, and air salinity affect PV performance. Studies by Attari et al.

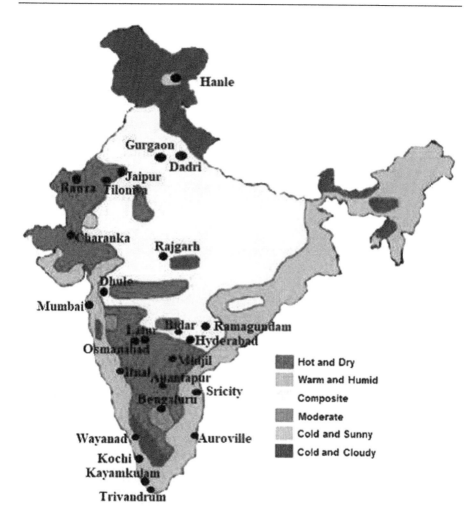

Figure 11.2 Climatic zone classifications of India (Chattopadhyay et al., 2015).

(2016) on a Moroccan solar plant conclude that in hot climates the losses in solar panels are higher due to soiling and dust. Studies performed by Ramanan et al. (2019) with polycrystalline and thin-film panels on small rooftop systems in warm and humid tropical regions concluded that thin films (with copper-indium-selenium) perform better. Studies by Boddapati and Daniel (2020) on a plant in warm and humid regions conclude that March has the maximum output, while July has the least due to rains. Studies by Ferrada et al. (2015) on different technologies in the same climatic region showed that thin film has a lesser drop when compared to monosilicon from winter to summer. Padmavathi and Daniel (2013) found that performance values varied by 60% in certain months; inverter failure and grid failure were pointed as reasons.

11.3 Defects in solar panels and cleaning mechanism

The defects in solar panels are due to interconnect breakage, encapsulant discoloration, delamination, snail trails, metallization, and backsheet degradation (Wohlgemuth et al., 2010). The interconnect breakage increases the series resistance that reduces the fill factor (FF), leading to a reduction in P_{max} (maximum power). These breakages can be point or line type, the severity of reduction in P_{max} having a direct correlation with the intensity of breakage (Chattopadhyay et al., 2015). The encapsulant discoloration is typically yellow or brown, which happens due to long-term exposure to humidity and high temperature, leading to metallization (Wohlgemuth et al., 2010). As the degree of encapsulant discoloration and affected area increases, the short circuit current (I_{sc}) is limited, leading to lower power output (Dubey et al., 2013). Delamination is defined as the loss of adhesion between the module laminates, usually occurring at the front side or on top of the solar cells, leading to an increase in the reflection losses that reduces the short circuit current I_{sc} and power output P_{max} (Mani & Pillai, 2010). Snail trails are curved black lines on top of solar cells due to the ingress of moisture and oxygen via microcracks (Richter et al., 2012). Metallization on PV panel fingers, bus bars, and interconnected ribbons conducts the current to the terminal box of the module. Corrosion is the cause of discoloration in the metallization of old modules, while in newer panels it can be due to manufacturing defects. This decolorization is additive in nature and can increase the series resistance, leading to loss in P_{max} (Dubey et al., 2013). The backsheet of the solar panel behaves as an insulation, and defects on the same due to chalking, cracking, burn marks, scratches, and delamination are termed backsheet degradation (Chattopadhyay et al., 2015). The front glass acts as a protection for solar cells, preventing environmental damage but allowing light. Degradation of the glass can be caused due to improper handling and cleaning practices, leading to haziness in the glass, spots, marks, scratches, or complete shattering. This can reduce short circuit current, allow scattering of rays, and affect transmittance. Degradation of the frame can be caused due to algae growth, corrosion, scratches, and joint separation. This will allow outside agents to pass through the frame, damaging the solar cells in the long run. Junction box degradation caused due to adverse environmental conditions opens the output terminals, while allowing ingress of water and moisture, affecting the overall output of the panel (Chattopadhyay et al., 2015).

Studies by Jordan et al. (2016) indicate that the solar panels in hot climates have more defects than in cold climates, as was the outcome of surveys conducted by Chattopadhyay et al. (2015), Chattopadhyay et al. (2016), and Dubey et al. (2013). Surveys conducted by Chattopadhyay et al. (2015) conclude that the degradation rate in young modules is largely due to the LID during the initial years of installation. This is more relevant in the crystalline panels that degrade faster during the initial years, while thin films degrade due to LID over a longer period of time (Jordan et al., 2018). Table 11.1 categorizes the climatic zones and the characteristics of the region. The table has considered hot and dry and warm and humid regions as hot regions, while composite, moderate, and cold and cloudy regions are

Table 11.1 Classification of climatic zones, common solar PV defects, and approximate reasons (Chattopadhyay et al., 2015; Chattopadhyay et al., 2016; Dubey et al., 2013).

Climatic zone	Regions/states	Mean monthly temperature (°C), humidity (%)	Common failures	Reasons
Zone A 1. Hot and Dry 2. Warm and Humid	Western Rajasthan, northern and central Gujarat, northern Karnataka, northwestern Telangana, northwestern Andhra Pradesh, central Maharashtra Southern Gujarat, coastal Maharashtra, coastal Karnataka, Kerala, coastal Tamil Nadu, coastal Andhra Pradesh, coastal Telangana, coastal Orissa, western Chhattisgarh, West Bengal, Assam, Arunachal Pradesh, Mizoram, Nagaland, Tripura, and Manipur	> 30°C, <55% > 30°C, >55%	Solder joint failure Encapsulant discoloration Delamination Metallization (discoloring and staining) Hot spots Snail trails Backsheet degradation Front-glass degradation	Manufacturing defect/difference in day-night temperatures Physical and chemical property changes of the EVA compound at high temperatures Found in warm and humid regions more than hot and dry. The reason being the ingress of moisture and high temperatures Corrosion due to ingress of water and improper soldering during manufacturing Accumulation of fine dust particles on glass cover/discoloration of the panels Microcracks during transportation Manufacturing defect Ingress of water and corrosion Careless installation practices Improper cleaning practices and handling during installation

(Continued)

Table 11.1 (Continued)

Climatic zone	Regions/states	Mean monthly temperature (°C), humidity (%)	Common failures	Reasons
Zone B 1. Moderate 2. Composite	Northeastern Tamil Nadu, Southeastern Karnataka Delhi, Haryana, Punjab, Uttar Pradesh, Bihar, Madhya Pradesh, Jharkhand, southern Chhattisgarh, and Uttarakhand	25°C–30°C, <75% None of the above or below in 6 months.	Snail trails Metallization of bus bars and interconnects Discoloration Backsheet degradation Interconnect breakages Delamination	Minute cracks during handling Scratches during cleaning practices Ingress of water, temperature variations, and manufacturing defects Physical and chemical reasons during manufacturing and exposure to the environment. Ingress of water and moisture, growth of algae, and corrosion Manufacturing defects Temperature variations Ingress of water and moisture Evaporation–Condensation process.
Zone C 1. Cold and Cloudy	Ladakh, eastern Kerala, and Sikkim	<25°C, >55%	Snail trails	Minute cracks during handling scratches during cleaning practices

considered as nonhot regions. The defects that occur in hot and nonhot regions are listed based on descending order from the most prevalent on the top. In hot and dry regions, which are predominantly northern Gujarat, western Rajasthan, and the Deccan Plateau, the mean temperatures are greater than 30°C, which go beyond 45°C during summer, and the humidity is less than 55%; the common defects are interconnect breakages due to solder joint failures and hot spots. LID, a common defect in all climatic zones, will have a higher degradation rate in hot and dry zones during the first year compared to other climatic zones.

The solder joint breakage and discoloration are two common defects in extreme temperatures (Jordan et al., 2017). Hot and dry and warm and humid regions have higher discoloration, leading to I_{sc} reduction. However, it is less in younger panels below 5 years, which conclude that discoloration starts after 5 years of life (Broek et al., 2012). Hot spots are common as solar irradiance increases beyond 700 W/m^2 (Spataru et al., 2015). The reason for hot spots is bypassed substrings, broken cells, and open solder joints. The stability of the backsheet plates plays an important role in preventing the ingress of moisture, leading to corrosion (Kempe et al., 2010). LID happens due to the presence of oxygen and takes a few months to stabilize, while thin film has higher degradation than silicon cells. The accepted variation of P_{max} per international standards of PV manufacturers is 0.6%−0.8% per year, which is usually considered a thumb rule, while the survey conducted (Chattopadhyay et al., 2015; Chattopadhyay et al., 2016; Dubey et al., 2013) has found the same varying between 0.6% and 4.0% per year. P_{max} degradation in young modules was largely due to FF, which throws light on the quality and handling of the panels in a price-competitive bidding process. For older modules, the I_{sc} losses due to encapsulant discoloration were the reason (Chattopadhyay et al., 2015; Chattopadhyay et al., 2016; Dubey et al., 2013). The roof-mounted solar panels were found to have a higher degradation ratio due to FF, which can be due to a careless installation procedure (Chattopadhyay et al., 2016). In general, encapsulant discoloration, metal and interconnect corrosion, and interconnect breakages are the common defects in hot and dry and warm and humid regions, while snail trails are typical in moderate and composite regions. Metallization (discoloration) is a common defect across all regions, while delamination is frequently found in warm and humid regions. The cold and cloudy regions have the minimum defects and are found to be the most suitable for solar PV plants. Table 11.2 and Fig. 11.3 describe the occurrence of various defects in different climatic zones. It is found that metallization is a common defect in all zones, while it is the single major defect in the composite region. Similarly, snail trails are a common defect in nonhot regions than in hot regions. Warm and humid regions have the maximum defects, while cold and cloudy regions have the minimum defects.

The P_{max} degradation rates per climatic zone due to I_{sc}, open voltage (V_{oc}), and FF are indicated in Fig. 11.4. It is found that in the hot and dry zone and warm and humid zone, the P_{max} reduction component is largely due to FF. Even in moderate regions, the FF factor is the main contributor, due to high metallization. The analysis is based on the survey conducted in 2013−14 by Chattopadhyay et al. (2015). Consecutive surveys by Chattopadhyay et al. (2016) and Dubey et al. (2013) have

Table 11.2 The occurrence of defects in various climatic zones with respect to a number of panels tested (inside brackets).

Defect	Interconnect breakage	Discoloration	Delamination	Snail trail	Metallization	Backsheet damage
Hot and dry	27 (142)	32 (104)	31 (142)	18 (142)	65 (112)	39 (104)
Warm and humid	53 (154)	63 (272)	33 (272)	6 (263)	73 (273)	63 (272)
Composite	28 (168)	45 (256)	13 (256)	3 (256)	67 (252)	60 (256)
Moderate	5 (135)	8 (135)	27 (125)	53 (135)	59 (131)	39 (135)
Cold and cloudy	—	0 (112)	2 (112)	45 (112)	16 (112)	—

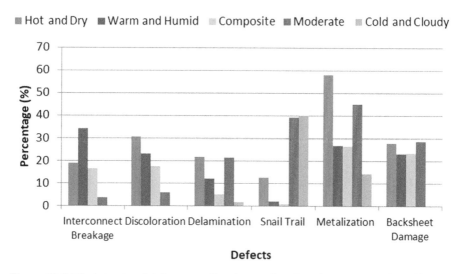

Figure 11.3 The intensity of defects per climatic zone based on the survey conducted by Dubey et al. (2013).

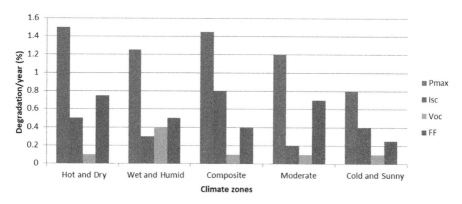

Figure 11.4 The degradation per year on P_{max} and corresponding contributors per climatic zone.

classified the degradation rate on technology-based solar panels, which is indicated in Fig. 11.5. Monocrystalline (Mono C-Si), multicrystalline/polycrystalline (Multi C-Si), amorphous silicon (a-Si), cadmium-telluride (Cd-Te), and a heterojunction with intrinsic thin-layer (HIT) panels were compared for P_{max} degradation. It is found that HIT panels are the best, while polycrystalline panels have the worst degradation pattern. The survey was based on the largest group of panels from monocrystalline and polycrystalline panels, while the least from the HIT panels.

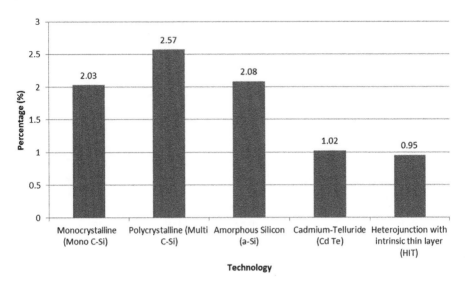

Figure 11.5 Degradation rate per year of solar PV panels based on technologies.

11.4   Results and discussion

The survey and research results show that there is a clear distinction between the degradation rates based on climatic patterns and the factors that contribute to the P_{max} reduction. Similarly, there is a clear distinction between the degradation rates and technologies with respect to time periods through the life of the panels. Thus, it will be critical to consider these degradation rates in the cost calculations, while arriving at the IRR, LCOE, and payback. The degradation rates by the manufacturers are between 0.6% and 0.8% per year with a higher rate during the first year due to LID.

Some of the factors that contribute to the degradation rates can be mitigated by integrated agriculture practices (agrivoltaics) or proper microclimate change mitigations (Thomas & Kumar, 2021). The mitigation for the accumulation of dust, hot spots, microcracks, corrosion, and ingress of water requires proper cleaning and maintenance measures. The defects, reasons for the same, mitigation management, type of maintenance, and impact on operational expenses are mentioned in Table 11.3.

Considering the climate-specific panel defects, mitigation strategies, the impact on operational cost, and degradation levels, the IRR, payback, LCOE, and carbon credit eligibility are calculated. A 1-MW ground-mounted solar plant is evaluated for different climatic zones. Hot and dry and wet and humid are clubbed as Zone A, while composite and moderate as Zone B and cold and cloudy as Zone C. The following considerations are considered as constants for all the sites. The irradiance levels of Jodhpur, Madurai, Patna, Bangalore, and Ooty are considered to be from hot and dry, wet and humid, composite, moderate, and cold and cloudy regions. The following factors are kept constant.

1. Technology: Polycrystalline
2. Wattage: 180 Wp

Table 11.3 Solar PV panel defects, mitigation mechanisms, and impact on operational cost.

Common defects	Reasons	Mitigation management	Type of maintenance	Impact on operational expenses
Solder joint failure	Manufacturing defect/difference in day-night temperatures	Integrated agriculture practices to cool the solar panels	Routine maintenance on additional fixed assets	High
Encapsulant discoloration	Physical and chemical property changes of the EVA compound at high temperatures	Cooling of solar panels, through agriculture, and intermediate cleaning practices	Routine maintenance	High
Delamination	Found in warm and humid regions more than hot and dry. The reason being the ingress of moisture and high temperatures	Water-resistant painting, monitoring of beading, sealant, joint conditions, and replacement	Preventive maintenance	High
Metallization (discoloring and staining)	Corrosion due to ingress of water and improper soldering during manufacturing	Anticorrosion and efficient sealant replacement	Preventive maintenance	High
Hot spots	Accumulation of fine dust particles on glass cover/discoloration of the panels.	Efficient cleaning mechanism	Routine maintenance	Low
Snail trails	Microcracks during transportation and manufacturing defect	Proper investigation and replacement at the initial operation period	Preventive maintenance	Low
Backsheet degradation	Ingress of water and corrosion and careless installation practices	Anticorrosion and efficient enclosure protection	Preventive maintenance	High
Front-glass degradation	Improper cleaning practices and handling during installation	Efficient cleaning practices with proper water management	Routine maintenance	Low

3. Open circuit voltage: 30 V
4. Operating voltage: 24 V
5. Plant capacity: 1 MW
6. Capital cost: 400 lakhs INR
7. Debt:equity ratio: 70:30
8. Interest on debt: 10%
9. Interest on working capital: 11%
10. Accelerated depreciation: 12 years
11. Bid price: 2.44 INR/kWh for 25 years through a reverse bidding process.

Fig. 11.6 shows the LCE is high in Zone A and Zone B, considering the higher degradation levels of panels and corresponding operational cost to mitigate. Zone B needs higher mitigation from water ingress along with interconnect breakages. Zone C has better LCOE and IRR compared to Zones A and B (Fig. 11.7), considering the lower operational cost and lower mitigation management strategies. The payback hence is better in Zone C (Fig. 11.8), while for Zone A and Zone B the higher degradation rates reduce the payback potential, even though there is a better irradiation.

The CO_2 arrest equivalent for the solar-generated and carbon emission reduction (CER) certificate for various climatic zones and under different degradation rates are represented in Fig. 11.9. It is seen that Zone C has a better CO_2 content and CER certificate eligibility considering lower degradation and better production.

Considering the different climatic zones and corresponding degradation/year, it is found that there is a considerable difference in IRR and LCOE. However, in a competitive reverse-bidding process, the policy and regulatory norms are the same irrespective of the climatic zones, even though the operational cost for the mitigation of defects against various degradation levels is different.

Considering most of the installations of solar PV happen in hot zones (hot and dry, wet and humid, and composite), a sensitivity analysis is performed on a 1-MW solar

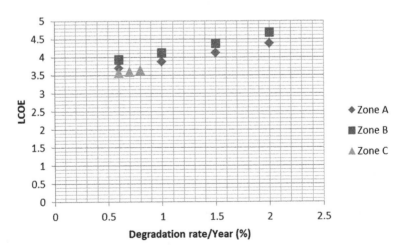

Figure 11.6 LCOE for ground-mounted solar PV projects for different climatic zones with varying annual degradation levels for solar panels.

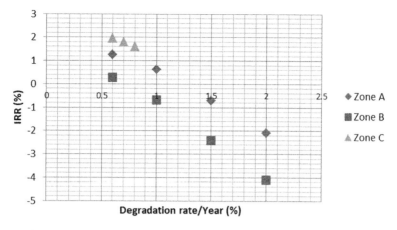

Figure 11.7 IRR for solar PV projects for different climatic zones under varying annual degradation levels for solar panels.

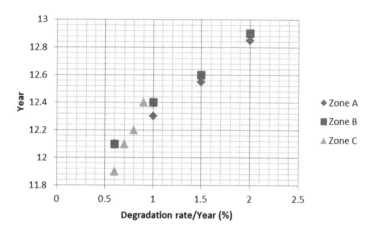

Figure 11.8 Payback for solar PV projects in number of years for different climatic zones at varied degradation levels for solar panels.

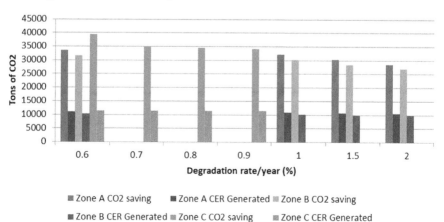

Figure 11.9 The CO_2 arrest and CER generated per climatic zone based on different degradation levels of solar panels.

plant with polycrystalline technology varying the capital cost, operational cost, yearly escalations, rate of interest on debt and working capital, and average yearly duration of solar panels. The objective of the sensitivity analysis is to bring the project IRR to 3% and above while matching the least reverse-bidding price of 2.44 INR/kWh.

11.5   The following scenarios are considered for LCOE, IRR, and payback

1. CASE 1: Capex 400 lakhs INR, O&M 5% with 3.85% escalation, panel yearly degradation rate 1.5%, IRR before tax, and no depreciation considered
2. CASE 2: Capex 400 lakhs INR, O&M 5% with 5% escalation, solar panel degradation rate 1.5%, IRR before tax, and no depreciation considered
3. CASE 3: Capex 450 lakhs INR, O&M 5% with 5% escalation, solar panel annual degradation rate 1.5%, IRR before tax, and no depreciation considered.
4. CASE 4: Capex 450 lakhs INR, O&M 5% with 5% escalation, panel degradation rate 1.5%, IRR before tax, depreciation not considered, and 0.5% escalation on bid price
5. CASE 5: Capex 450 lakhs INR, O&M 5% with 5% escalation, panel degradation rate 1.5%, IRR before tax, depreciation not considered, and 1.0% escalation on bid price
6. CASE 6: Capex 450 lakhs INR, O&M 5% with 5% escalation, panel degradation rate 1.5%, IRR before tax, depreciation not considered, and 1.5% escalation on bid price.
7. CASE 7: Capex 450 lakhs INR, O&M 5% with 5% escalation, panel degradation rate 1.5%, IRR before tax, depreciation not considered, and 2.0% escalation on bid price.
8. CASE 8: Capex 400 lakhs INR, O&M 5% with 5% escalation, panel degradation rate 1.5%, IRR before tax, depreciation not considered, and 2.0% escalation on bid price

Current reverse-bidding practices have to opt for an interest rate of 11.5% and 10% for loan and working capital, while offshore funding has an interest rate as close to 2%. The accelerated deprecation levels are kept for a 12-year period as per CERC guidelines, while the debt:equity ratio is kept at 70:30, and operational expenses are kept at approximately 11 lakhs/MW with a yearly escalation of 3.85% (Microsoft Word, May2020-Clean (Cercind.Gov.in)Accessed On, 2016). The capital expenses excluding the land vary between 4 and 5 crores INR/MW. The solar panels contribute 42% of the capital cost of the solar plant, while the annual escalation for maintenance cost is 3.85% according to CERC (Microsoft Word, May2020-Clean (Cercind.Gov.in). Accessed on, 2016).

Operational cost, depreciation, and debt repayment are the main contributors that decide the IRR and LCOE calculations (Probst et al., 2020). Fig. 11.10 shows the IRR and the LCOE under different case scenarios as above for the hot and dry zone, with annual degradation levels at 1.5%.

It is found that Case 8, where the Capex is 400 lakhs INR, with O&M cost of 5%, with 5% escalation at a debt interest rate of 2%, and annual solar feed rate of 2% per year, is the most feasible solution achieving an LCOE of 2.85 INR/kWh and an IRR of 3.8%. There can be a further reduction in LCOE if the O&M escalation can be brought down to 3.85% from 5% at a compromise of mitigation of climate-based degradation levels.

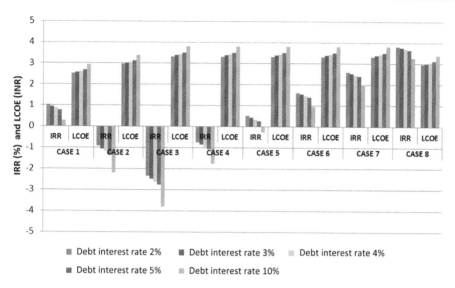

Figure 11.10 IRR and LCOE for different case scenarios (as above) under different interest rates for loans on debt.

11.6 Conclusion

The defects in solar PV panels based on different climatic conditions during their operational lifespan were explored in detail. It is found from earlier surveys and studies that the factors contributing to P_{max} reduction in hot and dry, and wet and humid zones are fill factor (FF), while for composite and moderate regions along with FF, short circuit current is also equally contributing. The defects in solar panels are climate specific and related to high temperature, humidity, intermittent cold and hot climates, dust accumulation, and cleaning methods. Based on the yearly deterioration rates in the relevant climatic zone, the LCOE, IRR, and payback figure were determined. However, the policy and regulation are uniform, and reverse bidding does not consider this variation in cost-competitive bidding, where the debt:equity ratio is 70:30 with high debt interest rates.

Sensitivity analysis is done on e-reverse bidding price with respect to changes in capital cost, O&M cost, annual O&M escalation rates, feed-in tariff escalation rates, panel degradation rates, and interest rates on debt and working capital. The objective was to analyze different case scenarios varying the Capex between 400 and 500 lakhs INR, interest rates on debt and working capital varied between 2% and 10%, O&M cost varied between 2.5% and 5%, and escalation varied between 3.85% and 5%. It is found that the annual O&M escalation is the key component along with the interest rate on debt that will be critical. The Capex below 400 lakhs INR/MW will reduce the LCOE, which can be possible only at a high-capacity plant ranging above 100 MW. The low debt and working capital interest of 2% is possible with foreign investments, while the interest rates as per CERC from Indian banks stand between 10% and 11.5%. It is found that a 2% interest rate for debt and working capital with an annual feed-in tariff escalation of 2% with a capital cost below 400 lakhs INR/MW and operational expenses at 5% with a 3.85% escalation shall bring justice to pricing close to 2.44 INR/kWh with an IRR of 3.84% and LCOE to closely match the first-year feed-in tariff. The aforementioned calculation was done for Zone A, which includes hot and dry, and wet and humid regions. It is thus recommended that reverse auction should consider the climatic zones as a factor during bidding as the O&M cost is a critical component to mitigate the climate-specific degradation rates.

References

Al Shehri, A., Parrott, B., Carrasco, P., Al Saiari, H., & Taie, I. (2016). Impact of dust deposition and brush-based dry cleaning on glass transmittance for PV modules applications. *Solar Energy, 135,* 317–324. Available from https://doi.org/10.1016/j.solener.2016.06.005.

Attari, K., Elyaakoubi, A., & Asselman, A. (2016). Performance analysis and investigation of a grid-connected photovoltaic installation in Morocco. *Energy Reports, 2,* 261–266.

Boddapati, V., & Daniel, S. A. (2020). Performance analysis and investigations of grid-connected solar power park in Kurnool, south India. *Energy for Sustainable Development, 55,* 161–169.

Bose, A. S., & Sarkar, S. (2019). India's e-reverse auctions (2017–2018) for allocating renewable energy capacity: An evaluation. *Renewable and Sustainable Energy Reviews, 112,* 762–774. Available from https://doi.org/10.1016/j.rser.2019.06.025.

Broek, M.K., Bennett, J.L., Jansen, J.M., van der Borg, N.J.C. (2012). 2012 proceedings of the 27th European PV Solar Energy Conference and Exhibition. Available from https://doi.org/10.4229/27thEUPVSEC2012-4DO.6.5.

Chattopadhyay, S., Dubey, R., Kuthanazhi, V., John, J.J., Solanki, C.S., Kottantharayil, A., Arora, B.M., Narasimhan, K.L., Vasi, J., Bora, B., Singh, Y.K., Sastry, O.S. (2015). 2015 IEEE 42nd Photovoltaic Specialist Conference, PVSC 2015 9781479979448, Institute of Electrical and Electronics Engineers Inc. India All India Survey of Photovoltaic Module Degradation 2014: Survey methodology and statistics. Available from https://doi.org/10.1109/PVSC.2015.7355712.

Chattopadhyay, S., Dubey, R., Kuthanazhi, V., Zachariah, S., Bhaduri, S., Mahapatra, Rambabu, S., Ansari, F., Bora, Kumar, G., Singh, Y.K., Bangar, M., Kumar, M., Haldkar, A.K., Singh, R., et al. (2016). All India survey of photovoltaic module reliability.

Dubey, R., Chattopadhyay, V. Kuthanazhi, J.J., John, M.B., Arora, A., Kottantharayil, L.K., Narasimhan, S.C., Solanki, V., Kuber, Vasi. (2013). All India Survey of Photovoltaic Module Degradation.

Ferrada, P., Araya, F., Marzo, A., & Fuentealba, E. (2015). Performance analysis of photovoltaic systems of two different technologies in a coastal desert climate zone of Chile. *Solar Energy, 114,* 356–363. Available from https://doi.org/10.1016/j.solener.2015.02.009.

Fundamentals and Innovations in Solar Energy. Energy Systems in Electrical Engineering. Available from https://doi.org/10.1007/978-981-33-6456-1_10.

Hammad, B., Al−Abed, M., Al−Ghandoor, A., Al−Sardeah, A., & Al−Bashir, A. (2018). Modeling and analysis of dust and temperature effects on photovoltaic systems' performance and optimal cleaning frequency: Jordan case study. *Renewable and Sustainable Energy Reviews, 82,* 2218–2234. Available from https://doi.org/10.1016/j.rser.2017.08.070.

Jamil, B., Siddiqui, A. T., & Akhtar, N. (2016). Estimation of solar radiation and optimum tilt angles for south-facing surfaces in Humid Subtropical Climatic Region of India. *Engineering Science and Technology, an International Journal, 19*(4), 1826–1835. Available from https://doi.org/10.1016/j.jestch.2016.10.004.

Jordan, D. C., Deline, C., Kurtz, S. R., Kimball, G. M., & Anderson, M. (2018). Robust PV degradation methodology and application. *IEEE Journal of Photovoltaics, 8*(2), 525–531. Available from https://doi.org/10.1109/JPHOTOV.2017.2779779.

Jordan, D. C., Kurtz, S. R., VanSant, K., & Newmiller, J. (2016). Compendium of photovoltaic degradation rates. *Progress in Photovoltaics: Research and Applications, 24*(7), 978–989. Available from https://doi.org/10.1002/pip.2744.

Jordan, D. C., Silverman, T. J., Sekulic, B., & Kurtz, S. R. (2017). PV degradation curves: non-linearities and failure modes. *Progress in Photovoltaics: Research and Applications, 25*(7), 583–591. Available from https://doi.org/10.1002/pip.2835.

Kempe, M., Reese, M., Dameron, A., Moricone, T. (2010) Types of Encapsulant Materials and Physical Differences Between Them. NREL Photovoltaic Module Reliability Workshop.

Khare, V., Nema, S., & Baredar, P. (2013). Status of solar wind renewable energy in India. *Renewable and Sustainable Energy Reviews, 27*, 1–10. Available from https://doi.org/10.1016/j.rser.2013.06.018.

Köppen, W. (1918). Classification of climates according to temperature, precipitation and seasonal cycle. *Petermanns Geographische Mitteilungen, 64*, 243–248.

Mani, M., & Pillai, R. (2010). Impact of dust on solar photovoltaic (PV) performance: Research status, challenges and recommendations. *Renewable and Sustainable Energy Reviews, 14*(9), 3124–3131. Available from https://doi.org/10.1016/j.rser.2010.07.065.

Microsoft Word—RE-Tariff-Regulations-EM-12May2020-clean (cercind.gov.in) Accessed on. CERC. (2016).

New Research Directions in Solar Energy Technologies. (n.d.). Energy, Environment, and Sustainability. Available from https://doi.org/10.1007/978-981-16-0594-9.

Padmavathi, K., & Daniel, S. A. (2013). Performance analysis of a 3MWp grid connected solar photovoltaic power plant in India. *Energy for Sustainable Development, 17*(6), 615–625. Available from https://doi.org/10.1016/j.esd.2013.09.002.

Panjawani, P. R., Jain, V. D., Bhandari, R. K., Gaikwad, U. S., & Deokar, U. (2020). Design and analysis of solar structural and mountings for solar panel. *International Journal of Future Generation Communication and Networking, 13*, 668–679.

Probst, B., Anatolitis, V., Kontoleon, A., & Anadón, L. D. (2020). The short-term costs of local content requirements in the Indian solar auctions. *Nature Energy, 5*(11), 842–850. Available from https://doi.org/10.1038/s41560-020-0677-7.

Rahman, M. M., Hasanuzzaman, M., & Rahim, N. A. (2015). Effects of various parameters on PV-module power and efficiency. *Energy Conversion and Management, 103*, 348–358. Available from https://doi.org/10.1016/j.enconman.2015.06.067.

Ramanan, K., Murugavel, K., & Karthick, A. (2019). Performance analysis and energy metrics of grid-connected photovoltaic systems. *Energy for Sustainable Development, 52*, 104–115.

Reis, A. M., Coleman, N. T., Marshall, M. W., Lehman, P. A., & Chamberlin, C. E. (2002). Comparison of PV module performance before and after 11-years of field exposure. *Conference Record of the IEEE Photovoltaic Specialists Conference*, 1432–1435. Available from https://doi.org/10.1109/PVSC.2002.1190878.

Richter, S., Werner, M., Swatek, S., & Hagendorf, C. (2012). Understanding the snail trail effect in silicon solar modules on microstructural scale. *Photovoltaic Solar Energy Conference and Exhibition*, 3439–3441.

Shah, A. H., Hassan, A., Laghari, M. S., & Alraeesi, A. (2020). The influence of cleaning frequency of photovoltaic modules on power losses in the desert climate. *Sustainability*, *12*(22). Available from https://doi.org/10.3390/su12229750.

Shenouda, R., Abd-Elhady, M. S., & Kandil, H. A. (2022). A review of dust accumulation on PV panels in the MENA and the Far East regions. *Journal of Engineering and Applied Science*, *69*(1). Available from https://doi.org/10.1186/s44147-021-00052-6.

Slamova, K., Glaser, R., Schill, C., Wiesmeier, S., & Köhl, M. (2012). Mapping atmospheric corrosion in coastal regions: Methods and results. *Journal of Photonics for Energy*, *2*(1). Available from https://doi.org/10.1117/1.JPE.2.022003.

Solar energy policies and guidelines. (n.d.). Available at MNRE - SOLAR POLICIES. MNRE.

Spataru, V. S., Sera, D., Hecke, P., Kerekes, T., & Teodorescu, R. (2015). Fault identification in crystalline silicon PV modules by complementary analysis of the light and dark current−voltage characteristics. *Progress in Photovoltaics: Research and Applications*, *2015*.

Thapar, S., Sharma, S., & Verma, A. (2018). Analyzing solar auctions in India: Identifying key determinants. *Energy for Sustainable Development*, *45*, 66−78. Available from https://doi.org/10.1016/j.esd.2018.05.003.

Thomas, S., Kumar, A., Sahoo, S. S., & Thomas, S. (2023). Sustainability of livelihood systems in bottom-up approach method: Energy-Water-Food nexus and potential of renewable energy integration. *International Journal of Energy for a Clean Environment*.

Thomas, S. J., Thomas, S., Sahoo, S. S., Gobinath, R., & Awad, M. M. (2022). Allotment of waste and degraded land parcels for PV based solar parks in India: Effects on power generation cost and influence on investment decision-making. *Sustainability*, *14*(3). Available from https://doi.org/10.3390/su14031786.

Thomas, S. S. Sahoo, & Kumar, A. (2021). *Renewable energy adaptation model for sustainable small islands in AI in manufacturing and green technology* (2021). CRC Press.

Windarta, J., Handoko, S., Sukmadi, T., Irfani, K. N., Masfuha, S. M., & Itsnareno, C. H. (2021). Technical and economic feasibility analysis of solar power plant design with off grid system for remote area MSME in Semarang City. *IOP Conference Series: Earth and Environmental Science*, *896*(1). Available from https://doi.org/10.1088/1755-1315/896/1/012007.

Wohlgemuth, J., Cunningham, W.D., Nguyen, A. (2010). NREL PV Reliability Workshop Failure Modes of Crystalline Silicon Modules.

Yuy, I. Z., Yu, Z., Huanxin, L., Yunjia, L., & Liang, L. (2013). Control system design for a surface cleaning robot. *International Journal of Advancements in Robotic Systems*, *10*.

Zaihidee, F. M., Mekhilef, S., Seyedmahmoudian, M., & Horan, B. (2016). Dust as an unalterable deteriorative factor affecting PV panel's efficiency: Why and how. *Renewable and Sustainable Energy Reviews*, *65*, 1267−1278. Available from https://doi.org/10.1016/j.rser.2016.06.068.

The emergence of digital metasurfaces: a new technology for enhancing photovoltaic systems

Mohammed Berka[1,2], Amina Bendaoudi[2], Kaddour Benkhallouk[2],
Zoubir Mahdjoub[2] and Ahmed Yacine Rouabhi[2]
[1]Department of Electrotechnic, University of Mustapha Stambouli Mascara, Mascara City,
Algeria, [2]Laboratory E.P.O, University of S.B.A., Sidi Bel Abbes City, Algeria

12.1 Introduction

The development of various economic and industrial sectors requires good control of electrical energy. The supply of this energy and its exploitation in a rational way becomes a condition and an indispensable factor for achieving the desired objectives. Due to the different infrastructures, the problem of distribution of electrical energy in rural areas and fields becomes a constraint for the architecture and installation of renewable energy networks. So the inaccessibility of a power grid is the challenge for rural and remote areas.

The contribution of solar electricity to the world's total electricity generation is currently a small percentage of the total world energy production (0.2%) (corresponding to 72 GWp capacity of solar plants as compared to 3600 GWp capacity of generated power plants). During the last several years, the average annual growth rates of renewable energy capacity have been 70% (Renewable Energy Policy Network for the 21st century, 2012).

The wavelength of the sun lies mainly between 300 and 3000 nm, which implies that the frequency is between 100 and 1000 THz by referring Fig. 12.1 (https://socratic.org/questions/radiation-from-the-sun-often-gets-trapped-in-the-earth-s-atmosphere). The working of usual solar cells operates when the light photons reach the solar panel and are absorbed by a semiconductor or some other kind of material.

The most used material for the design of PV cells is silicon (Si); this kind of material shows some efficiency, but improvements and design techniques have been made to optimize and amplify the absorption coefficients of this material with compact (Hassoune et al., 2017) and less-expensive architectures (Green et al., 2012). The appearance of a new class of materials called metamaterials has made a remarkable contribution to improving the absorption qualities of silicon cells. The electromagnetic characteristics of metamaterials such as negative permittivity,

Performance Enhancement and Control of Photovoltaic Systems. DOI: https://doi.org/10.1016/B978-0-443-13392-3.00013-X

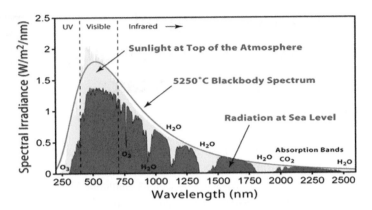

Figure 12.1 Solar spectrum Google images (https://socratic.org/questions/radiation-from-the-sun-often-gets-trapped-in-the-earth-s-atmosphere).

permeability (Estep et al., 2014; Liu et al., 2008), and therefore a negative refractive index (Zhu et al., 2018) have contributed in one way or another to the realization of high-quality electrical devices. Metamaterials (2-D) also called "metasurfaces" show great efficiency for a variety of applications (Hu et al., 2013, 2019; Xu et al., 2019; Zhang et al., 2018; Zhao et al., 2016). The encoded 2-D metasurfaces show their effectiveness for various applications in renewable energy systems such as the ease of controlling networks and power generation stations located in rural and remote areas. The physical characteristics of coded metasurfaces can also contribute to the optimization of PV cell energy storage systems (for long life), which is necessary for areas and places that do not have accessibility.

The rest of the chapter is organized as follows: In Section 12.2, the state of the art is presented. In Section 12.3, we provide a general overview of solar PV systems. Solar energy technologies are mentioned. In Section 12.4, we present the principles of coded metasurfaces with the necessary conditions and criteria. In Section 12.5, we present our metasurface-based absorber (MSA). This MSA is formed by diagonally square-shaped metasurface cells for the different configurations. In Section 12.6, simulation results are presented for each studied structure. Section 12.7 concludes this chapter.

12.2 State of the art

Since the first proposal to study and design them by the physicist V. Veselago in 1967 (Veselagi, 1968), metamaterials have opened up great fields to researchers and designers for the development of devices and systems of industry technology. The first metamaterial structure in 3-D was made by J. Pendry (Marquez et al., 2002); this structure shows a magnetic resonance for negative permeability and low dimensions. In the early 2000s, metamaterials (3-D) were used for the design of circuits evolving in microwave filters (Begaud et al., 2018; Kim et al., 2018), smart antennas (Hou et al., 2005; Hsiao et al., 2017; Hu et al., 2017), radar link (Hu et al., 2016; Huang et al., 2013), and biosensors (Ebrahimi et al., 2018; Gennarelli et al., 2013; Su et al., 2016; Yang et al., 2016). Recently, the trend toward light and infrared waves requires the use of metamaterials (2-D) called metasurfaces. Metasurfaces are thin layers that have an electromagnetic bandgap, allowing or preventing the propagation of electromagnetic waves in the desired directions, hence the existence of evanescent waves. Metasurfaces are characterized by unusual reflection and transmission properties in nature; they can be obtained by periodically disposing of many conductive tracks on dielectric substrates. According to the distribution of the tracks on the dielectric substrates, we distinguish two large classes of metasurfaces: penetrable and impenetrable. The study of penetrable metasurfaces can be adopted in the field of millimeter and submillimeter waves to solve the problem of losses in classical dielectric substrates (González-Ovejero et al., 2016). This kind of structure provides effective surface impedance, which can be specifically designed and manipulated for various applications, including absorption (Li et al., 2018a; Li et al., 2019), wave front engineering (Li et al., 2012; Liu et al., 2017), leaky-wave radiation (Pors et al., 2013; Principe et al., 2019), cloaking (Qian et al., 2020), polarization control (Rocco et al., 2020; Ruiz de Galarreta et al., 2020; Scheerlinck et al., 2009), as well as lenses and imaging structures (Ni et al., 2012; Person et al., 2013; Sirmaci et al., 2020). In Li et al. (2018b), a double-sided metasurface is employed in the design of a wideband GNSS antenna. The antenna's circularly slotted design gives it a circular polarization with a -10 dB reflection coefficient and 3 dB axial ratios of 34.3 and 38.1%, respectively. The antenna covers all the GNSS services' bands with a gain of 6.44 dBic at 1.6 GHz. As Faraday rotators, although transistor-loaded ring particles, nonreciprocal metasurfaces have been experimentally demonstrated in Kodera et al. (2010). These structures are applied to several guided-wave components. In Chen et al. (2013), the active and non-Foster lumped elements have been used to overcome the inherent bandwidth limitations of passive metasurfaces and improve their overall efficiency. Lee (2012) have experimentally demonstrated active modulations on both the amplitude (47%) and phase (32.2 degrees) of the transmitted THz wave for tunable metasurfaces. The gradient metasurfaces have been studied in Díaz-Rubio et al. (2017) for phase-controlling and bianisotropic devices. An adaptive metalens with the focal length controlled electrically was experimentally demonstrated in She et al. (2018) for an optical regime based on metasurfaces.

In the field of renewable energies, several studies and research based on metamaterials and metasurfaces have been carried out to improve and optimize the electrical qualities of PV cells. The main objective of these various studies is to increase the absorption coefficients of PV cells. In Houria and Mohammed (2018), a multilayer waveguide structure has been proposed to improve efficient silicon solar cells. For this study, the sunlight is guided by a metamaterial layer, and the transmission loss is eliminated by an aluminum back reflector. A layer of metasurface constituted by a (3×3) array of circular SRRs is used in Mohammed et al. (2019) to optimize the absorption qualities of PV cells. In Adnan et al. (2018), a novel metasurface absorber (MSA) based on a crossed-shaped resonator placed at the top surface is proposed; the absorption resonances and their corresponding spectral widths are modified and widened by adding extra nanorods to the crossed-shaped structure at different locations. The absorption levels obtained are of the order of 86.35%. Very recently, an optically switchable broadband metasurface absorber structure based on planar-patterned photoconductive silicon in the terahertz region has been numeri-cally demonstrated in Yongzhi et al. (2021). The designed metasurface is composed of a planar-square-ring-shaped structure photoconductive silicon array placed over a ground plane separated by a dielectric substrate. The proposed structure can realize a switching absorption from 2.8% to 99.9% and the relative bandwidth of the con-tinuous absorption of 90%.

In the present chapter, we intend to present a new approach to the design of digi-tal metasurfaces involved in THz metamaterial absorbers. The structures forming the metasurfaces are coded (in 0 and 1) according to the geometric shapes of the proposed SRRs. The dimensions of these resonators are optimized to have magnetic resonances in the solar spectrum. The absorption coefficients of the proposed meta-surfaces are calculated according to the reflections obtained from the studied SRRs. The locations of the metasurfaces are chosen relative to the silicon layers and ground planes to have the desired absorption coefficients. For validation purposes, this problem of interest was numerically modeled and simulated using ANSYS High-Frequency Structure Simulator (HFSS) (ANSYS Electromagnetics Suite, n.d.). ANSYS proposed structures that have a simple design are suitable for applica-tions in power generation stations using communication techniques (GSM, I0T, Ethernet, and GPRS). These techniques are now becoming essential for rural and remote installations.

12.3 Overview of solar photovoltaic systems

In nature, there are several ways to extract or obtain electrical energy; this energy can be primary or secondary. Energy captured or extracted directly (without trans-formation) from nature is called primary energy such as coal, crude oil, natural gas, nuclear fuel, hydroelectricity, biomass, solar energy, wind energy, geothermal energy, and ocean energy (référence sur charbon, n.d.). Primary energy is trans-formed into secondary energy in the form of electrical energy or fuel such as gaso-line, methanol, ethanol, or hydrogen. A secondary energy can be the result of the transformation of a primary energy or another secondary energy (Fig. 12.2).

12.3.1 Solar energy

Solar energy is by far the energy that currently concentrates the most development efforts (Fig. 12.3) and therefore registers the best annual growth rate since 1990 as well as the greatest cost reduction. In 2015, the global average cost for crystalline

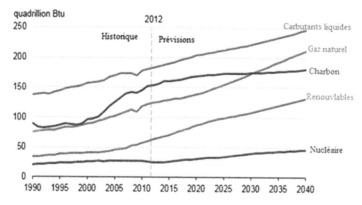

Figure 12.2 Evolution and forecasts of world energy consumption (International Energy Outlook, 2021).

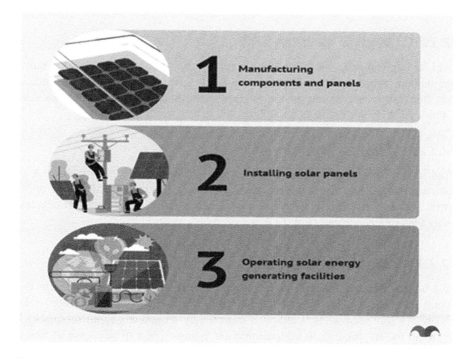

Figure 12.3 Best solar energy stocks to invest in 2022 (https://encrypted-tbn0.gstatic.com/image).

silicon photovoltaic cells was 122 $ per MWh compared to 143 $ in 2014. Specific projects will further reduce the costs of photovoltaic energy: the record is expected to be for a 200-MW installation, in the installation phase in Dubai by the company ACWA Power International, crowned by a contract at only 58.50 dollars per MWh (Pisco et al., 2017).

In the context of applications using solar energy, it is imperative to be able to measure the energy received, in the form of light, by a given surface during a given time. Two measurements are used for this purpose: solar radiation (also called sunshine) and solar irradiation (Smith et al., 2004). Solar radiation (sunshine) is the amount of energy, due to the flux of incident sunlight, received by the surface of the Earth (in kW/m^2). Not to be confused with the term "solar irradiation," which designates the energy, due to the sun, received by a given surface in a given time (in kWh or kWh/hour or kWh/day, etc.).

12.3.2 Solar spectrum

The solar spectrum refers to the distribution of electromagnetic radiation emitted by the sun as a function of incident wavelength on the outer surface of the atmosphere. Some of it is reflected by the Earth's atmosphere, while the rest is absorbed by the Earth's surface. Practically, we distinguish two different solar spectral distributions depending on the thickness of the atmosphere to cross.

12.3.2.1 Mass air spectrum AM0

Used for applications outside the Earth's atmosphere, it corresponds to solar radiation without atmospheric attenuation at an Earth-Sun distance of one astronomical unit (the astronomical unit is the average Earth-Sun distance and is equal to 1.496×10^8 km) (Valentine et al., 2008).

12.3.2.2 Mass air spectrum AM1.5

Used for terrestrial applications, it corresponds to a midday sun at an inclination of approximately 48 degrees (radiation passing through 1.5 times the thickness of the Earth's atmospheric layer). AM1 corresponds to a sun at the zenith (90 degrees) under the same conditions (Fig. 12.4).

12.3.3 Solar energy technologies

To produce electrical energy, solar energy can be exploited in two ways: thermodynamic solar energy and PV energy. One solution to using thermodynamic solar energy is tower solar where sunlight is reflected by thousands of mirrors surrounding a tall tower; this light is concentrated on a single point of the tower to heat a liquid, which in turn boils water to create steam to turn a turbine. Fig. 12.5 represents this kind of technology.

12.3.4 Photovoltaic energy

The basic element of a PV system is the solar cell that directly converts the energy received from sunlight into electric current. The operating principle is identical for all PV cell technologies. A PV cell always contains a junction between two

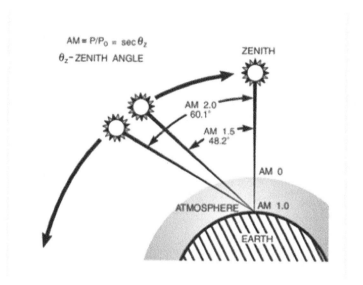

Figure 12.4 Air mass distribution (https://www.newport.com/mam/celum/celum_assets/AM-Definition-Sun_600w.jpg).

Figure 12.5 Solar energy on a tower.

different materials separated by an electrical potential barrier. The absorption of an amount of energy (photons) greater than the band gap of the semiconductor causes electrons to jump from the valence band to the conduction band, creating hole-electron pairs (Yang et al., 2014). The p–n junction forces electrons and holes to each go to an opposite side of the material. As a result, a potential difference

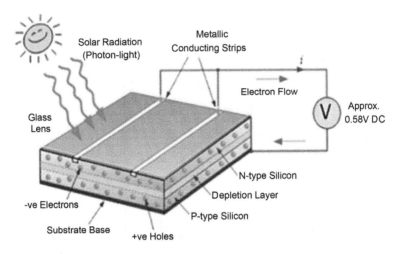

Figure 12.6 Basic diagram of a photovoltaic solar cell.

appears between the two faces. This effect, discovered by the French physicist Edmond Becquerel in 1839, is known as the PV effect. Fig. 12.6 shows the essential components of a monocrystalline silicon PV cell.

12.3.5 Digital technologies for rural and remote installations

For photovoltaic systems installed at rural locations, remote monitoring capabilities provide information in advance when system performance is degraded or is likely to fail. Based on this information, preventive maintenance can be carried out to improve the performance and life of the system, thereby reducing the overall operating cost. Monitoring systems for rural applications, based on communication techniques, have several advantages. For embedded computer systems (GSM), this technique has a low initial cost as well as a low operating cost. The GSM network is readily available in rural areas; this technique can be used easily (Peijiang & Xuehua, 2008). There are also other techniques, such as computer-to-computer communication (Ethernet), embedded system to embedded system (GSM and GPRS), and the Internet of Things technology (for supervising solar PV power generation and for the enhancement of the performance, monitoring, and maintenance of the plant). Today, the physical and electrical qualities of PV cells facilitate these control and maintenance operations. The effective maximum power point tracking (MPPT) has a direct relationship to PV cells. The functional diagram of the system proposed in VilasBhau (2021) is represented in Fig. 12.7.

12.4 Coded metasurfaces

Electromagnetic field is a physical behavior that is produced in a space due to time-varying electric charges and represents the interaction between electric and

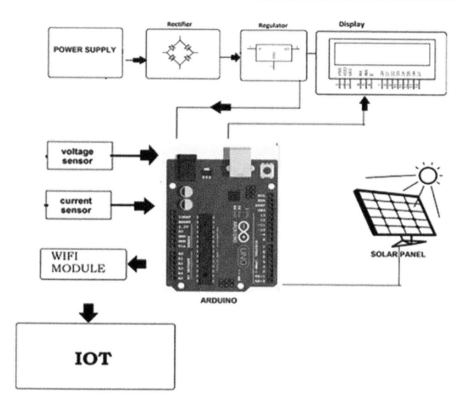

Figure 12.7 Functional diagram of the monitoring system based on IoT technique (VilasBhau, 2021).

magnetic fields. Unlike static charges that can only produce static electric fields in space, time-varying electric charges are one of the sources for the rise of magnetic fields, which in turn produce time-varying electric fields. This is summarized in the four time-varying Maxwell's equations given in differential form.

$$\nabla \times E = -\mu \frac{\partial H}{\partial t} \tag{12.1}$$

$$\nabla \times H = J(t) + \varepsilon \frac{\partial H}{\partial t} \tag{12.2}$$

$$\nabla . B = 0 \tag{12.3}$$

$$\nabla . E = \frac{\rho(t)}{\varepsilon} \tag{12.4}$$

where ρ is the time-varying volume charge density, ε and μ are the electric permittivity and magnetic permeability, respectively, J is the time-varying electric current density in a medium, D and B are the time-varying electric and magnetic flux densities, respectively, and E and H are the time-varying electric and magnetic field intensities, respectively.

A method for constructing "metamaterial bytes" through proper spatial mixtures of "digital metamaterial bits" has been proposed in Della Giovampaola and Engheta (2014). The "digital metamaterial bits" are some particles that possess distinct material properties. However, the resulting metamaterial bytes are still described by the effective medium parameters. Cui proposed the general concepts of "coding metamaterial," "digital metamaterial," and "programmable metamaterial," which means that a single metamaterial can be digitally controlled to obtain distinctly different functionalities (Cui et al., 2014; Moccia et al., 2017). Two types of unit cells with 0 and π phase responses to mimic the "0" and "1" elements were proposed for 1-bit digital metamaterials such that they can be controlled using the existing digital technology. By designing coding sequences of "0" and "1" elements in coding metamaterials, EM waves can be easily manipulated to obtain different functionalities. And this concept can also be extended to 2-bit (0, $\pi/2$, π, and $3\pi/2$) and 3-bit (0, $\pi/8$, $\pi/4$, $3\pi/8$, $\pi/2$, $5\pi/8$, $3\pi/4$, and $7\pi/8$) or more.

For coded metasurfaces, the absorption coefficients or levels of a microwave structure are obtained as a function of the structure's reflection and transmission. So, voltage is not a well-defined entity, and it is necessary to define the scattering parameters (S-parameter) in terms of the electric field. The computed electric field E_c on the port consists of the excitation plus the reflected field. The scattering parameters can be defined as following equations (Liu et al., 2012):

$$S_{11} = \frac{\int_{Port1}((E_c - E_1)E_1^*)dA_1}{\int_{Port1}(E_1.E_1^*)dA_1} \tag{12.5}$$

$$S_{21} = \frac{\int_{Port1}(E_c.E_2^*)dA_2}{\int_{Port1}(E_2.E_2^*)dA_2} \tag{12.6}$$

where E_1 and E_2 are the electric patterns on ports 1 and 2. The higher the ability of a cell to absorb the radiations, the more power it gains. However, the reflection and transmission reduce the power gain as can be observed in the following equation:

$$A(\omega) = 1 - |S_{11}(\omega)|^2 - |S_{21}(\omega)|^2 \tag{12.7}$$

It is also possible to calculate the S-parameters from the power flow through the ports in terms of the power flow.

$$S_{11} = \frac{\sqrt{\text{Power reflected from port 1}}}{\sqrt{\text{Power incident on port 1}}} \tag{12.8}$$

$$S_{21} = \frac{\sqrt{\text{Power delivred to port 2}}}{\sqrt{\text{Power incident on port 1}}} \tag{12.9}$$

where $A(\omega)$ is the absorption and $S_{11}(\omega)$ and $S_{21}(\omega)$ are the scattering parameters that are directly related to the reflected and transmitted radiations. Therefore, in order to optimize the absorption, the reflected and transmitted radiations need to be minimized as much as possible. Fortunately, in the proposed design, there is no transmission due to the fact that the back metal substrate prevents transmission losses. As a result, Eq. (12.9) becomes ($S_{21} = 0$) and Eq. (12.7) is used in the present study for the calculation of absorption.

$$A(\omega) = 1 - |S_{11}(\omega)|^2 \tag{12.10}$$

For light absorption based on metasurfaces, we apply Fermat's principle. This principle states that light travels along an extremum path. The derivative of the general relationship between an incident wave and the scattered waves is given by (Yu et al., 2011)

$$\begin{cases} n_t\sin(\theta_t) - n_i\sin(\theta_i) = \dfrac{\lambda_0}{2\pi}\dfrac{d\varnothing\,(x)}{dx} \\[3mm] n_t\sin(\theta_r) - n_i\sin(\theta_i) = \dfrac{\lambda_0}{2\pi}\dfrac{d\varnothing\,(x)}{dx} \end{cases} \tag{12.11}$$

where n_i and n_t are the refractive indices on the two sides of the interface, λ_0 is the free space wavelength, and θ_i, θ_r, and θ_t are the incident, reflected, and transmitted angles, respectively.

12.5   Proposed metasurface absorber

Our MSA is developed using unit cells made up of SRRs. These cells are binary coded (0 and 1). Using the combination of these encoded unit cells, our MSA resonates in excellent polarization insensitivity. As with the binary patterns using the coded arrangements "0" and "1" [00, 01, 10, and 11], the "1" position indicates the numerical characteristics by electric current and high surface concentration and vice versa for "0." Our proposed MSA consists of two layers, namely a metal ground plane (designed by copper) to prevent transmission and a dielectric substrate of FR4_Epoxy, which is chosen according to its absorption qualities (Berka et al., 2021). This substrate of height ($h = 2.5\,\mu m$) has the physical characteristics ($\varepsilon_r = 4.4$; $tg\delta = 0.02$).

12.5.1   Metasurface unit cell

The basic cell of the proposed metasurface consists of four diagonally rectangular-shaped SRRs. The structural configuration of each SRR is represented as binary models [00, 01, 10, 11]. The inclusion associated with each

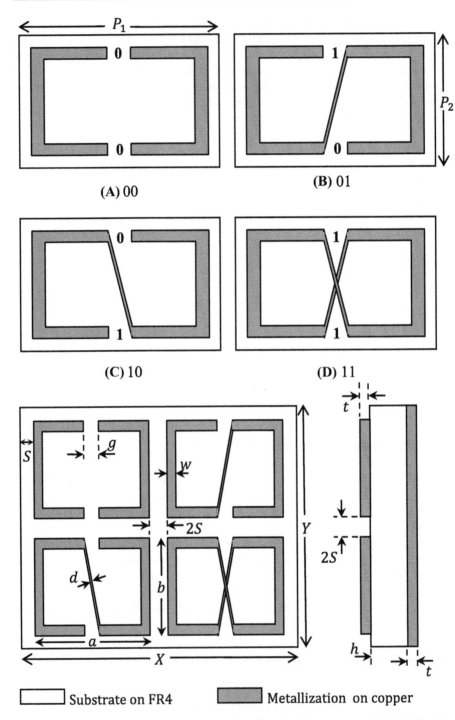

Figure 12.8 (A–D) Individual structural configurations as binary patterns [00, 01, 10, 11] with top and side view of the metasurface single unit cell.

code is engraved on the upper side of the chosen substrate. For 00 combination, the resonator is represented by a square loop, which has two symmetrical splits on both sides because the splits were represented by "0". For 01 combination, we add a diagonal arm connecting the two splits from bottom to top (from left to right) to the resonator of the previous shape. Therefore, for 10 combination, the add arm is directed from top to bottom (in the same direction). For the last combination 11, the resonator contains the two diagonal arms. Thus, four individual unit cells for the period $(P_1 \times P_2)$ were designed as shown in Fig. 12.8.

12.5.2 Dimensions of the proposed MSA unit cell

For our MSA unit cell, we have chosen dimensions on the micrometric scale to be able to exploit the THz resonances of these cells formed by the SRRs in the optimization of the electromagnetic qualities of the silicon layers. Our MSA cell with $h = 1.5$ μm and $t = 0.35$ μm has the dimensions summarized in Table 12.1.

For these dimensions, the area of the unit metasurface cell will be $(X \times Y) = (21.6 \times 17.6)$ μm^2 and $(P_1 \times P_2) = (10.8 \times 8.8)$ μm^2 for individual resonators.

Table 12.1 Design parameters of the proposed MSA unit cell.

Parameter	g	w	S	d	a	b
Value (μm)	0.8	0.4	0.4	0.1	10	8

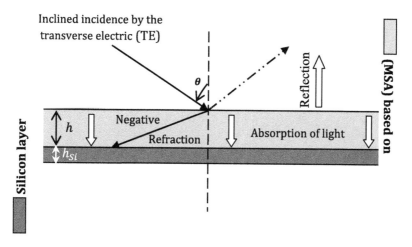

Figure 12.9 Incidence in the global cell.

12.5.3 MSA for optimized array

To improve the absorption qualities of silicon cells, we associate the metasurface layer for an optimized array with the silicon layer for thickness ($h_{si} = 1.5 \ \mu m$) and for a surface identical to that of the metasurface layer. Our optimized MSA consists of a (4×4) and (8×8) array of individual cells coded in four and eight different configurations. The study of the global cell (metasurface + silicon layer) is done for electrical transverse propagation modes (TE) according to the angle of incidence $\left(0 < \theta < \pi/2 \right)$. Fig. 12.8 shows the propagation of light for TE modes as a function of angles θ (Fig. 12.9).

12.6 Simulation results and discussion

12.6.1 Individual unit cells

The individual unit cell is represented for four different configurations according to the four codes. With the dimensions given in Table 12.1, these structures on the 3-D Modeler of the HFSS simulator are shown in Fig. 12.10.

To simulate each coded structure in HFSS, we set the band conditions for the E and H electromagnetic field. Fig. 12.11 shows these conditions for the cell coded as "00."

After simulating the structure, we obtain the following transmission and reflection coefficients represented in Fig. 12.12.

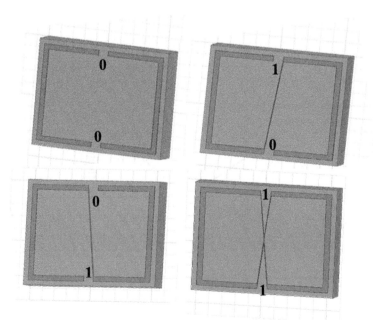

Figure 12.10 Different configurations of coded individual cells.

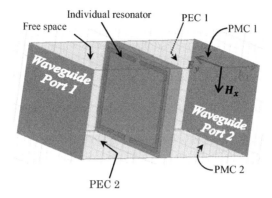

Figure 12.11 Simulation setup of the first individual cell coded in "00."

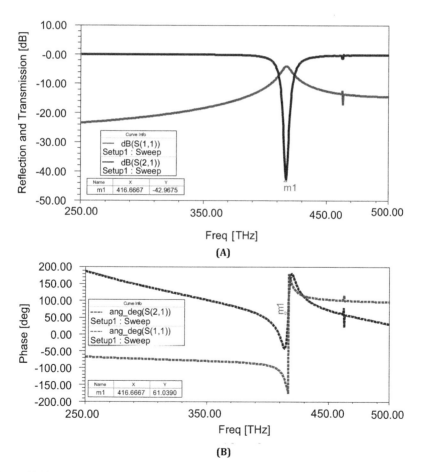

Figure 12.12 Reflection and transmission of the individual cell coded in "00" (A) amplitude (B) phase.

Fig. 12.12 represents the two coefficients of reflection and transmission of the individual cell coded in 00. In Fig. 12.12A, we observe a stope-band behavior at the resonant frequency of 416.66 THz. The transmission of the cell is −42.96 dB. In Fig. 12.12B, the two phases of the individual cell are between −π and +π.

For the same simulation conditions, the frequency responses of the four coded individual cells are shown in Fig. 12.13.

In Fig. 12.13, it is noted that the transmissions of each individual cell are different from one code to another. For code 00, we have a single resonance for a single transmission. For code 01, we have two resonances at 369.23 THz and 403.12 THz for the two transmissions of −26.24 dB and −66.52 dB, respectively. For code 10, we have three different resonances at frequencies 312.36 THz, 362.14 THz, and 381.78 THz of −43.51 dB, −33.99 dB, and −16.34 dB transmissions, respectively. For the last code 11, we observe four resonances 332.66 THz, 348.34 THz, 426.42 THz, and 491.25 THz of −38.85 dB, −31.16 dB, −59.34 dB, and −49.62 dB transmissions, respectively. The confinement of the electric field is represented by Fig. 12.14.

12.6.2 Metasurface response

For the four configurations obtained according to the possible coding matrices, the metasurface cells are represented in Fig. 12.15.

The reflection coefficients of the metasurface cells corresponding to each coded configuration are represented in Fig. 12.16.

The reflection coefficients of the four possible configurations of the metasurface cells are shown in Fig. 12.16. We note that the reflection of the first case of code 00 is identical to the reflection of the fourth configuration of code 11. We also observe multiple reflections for different resonances, the largest of which is around −48.56 db. For the two other codes 01 and 10, we notice that they also have the

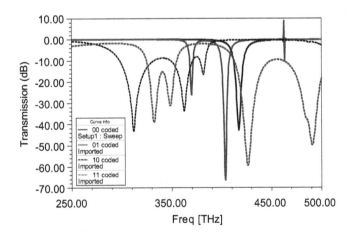

Figure 12.13 Transmission of the four individual cells for different codes.

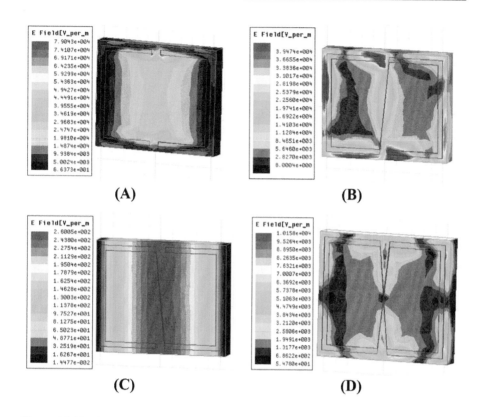

Figure 12.14 Electric field confinement at the first resonant frequency for each cell.

same multiple reflections, the most important of which is of the order of −34.88 dB. This behavior of identical multiple reflections is due to the symmetry of the metasurface cells for the four configurations.

12.6.3 Optimized MSA

Our optimized metasurface absorber consists of an individual cell array. The first designed array consists of (4×4) cells, and the second network contains (8×8) cells. The optimized MSA for both networks is shown in Fig. 12.17.

The absorption of the overall structure (MSA and silicon) is shown in Fig. 12.18.

Fig. 12.18 shows the absorption coefficient of the silicon layer for different cases. It is observed that this layer without the proposed MSA has the absorption of the order of 51.16% at the resonant frequency of 406 THz. When we add our optimized metasurface absorber to this silicon layer, the structure for the network (4×4) has the absorption of the order of 73.58% at the first resonant frequency of 344 THz, while for the network (8×8), we observe a near perfect absorption of the order 99.37% at the resonant frequency of 344.33 THz.

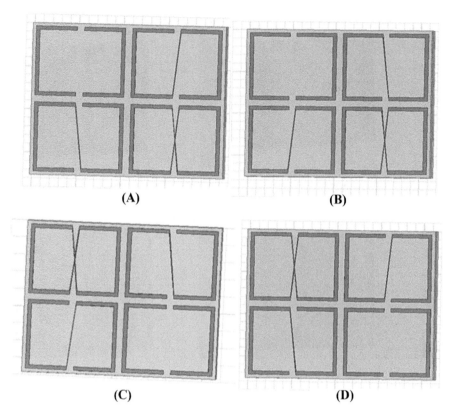

Figure 12.15 The four configurations of the proposed metasurface according to the possible coding matrices. (A) $\begin{bmatrix} 00 & 01 \\ 10 & 11 \end{bmatrix}$, (B) $\begin{bmatrix} 00 & 10 \\ 01 & 11 \end{bmatrix}$, (C) $\begin{bmatrix} 11 & 10 \\ 01 & 00 \end{bmatrix}$, and (D) $\begin{bmatrix} 11 & 01 \\ 10 & 00 \end{bmatrix}$.

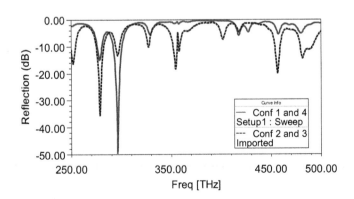

Figure 12.16 The reflection coefficients of the four configurations.

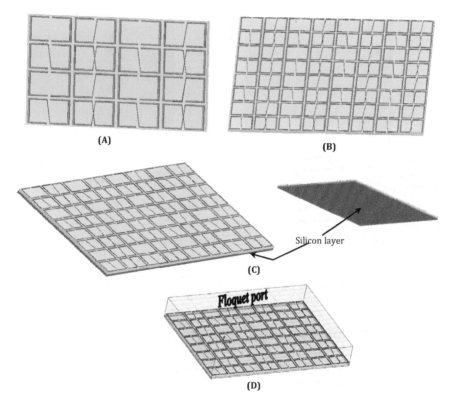

Figure 12.17 Proposed MSA (A) for (4 × 4) array, (B) for (8 × 8) array, (C) MSA with a silicon layer, and (D) simulation setup.

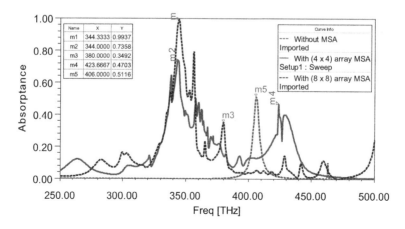

Figure 12.18 Absorptance of the silicon layer with and without MSA.

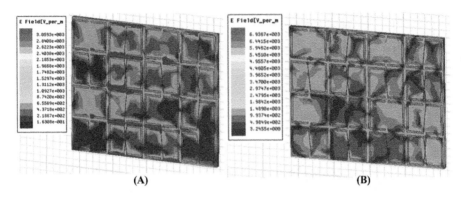

Figure 12.19 Mapping of the electrical field at (A) 344 THz and (B) 423.66 THz.

We have chosen to represent the distribution of the electric field based on the (4 × 4) array MSA at the two resonant frequencies. Fig. 12.19 represents this distribution.

12.7 Conclusion and future work

In this chapter, a new technology for improving solar PV systems is provided. We presented a general study of solar systems with an overview of PV systems. Then, we presented our contribution based on the use of digital metasurfaces to optimize the absorption qualities of PV cells. Our proposed metasurface is based on the diagonally square-shaped split-ring metamaterial resonators. The dimensions of each cell are chosen for a micrometric scale to be able to associate our metasurface absorber with the silicon layers. The electromagnetic characteristics of the proposed metasurfaces are studied. Two networks of encoded metasurface cells are optimized for different configurations to have the desired absorption levels. These coefficients obtained after simulations were carried out show the impact of the coded metasurfaces for each configuration and each network on the performance of the solar panels. The outlooks of this study are numerous, but the most important is the trend toward more miniaturized metasurface structures without increasing the losses and design costs for solar systems. It is anticipated that more novel fully balanced solutions and innovations will be seen in the near future; a major driver for innovation has been the push for higher efficiency. This is reflected by the expansion of passivated emitter and rear cell/contact (PERC) technology, which offers more efficient solar cells and as such increases the performance of solar panels. Tandem and perovskite technologies also offer interesting perspectives, but in the longer term due to barriers that still need to be addressed and overcome such as durability and price. Concerning the solutions proposed for monitoring rural and remote areas, the process of monitoring is done by conducting manual inspections; however, these can be replaced by intelligent systems, such as drones.

Acknowledgment

This work was supported by the Algerian Ministry of Higher Education and Scientific Research and the General Directorate of Scientific Research and Technological Development (DGRSDT) via funding through the PRFU Project No. A25N01UN220120200001.

References

Adnan, D. K., Aimal, D., Sultan, D., & Muhammad, N. (2018). Light absorption enhancement in tri-layered composite metasurface absorber for solar cell applications. *Optical Materials*, *84*, 195−198. Available from https://doi.org/10.1016/j.optmat.2018.07.009.

ANSYS® Electromagnetics Suite. (n.d.). 17.2 [Online], http://www.ansys.com.

Begaud, X., Lepage, A. C., Varault, S., Soiron, M., & Barka, A. (2018). Ultra-wideband and wideangle microwave metamaterial absorber. *Materials*, *11*(10), 2045.

Berka, M., Azzeddine, H., Bendaoudi, A., Mahdjoub, Z., & Rouabhi, A. Y. (2021). Dual-band bandpass filter based on electro-magnetic coupling of twin square metamaterial resonators (SRRs) and complementary resonator (CSRR) for wireless communications. *Journal of Electronic Materials*, *50*, 4887.

Chen, P. Y., Argyropoulos, C., & Alù, A. (2013). Broadening the cloaking bandwidth with non-Foster metasurfaces. *Physical Review Letters*, *111*, 233001.

Cui, T. J., Qi, M. Q., Wan, X., Zhao, J., & Cheng, Q. (2014). Coding metamaterials, digital metamaterials and programmable metamaterials. *Light: Science & Applications*, *3*, e218. Available from https://doi.org/10.1038/lsa.2014.99.

Della Giovampaola, C., & Engheta, N. (2014). Digital metamaterials. *Nature Materials*, *13*, 1115. Available from https://doi.org/10.1038/nmat4082.

Díaz-Rubio, A., Asadchy, V., Elsakka, A., & Tretyakov, S. (2017). From the generalized reflection law to the realization of perfect anomalous reflectors. *Science Advances*, *3*, e1602714.

Ebrahimi, A., Scott, J., & Ghorbani, K. (2018). Differential sensors using microstrip lines loaded with two split-ring resonators. *IEEE Sensors Journal*, *18*(14), 5786−5793.

Estep, N., Askarpour, A., & Shvets, S. (2014). Transmission-line model and propagation in a negative-index, parallel-plate metamaterial to boost electron-beam interaction. *Antenna and Propagation, IEEE Transactions*, *62*, 3212−3221.

Gennarelli, G., Romeo, S., Scarfi, M. R., & Soldovieri, F. (2013). A microwave resonant sensor for concentration measurements of liquid solutions. *IEEE Sensors Journal*, *13*(5), 1857−1864.

González-Ovejero, D., Rcck, J., Jung-Kubiak, C., & Chattopadhyay, G. (2016). A class of silicon mi-cromachined metasurface for the design of high-gain terahertz antennas. IEEE Int. Symp. Antennas Propag. (APSURSI), Fajardo, 1191−1192.

Green, M., Emery, K., & Dunlop, W. (2012). Solar cell efficiency tables (version39). *Progress in Photovoltaics: Research and Applications*, *20*, 12−20.

Hassoune, A., Khalf Allah, M., Mesbahi, A., & Breuil, D. (2017). "Electrical Design of a Photovoltaic-Grid System for Electric Vehicles Charging Station," 14th International Multi-Conference on Systems, Signals & Devices (SSD).

Hou, B., Xu, G., Wong, H., & Wen, W. (2005). Tuning of photonic bandgaps by a fieldinduced structural change of fractal metamaterials. *Optics Express*, *13*(23), 9149−9154. Available from https://doi.org/10.1364/opex.13.009149.

Houria, H., & Mohammed, M. S. (2018). Computational analysis of metamaterial−aluminum−silicon solar cell model. *Optical and Quantum Electronics*, *50*(448), 1−15. Available from https://doi.org/10.1007/s11082-018-1705-8.

Hsiao, H.-H., Chu, C. H., & Tsai, D. P. (2017). Fundamentals and applications of metasurfaces. *Small Methods*, *1*(4). Available from https://doi.org/10.1002/smtd.201600064.

Hu, J., Lang, T., & Shi, G. H. (2017). Simultaneous measurement of refractive index and temperature based on all-dielectric metasurface. *Optics Express*, *25*(13), 15241−15251. Available from https://doi.org/10.1364/OE.25.015241.

Hu, J., Lang, T., Xu, W., Liu, J., & Hong, Z. (2019). Experimental demonstration of electromagnetically induced transparency in a conductively coupled flexible metamaterial with cheap aluminum foil. *Nanoscale Research Letters*, *14*(1), 359. Available from https://doi.org/10.1186/s11671-019-3180-y.

Hu, J., Li, L., Lin, H., Zhang, P., Zhou, W., & Ma, Z. (2013). Flexible integrated photonics: Where materials, mechanics and optics meet [Invited]. *Optical Materials Express*, *3*(9), 1313. Available from https://doi.org/10.1364/ome.3.001313.

Hu, X., Xu, G., Wen, L., Wang, H., Zhao, Y., Zhang, Y., et al. (2016). Metamaterial absorber integrated microfluidic terahertz sensors. *Laser & Photonics Reviews*, *10*(6), 962−969. Available from https://doi.org/10.1002/lpor.201600064.

Huang, L., Chen, X., Mühlenbernd, H., Zhang, H., Chen, S., Bai, B., et al. (2013). Three dimensional optical holography using a plasmonic metasurface. *Nature Communications*, *4*(1), 2808. Available from https://doi.org/10.1038/ncomms3808.

International Energy Outlook. (2021). U.S. Energy Information Administration. https://www.eia.gov/outlooks/ieo/.

Kim, Y., Yuk, H., Zhao, R., Chester, S. A., & Zhao, X. (2018). Printing ferromagnetic domains for untethered fast-transforming soft materials. *Nature*, *558*(7709), 274−279. Available from https://doi.org/10.1038/s41586-018-0185-0.

Kodera, T., Sounas, D. L., & Caloz, C. (2010). Artificial Faraday rotation using a ring metamaterial structure without static magnetic field. *Applied Physics Letters*, *99*, 031114.

Lee, S. H., et al. (2012). Switching terahertz waves with gate controlled active graphene metamaterials. *Nature Materials*, *11*, 936−941.

Li, P., Dolado, I., Alfaro-Mozaz, F. J., Casanova, F., Hueso, L. E., Liu, S., et al. (2018a). Infrared hyperbolic metasurface based on nanostructured van der Waals materials. *Science (New York, N.Y.)*, *359*(6378), 892−896. Available from https://doi.org/10.1126/science.aaq1704.

Li, S., Zhou, C., Ban, G., Wang, H., Lu, H., & Wang, Y. (2019). Active all-dielectric bifocal metalens assisted by germanium antimony telluride. *Journal of Physics D: Applied Physics*, *52*(9), 9. Available from https://doi.org/10.1088/1361-6463/aaf7f3.

Li, X., Xiao, S., Cai, B., He, Q., Cui, T. J., & Zhou, L. (2012). Flat metasurfaces to focus electromagnetic waves in reflection geometry. *Optics Letters*, *37*(23), 4940−4942. Available from https://doi.org/10.1364/OL.37.004940.

Li, X.-S., Cheng, L.-L., Liu, X.-Y., & Liu, Q. H. (2018b). Wideband GNSS antenna covered by a double sided metasurface. *AEU— International Journal of Electronics and Communications*, *96*, 170−177.

Liu, N., Guo, H., Fu, L., & Kaiser, S. (2008). Three dimensional photonic metamaterials at optical frequencies. *Nature Materials*, *7*, 31−37.

Liu, X., Fan, K., Shadrivov, I. V., & Padilla, W. J. (2017). Experimental realization of a terahertz all-dielectric metasurface absorber. *Optics Express*, *25*(1), 191−201. Available from https://doi.org/10.1364/OE.25.000191.

Liu, Y., Chen, Y., Liand, J., Hung, T. C., & Jianping, L. (2012). Study of energy absorption on solar cell using metamaterials. *Solar Energy*, *86*, 1586e99.

Marquez, R., Medina, F., & Raffi, R. (2002). Role of bianisotropy in negative permeability and left-handed metamaterials. *Physical Review B*, *65*.

Moccia, M., Liu, S., Wu, R. Y., et al. (2017). Coding metasurfaces for diffuse scattering: Scaling Laws, bounds, and suboptimal design. *Advanced Optical Materials*, *5*(19). Available from https://doi.org/10.1002/adom.201700455.

Mohammed, B., Zoubir, M., & Baghdad, B. A. (2019). Improvement of the absorption qualities of photovoltaic cells based on the Metasurface for magnetic activity, 5th International Conference on Advances in Mechanical Engineering, ISTANBUL.

Ni, X., Emani, N. K., Kildishev, A. V., Boltasseva, A., & Shalaev, V. M. (2012). Broadband light bending with plasmonic nanoantennas. *Science (New York, N.Y.)*, *335*(6067), 427. Available from https://doi.org/10.1126/science.1214686.

Peijiang, C., Xuehua, J. (2008). Design and Implementation of Remote Monitoring System Based on GSM. IEEE Pacific-Asia Workshop on Computational Intelligence and Industrial Apllication.

Person, S., Jain, M., Lapin, Z., Saenz, J. J., Wicks, G., & Novotny, L. (2013). Demonstration of zero optical backscattering from single nanoparticles. *Nano Letters*, *13*(4), 1806−1809. Available from https://doi.org/10.1021/nl4005018.

Pisco, M., Galeotti, F., Quero, G., Grisci, G., Micco, A., Mercaldo, L. V., et al. (2017). Nanosphere lithography for optical fiber tip nanoprobes. *Light: Science & Applications*, *6*(5), e16229. Available from https://doi.org/10.1038/lsa.2016.229.

Pors, A., Albrektsen, O., Radko, I. P., & Bozhevolnyi, S. I. (2013). Gap plasmon-based metasurfaces for total control of reflected light. *Scientific Reports*, *3*, 2155. Available from https://doi.org/10.1038/srep02155.

Principe, M., Consales, M., Castaldi, G., Galdi, V., & Cusano, A. (2019). Evaluation of fiber-optic phase-gradient meta-tips for sensing applications. *Nanomaterials and Nanotechnology*, *9*. Available from https://doi.org/10.1177/1847980419832724.

Qian, C., Zheng, B., Shen, Y., Jing, L., Li, E., Shen, L., et al. (2020). Deep-learning enabled self-adaptive microwave cloak without human intervention. *Nature Photonics*, *14*(6), 383−390. Available from https://doi.org/10.1038/s41566-020-0604-2.

Référence sur charbon (n.d.), pétrole brut, gaz naturel, combustible nucléaire, hydroélectricité, biomasse, énergie solaire, énergie éolienne, géothermique, et énergie des océans.

Renewable Energy Policy Network for the 21st century (2012). "Renewables 2012, Global Status report."

Rocco, D., Carletti, L., Caputo, R., Finazzi, M., Celebrano, M., & De Angelis, C. (2020). Switching the second harmonic generation by a dielectric metasurface via tunable liquid crystal. *Optics Express*, *28*(8), 12037−12046. Available from https://doi.org/10.1364/OE.386776.

Ruiz de Galarreta, C., Sinev, I., Alexeev, A. M., Trofimov, P., Ladutenko, K., Garcia- Cuevas Carrillo, S., et al. (2020). Reconfigurable multilevel control of hybrid alldielectric phase-change metasurfaces. *Optica*, *7*(5), 476. Available from https://doi.org/10.1364/optica.384138.

Scheerlinck, S., Dubruel, P., Bienstman, P., Schacht, E., Van Thourhout, D., & Baets, R. (2009). Metal grating patterning on fiber facets by UV-based nano imprint and transfer lithography using optical alignment. *Journal of Lightwave Technology*, *27*(10), 1415−1420. Available from https://doi.org/10.1109/jlt.2008.2004955.

She, A., Zhang, S., Shian, S., Clarke, D. R., & Capasso, F. (2018). Adaptive metalenses with simultaneous electrical control of focal length, astigmatism, and shift. *Science Advances*, *4*, eaap9957.

Sirmaci, Y. D., Tang, Z., Fasold, S., Neumann, C., Pertsch, T., Turchanin, A., et al. (2020). Plasmonic metasurfaces situated on ultrathin carbon nanomembranes. *ACS Photonics, 7* (4), 1060−1066. Available from https://doi.org/10.1021/acsphotonics.0c00073.

Smith, D. R., Pendry, J. B., & Wiltshire, M. C. (2004). Metamaterials and negative refractive index. *Science (New York, N.Y.), 305*(5685), 788−792. Available from https://doi.org/10.1126/science.1096796.

Su, L., Contreras, M. J., Velez, P., & Martin, F. (2016). Configurations of splitter/combiner microstrip sections loaded with stepped impedance resonators for sensing applications. *Sensors, 16*(12), 2195.

Valentine, J., Zhang, S., Zentgraf, T., Ulin-Avila, E., Genov, D. A., Bartal, G., et al. (2008). Three-dimensional optical metamaterial with a negative refractive index. *Nature, 455* (7211), 376−379. Available from https://doi.org/10.1038/nature07247.

Veselagi, V. G. (1968). The electrodynamics of substances with simultaneously negative values of ε and μ. *Soviet Physics Uspekhi, 10*, 509−514.

Vilas Bhau, G., et al. (2021). IoT based solar energy monitoring system. *Materials Today: Proceedings, 80*(3), 3697−3701. Available from https://doi.org/10.1016/j.matpr.2021.07.364.

Xu, J., Fan, Y., Yang, R., Fu, Q., & Zhang, F. (2019). Realization of switchable EIT metamaterial by exploiting fluidity of liquid metal. *Optics Express, 27*(3), 2837−2843. Available from https://doi.org/10.1364/OE.27.002837.

Yang, C.-L., Lee, C.-S., Chen, K.-W., & Chen, K.-Z. (2016). "Noncontact measurement of complex permittivity and thickness by using planar resonators,". *IEEE Transactions on Microwave Theory and Techniques, 64*(1), 247−257.

Yang, Y., KravchenkoII., Briggs, D. P., & Valentine, J. (2014). All-dielectric metasurface analogue of electromagnetically induced transparency. *Nature Communications, 5*, 5753. Available from https://doi.org/10.1038/ncomms6753.

Yongzhi, C., Jiaqi, L., Fu, C., Hui, L., & Xiangcheng, L. (2021). Optically switchable broadband metasurface absorber based on square ring shaped photoconductive silicon for terahertz waves. *Physics Letters A, 402*(28), 127345. Available from https://doi.org/10.1016/j.physleta.2021.127345.

Yu, N., Genevet, P., Kats, M. A., Aieta, F., Tetienne, J. P., Capasso, F., et al. (2011). Light propagation with phase discontinuities: generalized laws of reflection and refraction. *Science (New York, N.Y.), 334*(6054), 333−337. Available from https://doi.org/10.1126/science.1210713.

Zhang, W., Song, Q., Zhu, W., Shen, Z., Chong, P., Tsai, D. P., et al. (2018). Metafluidic metamaterial: A review. *Advances in Physics: X, 3*(1), 1417055. Available from https://doi.org/10.1080/23746149.2017.1417055.

Zhao, W., Jiang, H., Liu, B., Song, J., Jiang, Y., Tang, C., et al. (2016). Dielectric Huygens' metasurface for high-efficiency hologram operating in transmission mode. *Scientific Reports, 6*, 30613. Available from https://doi.org/10.1038/srep30613.

Zhu, Y., Li, Z., Hao, Z., DiMarco, C., Maturavongsadit, P., Hao, Y., et al. (2018). Optical conductivity-based ultrasensitive mid-infrared biosensing on a hybrid metasurface. *Light: Science & Applications, 7*, 67. Available from https://doi.org/10.1038/s41377-018-0066-1.

Challenge of modern photovoltaic systems under large-scale forms to distribution grid in preinstallation and operation: Vietnam study case evaluation

13

Minh Quan Duong[1], Tuan Le[2], The Hoang Tran[3], Thi Minh Chau Le[4]
and Ngoc Thien Nam Tran[5]
[1]The University of Danang-University of Science and Technology, Danang, Vietnam,
[2]Roberval Laboratory, University of Technology of Compiegne, Compiegne, France,
[3]The University of Auckland, Auckland, New Zealand, [4]Hanoi University of Science and
Technology-School of Electrical and Electronic Engineering, Hanoi, Vietnam, [5]Delta
Electronic Inc., Tainan, Taiwan

13.1 Introduction

The development of socioeconomic development requires a growing electricity demand. As a result, power companies face many grid planning and power generation problems to suffice the load and ensure power quality and energy security. On the other hand, after the Paris Agreement on climate change in 2015, reducing CO_2 emissions is a top priority that requires countries to make urgent changes in industries, especially in the energy industry (Kemp, 2018; Lacal Arantegui & Jäger-Waldau, 2018). In which the main strategies for future power system development identify the limitations of conventional generators and research alternatives to power sources such as fossil power plants, nuclear power plants, etc. Furthermore, renewable energy sources (RESs) play an essential role in the modern power grid, especially wind power and photovoltaic (PV).

Due to low investment rates and short construction periods compared to other RESs, PV is increasing rapidly in quantity and generation capacity, exceeding renewable portfolio standard (RPS) expectations, especially in tropical and subtropical countries (Barbose, 2019; Barragán-Escandón & Zalamea-León, 2019; D'Adamo et al., 2018; Gagnon et al., 2018; The Vietnamese Prime Minister, 2016). However, the nature of PV or, by extension, RESs is unstable because of environmental uncertainty factors such as weather, temperature, etc. The centralized distribution of these sources causes many potential dangers to the power system. Connecting these sources as distributed energy sources (DESs) provides higher

Performance Enhancement and Control of Photovoltaic Systems. DOI: https://doi.org/10.1016/B978-0-443-13392-3.00014-1

input power stability (Duong et al., 2017a; Joshi et al., 2009). As a result, technologies for DESs and power systems have been raised for mining efficiencies, such as the application of Internet of Things (IoT) and distribution automation systems (DAS) for improving controllability and monitoring. Moreover, modern PV systems with integrated inverters with advanced maximum power point tracking algorithms can capture the maximum amount of input power as well as improve the stability of PVs (Nguyen et al., 2020; Tran & Duong, 2019; Van Tan et al., 2020).

Photovoltaic penetration as a DES can significantly affect the steady-state and the fault ride-through capability of the power system, leading to significant impacts on reliability and safety during operation. Following the IEEE standard for distributed energy resources (IEEE, 2003; UL, 1741), PV should not operate in the voltage-regulating mode (P−V mode) at the point of common coupling (PCC), which means PV is considered an active power generator. The different penetration levels of PV into the distribution grid (DG) have been the subject of many studies (Ahmadi et al., 2018; Bawazir & Cetin, 2020; Jamil & Anees, 2016; Prakash & Khatod, 2016; Ullah et al., 2018). These studies focus on optimizing power flow in the grid, finding optimal connection locations, or control solutions to improve operating efficiency, but the PV power level is assumed to be relatively low in the grid or only considers the steady state of the grid. Other research focuses on the grid with only one PV plant coordinating with other generators to investigate the impact (Duong et al., 2017b, 2019).

Some works show that depending on the PV penetration level, the load condition can cause many voltage effects and grid problems and offer necessary adjustment solutions, especially inverter control in PV (Liu et al., 2008; Smith et al., 2011; Turitsyn et al., 2010). The studies (Bawazir & Cetin, 2020; Haifeng et al., 2010; Tran et al., 2020), along with a common point raised, show that PV penetration causes a diverse impact on the grid in dynamic mode, but the systems used in these studies are relatively small or use a typical grid. Specifically, in the study (Achilles et al., 2008; Refaat et al., 2018), the IEEE 39 Bus and the 30 IEEE Bus systems, respectively, were used to consider the grid response to many simulations of incidents and penetration levels of PV. The power flow change in the actual DG with PV is connected in Castro et al. (2020) and Guerrero et al. (2020) but does not consider the dynamic mode analysis process. Necessarily, the impact of PV needs to be considered in actual grid cases when PV capacity increases, which can clearly and accurately reflect the risks.

Reports from IRENA show that the development trend of smart grid and clean energy is strongly exploding. Following this direction, PV in the form of a large-scale plant is rising and connected to the national power systems in both installation and generation capacity. In fact, the power system planning issue is considered as a long-term plan for 5 or 10 years, while LSPVs and related technologies develop year by year. This leads to out-of-sync in operation when LVPS connects more than expected. As mentioned, DG has to face reverse power flow, local congestion, LSPV power fluctuations, and difficult conditions to recover after incidents in high-penetration cases (Gope et al., 2019; Paul, 2022). Power companies had recommendations to limit the generating capacity from LVPVs, but it causes unfairness to

LSPV investors when it reduces their revenue. Therefore, prefeasibility studies need to be conducted to determine the effect of LSPVs on DG so that optimization can be achieved.

For all the rationale of aforementioned, this chapter comprehensively investigates the penetration levels of LSPVs into the distributed grid, considering the transmission system. The impacts of LSPVs are reviewed in detail regarding power flow, voltage fluctuation, fault ride-through capability, and the effect on nearby generators in the grid. Mathematical model of the whole system will be provided to easily analyze the responses occurring as well as provide a comprehensive overview of the DG operating modes with integrated LSPVs. After that, ETAP software will simulate the grid elements with various operations in steady state and transient state to show LSPVs positive and negative impacts when connected to the grid in different scenarios. This software can easily exchange operational data through specialized software used in DG, such as PSSE or DAS. In addition, programming software such as Matlab can be combined to improve system controllability.

To carry out this study, the distribution grid of Quang Ngai province, central Vietnam, was selected, which is an area with an average solar irradiance of about 5057 kW/m^2/day with a total sunshine of about 2131 h/year. This DG contains a full range of load models and conventional generator types such as thermoelectricity and hydroelectricity along with medium- and high-voltage transmission systems, powered by the Vietnam national grid through extra-high voltage systems: 220 kV and 500 kV. Moreover, because of favorable conditions, the number of LSPV projects in this locality is increasing and promises to account for a large proportion of the total electricity production from RESs. Thus, it can be considered a typical model for performing research.

13.2 Analytical problems and system model under study

13.2.1 Steady-state stability

Steady-state stability is the ability of a system to maintain equilibrium under constrained operating conditions (Eftekharnejad et al., 2013). These constraints include the limited voltage at the nodes; the limited capacity of the line to exchange power; and the balance of power between generators, loads, and losses. The most significant difference between PV and conventional or wind generators commonly used is the movement or motor constructs that can generate direct alternating power. In addition, PV needs to use an inverter to create the AC power and face the reactive power limit.

The inverter can generate low reactive power due to limited sizing. So, the PCC can operate in the PV bus mode with Q-limited; then plants can still adjust the voltage but cannot be flexible. Most PVs operate in the unity power factor mode without generating reactive power. In other words, the coupling point for plants is referred to as PQ bus with $Q = 0$. This situation will cause more influence when the grid needs to provide more reactive power so that the devices in these plants can operate, especially transformers. In this study, the degree of penetration of LSPVs can be calculated based on the following equation:

$$\%\mathrm{LSPV}_{\mathrm{penetration}} = \frac{\sum \mathrm{LSPV\ generator\ capacity}}{\sum \mathrm{Load\ demand}} \times 100 \qquad (13.1)$$

13.2.2 Transient stability (dynamic stability)

The disturbance in the power system is indispensable in the face of faults on the lines, at buses, or primarily the shading phenomenon, which is the main reason causing the LSPV output power to fluctuate and be lost entirely. Therefore, transient stability will show the ability to maintain the operation and synchronization of the power system when disturbances occur.

The inertia of the generators has a critical role in controlling synchronization speed in the power system whenever there is a disturbance leading to an imbalance between the input mechanical power and generator output power. The voltage angle can also affect the synchronization process when injecting too much LSPV power. Moreover, the system inertia, including LSPVs, will be changed more than the wind energy source, leading to higher potential disturbances because of low system inertia.

Almost all LSPVs are equipped with full-size inverters. The active and reactive power can be independently adjusted based on the control model by transforming from three-phase frame to d–q frame and vice versa as in Fig. 13.1, similar to the existing wind turbine generator type 4 (Huang & Wang, 2018; Xu, 2013; Zhu et al., 2013). Because of that, the inverter of LSPVs can control the voltage in the connected area but to a very limited extent. So, considering the response at PCC in transient mode, especially after the fault, is necessary.

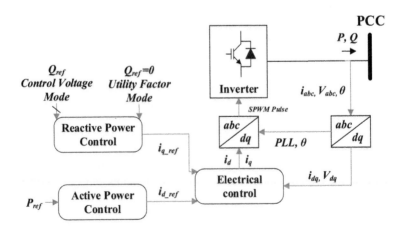

Figure 13.1 Independent control mode of active and reactive power in LSPV plants.

13.2.3 Real system model—Quang Ngai distribution grid

The analysis cases are performed on the actual DC in Vietnam to consider the overall impact of LSPVs when changing penetration levels. Quang Ngai grid is selected with many different voltage levels, including conventional generation sources, transmission and distribution systems, and additional support from the Vietnamese power system. PV connects with the grid based on the large-scale plant, adapts to local load demand, and reduces the transmission power burden of the Vietnamese power system. Single-line diagram and the exchange power are shown in Fig. 13.2 and Table 13.1.

Strengthening distributed generators is essential because the load demand is not supplied enough, which is a stepping stone for the penetration of LSPV plants. Construction work of three LSPVs, MoDuc PV, BinhNguyen PV, and PhoAn PV, was completed at the end of 2021, and they were injected into the grid with a capacity equivalent to 50% of the installed capacity. The power company sets this limitation to ensure the safety of the system when the study to assess the impact of these plants at 100% generating capacity has not been completed. It is expected that by early 2023, when the mechanisms are completed, all three LSPVs will be able to generate maximum power for DG, reaching 210 kW. Due to geographical and climatic features, LSPV plants can be built more in this area, not just stop at three. The study on this grid can be used as reference results for power system operators with similar grids, which have an increasing level of similar PV connection in many countries. These plants will be modeled in steady-state and dynamic modes to show the overall influence of the parameters shown in Table 13.2.

For steady-state stability analysis, different penetration scenarios are divided as in Table 13.3 for Quang Ngai DG to investigate the effect on the whole grid and at the coupling points with the PV penetration level defined by Eq. (13.1).

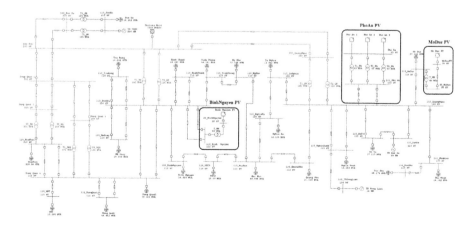

Figure 13.2 Single-line diagram of Quang Ngai distribution grid in ETAP.

Table 13.1 DG configuration.

Load demand	P (MW)	748
	Q (Mvar)	190
Total generation capacity in DG (without LSPVs)	P (MW)	333
	Q (Mvar)	131
Import from transmission system (without LSPVs)	P (MW)	415
	Q (Mvar)	60
LSPV plans	P (MW)	210

Table 13.2 Operating parameters of LSPVs.

LSPVs	Capacity (MW)	Inverter operation mode	Connected bus
MoDuc PV	19.2	Utility factor/PF control	110_MoDuc
Binh Nguyen PV	40.8	Utility factor/PF control	110_DocSoi (PCC1)
PhoAn PV	150	Utility factor/PF control	110_QuangNgai (PCC2)

Table 13.3 Penetration scenarios of PV in steady-state mode of real case.

Scenario	Description	P (MW)	% penetration
1	No PVs connected	0	0.0
2	9.6 MW from MoDuc PV and 10.2 MW from Binh Nguyen PV	19.8	2.6
3	9.6 MW from MoDuc PV, 10.2 MW from Binh Nguyen PV, and 50 MW from PhoAn PV	69.8	9.3
4	9.6 MW from MoDuc PV, 30.6 MW from Binh Nguyen PV, and 50 MW power from PhoAn PV	90.2	12.1
5	9.6 MW power from MoDuc PV, 20.4 MW from Binh Nguyen PV, and 100 MW from PhoAn PV	130	17.4
6	19.2 MW from MoDuc PV, 30.6 MW from Binh Nguyen PV, and 100 MW from PhoAn PV	149.8	20.0
7	19.2 MW from MoDuc PV, 40.8 MW from Binh Nguyen PV, and 100 MW from PhoAn PV	160	21.4
8	All three PVs generate full power	210	28.1

The transient stability analysis is proposed to check the grid recovery when different faults occur. Simulations are carried out in transient three-phase short circuit faults with transmission lines and LSPV outage, according to Table 13.4. Responses of voltage and frequency are considered besides the level of influence

Table 13.4 Transient stability analysis scenarios.

Scenario	Case 1—Three-phase short circuit faults		Case 2—LSPV outage faults	
	Connected LSPVs	Penetration (%)	Loss level of LSPVs	Outage power (%)
1	No LSPV plants	0.00	MoDuc PV	3.54
2	MoDuc PV	3.54	Binh Nguyen PV	7.51
3	Binh Nguyen PV	7.51	PhoAn PV	27.62
4	PhoAn PV	27.62	All three LSPVs	38.76
5	All three LSPVs	38.67		

on conventional generators nearby. The control model of modern inverter is built using the UDM tool in ETAP (Chapter 25, 2020), a tool similar to the Simulink function of Matlab based on the control model shown in Fig. 13.1.

13.3 Proposed evaluation model

13.3.1 Proposed steady-state mathematical model

The mathematical model is proposed based on the equivalent circuit, as shown in Fig. 13.3, and used to analyze the responses of power and voltage with different penetration levels of LSPVs in steady state. The proposed model was developed from several previous models and considered the effect of increasing reactive power in the grid when LSPVs penetrate too much. Previous research assumed this reactive power was zero. The model gives steady-state equations that clarify the relationship between load voltage, the total power of LSPVs, and reactive power consumed by LSPVs.

Suppose the conventional generators are referred to the equivalent terminal voltage of E_G with the deviation angle δ, power transmits through the line with impedance $Z_T = R_T + jX_T$, and total power of the generators in the grid is shown as follows:

$$\dot{S}_G = \dot{E}_G \dot{i}^* = \frac{|E_G| \quad (\cos\delta + j\sin\delta)(|E_G| \quad (\cos\delta \quad - j\sin\delta) - |V_L|)}{R_T - jX_T} = \dot{P}_G + j\dot{Q}_G$$

(13.2)

$$\dot{P}_G = \frac{|E_G|}{R_T^2 + X_T^2}(R_T|E_G| + |V_L|(X_T\sin\delta - R_T\cos\delta))$$

(13.3)

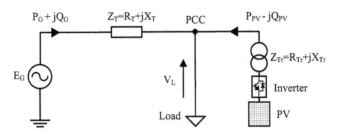

Figure 13.3 The mathematical model based on the equivalent circuit.

$$\dot{Q}_G = \frac{|E_G|(X_T \quad |E_G| - |V_L|R_T\sin\delta - |V_L| \quad X_T \quad \cos \quad \delta)}{R_T^2 + X_T^2} \tag{13.4}$$

The power dissipated in the grid is converted to losses on the transmission line, which can be calculated:

$$\dot{S}_T = \dot{V}_T \dot{i}^* = \frac{[|E_G|(\cos\delta + j\sin\delta) - |V_L|] \quad [|E_G|(\cos\delta - j\sin\delta) - |V_L|]}{R_T - jX_T} = \dot{P}_T + j\dot{Q}_T \tag{13.5}$$

$$P_T = \frac{|E_G|^2 R_T - 2|E_G||V_L|R_T\cos\delta + |V_L|^2 R_T}{R_T^2 + X_T^2} \tag{13.6}$$

$$Q_T = \frac{|E_G|^2 X_T - 2|E_G||V_L|X_T\cos\delta + |V_L|^2 X_T}{R_T^2 + X_T^2} \tag{13.7}$$

Balanced constraints of transaction power in the grid:

$$P_G = P_T + P_L - P_{LSPV} \tag{13.8}$$

$$Q_G = Q_T + Q_L + Q_{LSPV} \tag{13.9}$$

The relationship between reactive power and the effect can be calculated using the power factor at PCC; in this study, it can be assumed as follows:

$$Q_L = aP_L \quad and \quad Q_{LSPV} = bP_{LSPV} \tag{13.10}$$

Combine Eqs. (13.3), (13.6), and (13.8):

$$-|E_G| \quad |V_L|(X_T\sin\delta + R_T\cos\delta) = (P_{LSPV} - P_L)(R_T^2 + X_T^2) - |V_L|^2 R_T \tag{13.11}$$

Combine Eqs. (13.4), (13.7), and (13.9):

$$|E_G| \quad |V_L|(X_T\cos\delta - R_T\sin\delta) = aP_L\left(R_T^2 + X_T^2\right) + |V_L|^2 X_T + bP_{LSPV}\left(R_T^2 + X_T^2\right)$$
$$(13.12)$$

Eqs. (13.11) and (13.12) combined with condition (13.10) shows the quadratic characteristic of the load voltage and power generated from LSPVs:

$$|V_L|^4 + |V_L|^2\left[2X_T(aP_L + bP_{LSPV}) - 2R_T(P_{LSPV} - P_L) - |E_G|^2\right]$$
$$+ (aP_L + bP_{LSPV})^2\left(R_T^2 + X_T^2\right) + (P_{LSPV} - P_L)^2\left(R_T^2 + X_T^2\right) = 0$$
$$(13.13)$$

Solve Eq. (13.13) with an open circuit solution with EG = 1; the relationship between the voltage at the coupling point and the penetration level of LSPVs when changing variables a, b, R_T, X_T, and P_L is shown by the characteristics in Fig. 13.4.

With the obtained results, the quadratic voltage response did not change when changing the system parameters and reactive consumption ratio, considering that reactive power is supplied and not supplied to LSPVs. The responses show that only one voltage peak can be found when LSPVs penetrate at a determined value. Voltage can be markedly improved if the penetration level is limited between zero and that determined value. However, if the increased level exceeds the limit, the grid will face a voltage drop. The simulation progresses in existing systems to determine problems even more clearly. The results can demonstrate the feasibility of the proposed method and the impact of LSPV penetration.

13.3.2 Transient stability mathematical model

The responses of voltage, frequency, and conventional generators nearby are essential factors in evaluating the distribution system stability when LSPV penetration

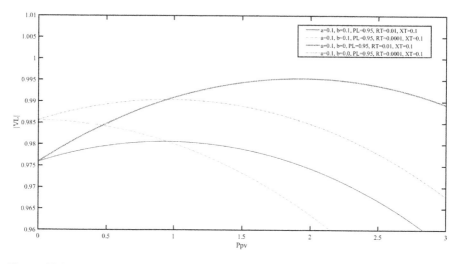

Figure 13.4 Relationship between load voltage and penetration level of LSPVs.

causes the fault. Furthermore, the system's dynamic structure will change when too many LSVPs are connected, such as system inertia and voltage angle. Therefore, investigating a comprehensive model is necessary to observe potential risks of disturbances inside the system. Based on the kinetic equations, the model is synthesized and proposed to assess responses in the system when errors arise due to different penetration levels of LSPVs.

The unbalanced mechanical-electrical power in the generator after faults occur causes each generator to behave differently and leads to disturbances in system synchronism. The unbalance and inertia deviation inside generators can be determined through the electromechanical equation (Glover et al., 2015):

$$J \frac{d^2\theta_m}{dt^2} = T_{in} - T_{out} \tag{13.14}$$

$$J\omega \frac{d^2\theta_m}{dt^2} = P_{in} - P_{out} \tag{13.15}$$

where $T_{mechanical}$ and $T_{electrical}$ are mechanical and electrical torque at total moment of inertia J and angular displacement of the rotor θ_m; $P_{mechanical}$ and $P_{electrical}$ refer to input mechanical power and output electrical power, respectively, with angular velocity ω.

The dissimilarity in technology and installed capacity causes generators to have different inertia. Thus, generators must obey the synchronous inertia J_{sys} and operate at the synchronous speed ωsys of the system. With inertia constant M and per unit inertia constant H, the generator dynamic model in synchronous mode can be considered as follows:

$$\frac{2}{p} M \frac{d^2\theta_m}{dt^2} = P_{in} - P_{out} \tag{13.16}$$

$$\frac{2H}{\omega_{sys}} \frac{d^2\theta_m}{dt^2} = P_{in} - P_{out} \tag{13.17}$$

where the relationship between mechanical angle θ_m and electric angle θ_e of the generator that has p poles can be seen:

$$\theta_e = \frac{p}{2} \theta_m \tag{13.18}$$

Considering the power system connected to multiple generators, generated power P_G provides for the demand P_L; the system swing equation is rewritten from Eq. (13.14) as follows (Duong et al., 2017a; Nguyen et al., 2020):

$$\frac{d}{dt}\left(\frac{1}{2} j_{sys}\omega_{sys}^2\right) = P_G - P_L \tag{13.19}$$

And the output voltage of generators can represent by

$$V_{tRMS} = \sqrt{2}\pi N\Phi f_{sys} \tag{13.20}$$

where N is the number of one-phase stator windings turns; Φ is the magnetic flux through stator; and f_{sys} is the system frequency.

Following model construction from Eqs. (13.17), (13.19), and (13.20), system responses and effects on conventional generators can be determined when LSPVs penetrate.

13.3.3 Inverter model integrated into ETAP

Similar to the actual mechanism of the actual inverter and in previous studies, the inverter model in ETAP goes through two control loops: inner current control loop and outer voltage control loop, as in Fig. 13.5.

The outer loop takes responsibility for tracking the reference DC voltage (V_{DCref}) from the MPPT system by regulating the DC voltage. This allows DC power from the PV panels to deliver maximum power before being converted to AC. Meanwhile, the voltage of the DC-link through charge–discharge process of integrated capacitor maintains reference values and reduces the loss for IGBT/ Mosfet. The equation shows the process of voltage balancing on the DC side:

$$\frac{d}{dt}\frac{CV_{DC}^2}{2} = P_{PV} - P_{inv\_in} \tag{13.21}$$

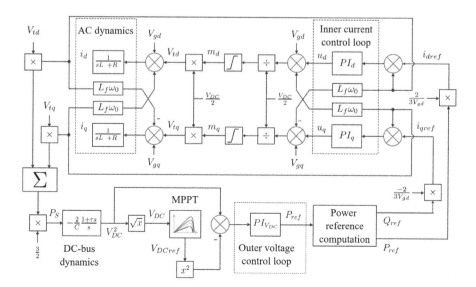

Figure 13.5 Schematic diagram of the PV inverter control applied in ETAP.

where P_{PV} is the PV power output at DC side and P_{inv_in} is the DC input power of the PV inverter.

Eq. (13.21) demonstrates that with probable DC output power of the considered PV panel. By adjusting the inverter input power, which can change the AC current in turn, DC-link capacitor voltage is controlled. During operation, the compensation for errors between V_{DC} and V_{ref} is handled by a PI controller.

Different from the mechanism of outer control loop, which regulates voltage, inner loop controls the AC side of inverter based on the current. Two AC current components of the inverter, i_d and i_q, are controlled based on their respective reference currents generated by the outer loop, including i_{dref} and i_{qref}. These reference signals will be compared with the signal from the grid and compensated for errors through PI controller similar to voltage control loop. Finally, the pulse modulation signals are fed to the IGBT/Mosfet of the inverter to perform switching operations.

From that, i_d and i_q can be observed from the following equation:

$$\begin{cases} L_f \dfrac{di_d}{dt} = L_f \omega_0 i_q - R_f i_d + V_{td} - V_{gd} \\ \\ L_f \dfrac{di_q}{dt} = L_f \omega_0 i_d - R_f i_q + V_{tq} - V_{gq} \end{cases} \qquad (13.22)$$

where L_f and R_f are filter inductance and resistance, V_{gd} and V_{gq} are voltages of grid side, and V_{td} and V_{tq} are inverter terminal voltages, which can be calculated from modulation indices m_{dq}:

$$\begin{cases} V_{tdq} = \dfrac{V_{DC}}{2} m_{dq} \\ \\ V_{tdq} = \dfrac{V_{DC}}{2} m_d \\ \\ V_{tdq} = \dfrac{V_{DC}}{2} m_q \end{cases} \qquad (13.23)$$

As presented, the PI controller will be used to compensate for the errors that occur between the D-axis and Q-axis components. Variables u_d and u_q are added to determine modulation indices on the two axes, m_d and m_q:

$$\begin{cases} m_d = \dfrac{2}{V_{DC}} \left(u_d - L_f \omega_0 i_q + V_{gd} \right) \\ \\ m_q = \dfrac{2}{V_{DC}} \left(u_q - L_f \omega_0 i_d + V_{gq} \right) \end{cases} \qquad (13.24)$$

So:

$$\begin{cases} L_f \dfrac{di_d}{dt} = V_{gd} - R_f i_d + u_d \\ \\ L_f \dfrac{di_q}{dt} = V_{gq} - R_f i_q + u_q \end{cases} \qquad (13.25)$$

From the aforementioned equations, it is easy to see that the components d and q are linear and can be controlled via two control loops independently. For PI controller, the control responses are established based on proportional coefficients, k_{pd} and k_{pq}, along with integral coefficients, k_{id} and k_{id}:

$$\begin{cases} PI_d = k_{pd} + \dfrac{k_{id}}{s} \\ \\ PI_q = k_{pq} + \dfrac{k_{iq}}{s} \end{cases} \qquad (13.26)$$

13.4 Simulation result and discussion

13.4.1 Steady state

The proposed analytical model shows the voltage magnitude points in Fig. 13.6 with different penetrations of LSPVs in Table 13.3. In addition, the quadratic curve

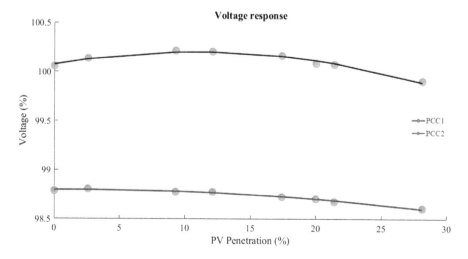

Figure 13.6 The steady-state voltage response.

representing the voltage characteristics is also determined asymptotically with the obtained voltage results by the polynomial regression method (Ostertagová, 2012). Compared to responses in Fig. 13.4, voltage behaves similarly when LSPV penetration increases. Therefore, the results analyzed from Eq. (13.2) to (13.13) can attest the accuracy of the simulation progress.

LSPVs raise the voltage at the nodes when penetrating to a certain extent. Nevertheless, if the penetration exceeds this limit, the voltage decrease and nearby buses respond similarly. The steady-state voltage of the system increases when the penetration level of the LSPVs reaches 5% at PCC1 and nearby locations, while in other areas, such as PCC2, it is 15%. In worse cases, the local overvoltage due to high LSPV penetration may occur, exceeding the permissible limit. Therefore, it is necessary to prevent buses from reaching this value. Adjusting the reactive power through capacitive shunts, SVC, or alternating the terminal voltage of conventional generators is a possible solution to minimize the adverse effects of LSPVs in the grid.

Fig. 13.7 shows power losses. In terms of benefits, losses can be reduced in the grid when the LSPV penetration reaches a limit similar to the voltage because of the shorter transmission distance and adjacent loads, and the power transmission in that scenario may be lower than in the traditional case (0% of LSPVs). At the same time, LSPV penetration reduces the voltage loss, as can be seen clearly in scenarios 2, 3, and 4 compared to the remaining scenarios. On the other hand, when LSPVs generate too much power, exceeding the demand for neighboring loads, a reverse power surge occurs in the grid. Surplus energy is supplied to loads with a longer transmission distance, so the loss of both power and voltage is increased in scenarios 5, 6, 7, and 8. The higher penetration level of LSPVs and the operation in utility factor mode make conventional generators required to generate a higher amount of reactive power so that devices in plants or nearby have the most stable operation. Thus, the response of reactive power loss has shown signs of increasing.

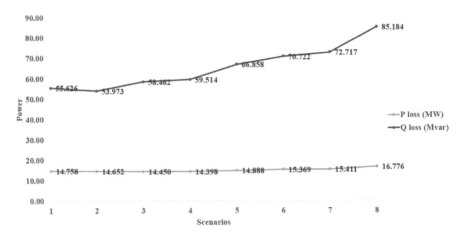

Figure 13.7 Power losses in the grid.

To summarize, LSPVs injecting capacity into the grid within a certain limit will reduce transmission power on the line, especially those located near the load. The losses in the grid also decrease for that reason, from which the voltage is improved. Alternatively, LSPVs supply power to nearby loads, limiting energy transmitted from conventional generators to loads at longer distances. However, suppose that supplied power exceeds the load demand. In that case, reverse power flow is generated and transmitted to a remote area, and the voltage decreases gradually as this flow goes up.

13.4.2 Transient stability

13.4.2.1 Case 1—short circuit fault

In this case, the short circuit is assumed at the critical location and has a higher probability in the system. Therefore, the three-phase short circuit fault is established on the transmission line, as shown in Fig. 13.8, which connects the LSPV area with the large system. The fault is generated at the 2nd second and cleared by the protection system after 150 Ms. System recovery is evaluated according to Grid Code (Circular No. 39/2015/TT-BCT of the Government of Vietnam; Onshore Grid Code, 2015) and is investigated for LSPV effects on grid recovery. In addition, the relative rotor speed and angle of generators near the fault area will be surveyed to assess the operating condition, primary frequency, and voltage.

The relative rotor speed and angle of a nearby conventional generator are shown in Fig. 13.9A and Fig. 13.9B. Generators are affected more at higher LSPV levels. However, this influence is positive for the distribution grid. The high level of penetration by LSPVs reduces the inertia and synchronous speed of the power system, which is shown in Eq. (13.19). Generators can more easily change the speed at low inertia. In detail, scenario 5, with the highest level of connected LSPVs, causes the

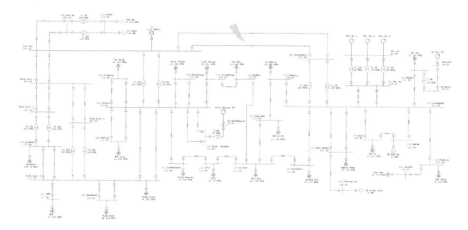

Figure 13.8 Fault location in a simulation.

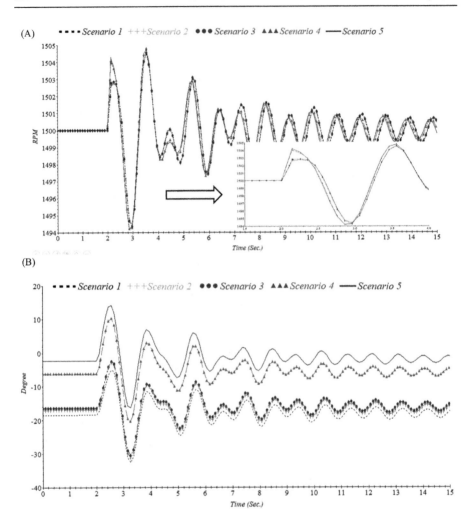

Figure 13.9 Case 1—Responses of conventional generator nearby.

velocity deviation to change more and makes the system recover faster when an incident occurs. Meanwhile, the rotor deflection angle increases and is more turbulent because of the sizeable reactive power transmitted to the fault site, which increases the voltage. This change can be explained through the vector diagram in Fig. 13.10 and the formula (13.27) (Bianchi, 2015; Kassakian & Schmalensee, 2011; Yahia Baghzouz, 2011):

$$Q = \frac{|E_G|}{X_d}(|E_a|\cos\delta - |V_t|) \tag{13.27}$$

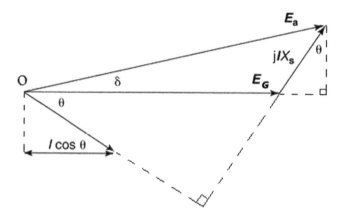

Figure 13.10 The relationship between quantities and reactive power in the generator.

where E_G is the terminal generator voltage (on the load side), E_a is the no-load electromotive force of the generator, X_d is the synchronous reactance of the generator, and δ is the phase angle between E_a and E_G.

The dynamic response of voltage and frequency is also assessed in this incident via the results in Fig. 13.11A and Fig. 13.11B. When LSPVs are connected to the distribution grid, system inertia is reduced because the total generated power of the grid increases while the load power remains constant. Time for fault occurs, and the low inertia makes the speed of the conventional generators flexible to change, as shown in Fig. 13.9A. From there, the system frequency is quickly adjusted, adapting to the disturbance and recovering more stably with the maximum intrusion scenario of the LSPVs. Moreover, the voltage in the fault event is also kept high and restored quickly after the fault is cleared.

13.4.2.2 Case 2—LSPV outage

The disadvantage of LSPVs is that the operation is unstable due to weather uncertainty, which may partially or entirely lose power if it encounters unknown causes by rain or cloudy. More importantly, this power disturbance occurs suddenly with responses that can be as small as one-tenth of a second. This result is one of the biggest challenges in the operation of LSPVs and makes power systems complicated to stabilize.

In this section, all LSPVs generate maximum power to get the worst scenarios, and a survey work is conducted similarly in the case of the short circuit. Power outage from LSPVs is started at the 2nd second with the assumption scenario presented in Table 13.4. The generator, voltage, and grid frequency responses are obtained in Figs. 13.12 and 13.13.

In the case of LSPV power failure, the results show that the small amount of LSPV output does not cause any problems to the grid and the generators. The system can still operate stably in incident scenarios 1 and 2 after the fault clearance. Meanwhile, with more serious incidents, the loss of PhoAn PV or the loss of power from all three plants puts the grid facing break out. For scenarios 3 and 4, the grid

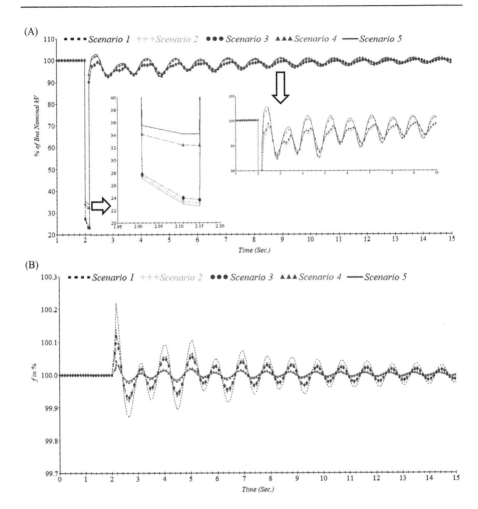

Figure 13.11 Case 1—Dynamic response of distribution grid.

loses a large amount of capacity, which causes conventional generators in the grid to increase speed rapidly to supply the shortfall. However, the acceleration process and the increased power of the generators are limited, making the shortfall in energy unable to be met in time. As a result, the system frequency in Fig. 13.13A and voltage in Fig. 13.13B fluctuate enormously and cannot recover after clearing the fault, according to Eqs. (13.19) and (13.20). This is the same reason the generators operate uncontrollably, like the responses observed in Fig. 13.12.

The probability of losing all the power from LSPV plants at once is very low since they are not centrally planned in one location. However, the loss of a large power plant is also a threat to grid security, as in this case, just losing power from PhoAn PV will make it difficult for the grid to overcome the fault. Therefore, supportive measures must be considered to protect against such adverse conditions.

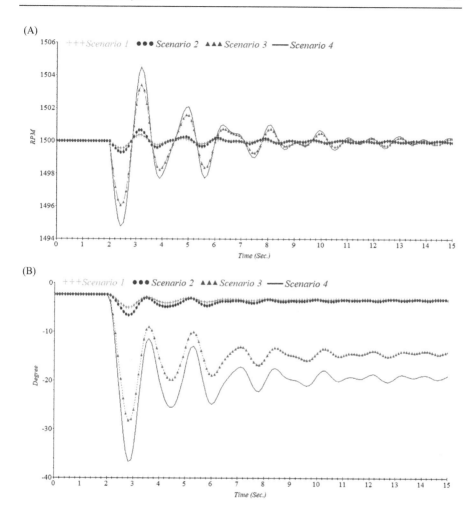

Figure 13.12 Case 2—Responses of conventional generator nearby.

13.4.3 Coordinate SCV for supporting grid

A static var compensator (SVC) is a device whose output is adjusted to exchange capacitive or inductive current for maintaining or controlling specific parameters of the electrical power system (Zhu et al., 2013). Following the dynamic responses, the outage power failure of PhoAn PV is the leading cause of the system breakout. Thus, SVC is installed at the coupling point of PhoAn PV to the grid (PCC2), which corresponds to a thyristor-controlled reactor. That can help to consume reactive power in case of high grid voltage due to load fluctuation or provide a significant amount of reactive power to assist in raising the grid voltage in the event of a failure to avoid grid breakout while implementing troubleshooting measures.

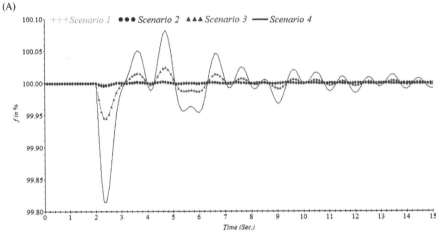

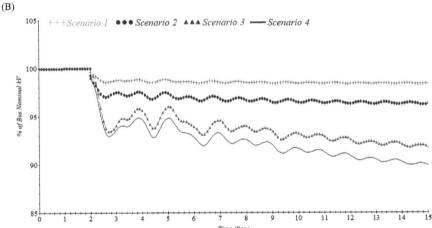

Figure 13.13 Case 2—Dynamic response of distribution grid.

In case of losing the PhoAn PV or all LSPVs, the voltage drops deeply, so placing SVC at PCC2 can support the grid immediately. To avoid unnecessary losses,
the power will not take the transmission time to the fault location. This SVC system
must be designed to keep voltage fluctuations within ± 0.1 p.u. (or 10%) from the
norm during incident and about ± 0.05 p.u. (or 5%) during recovery. With SVC
controlled by slow-firing angle α, the reactive power $Q_{SVC,i}$ injected to bus i^{th}
(Abdullah & Ibrahim, 2015; Savić & Đurišić, 2014) can be determined by

$$Q_{SVC,i} = \frac{U_i^2}{X_c} - U_i^2 \frac{2\pi - \alpha + \sin 2\alpha}{\pi X_L} \tag{13.28}$$

where U_i is the controlled voltage at SVC connection point; X_L and X_C are total inductance and capacitor reactance, respectively, inside SVC.

Maximum inductive and capacitive injected power of SVC is calculated:

$$\frac{U_i^2}{X_c} - \frac{U_i^2}{X_L} \leq Q_{SVC,i} \leq \frac{U_i^2}{X_c} \tag{13.29}$$

The simulation results when using SVC support show the effect in Figs. 13.14 and 13.15. The voltage response has shown promising resilience when supported by

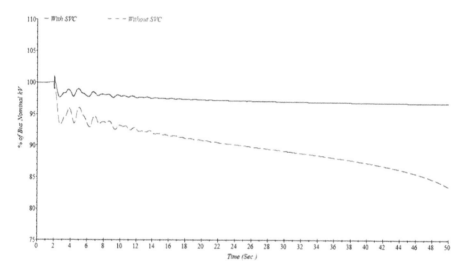

Figure 13.14 Voltage response in case of loss of power generated from PhoAn PV plant.

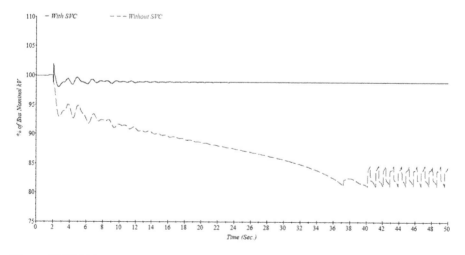

Figure 13.15 Voltage response in case of power loss from all LSPVs.

SVC in incidents with a voltage difference from the norm of only 3%. Copper also reduces the level of voltage fluctuations. Connecting SVC can ensure grid operation but will also increase operating costs due to the price of technology and equipment.

13.5 Conclusion

The impacts of the penetration levels of PVs into the distribution grid under the large-scale plant have been analyzed in detail in different operating modes. On the advantage, LSPVs contribute to renewable energy development, reducing the electricity burden on the national power system. Higher penetration of solar power plants contributes to voltage and frequency stability for faults outside the plant area. However, on the disadvantage, the sudden outage of power plants due to obscure phenomena and internal faults at high penetration will cause instability in the grid due to a significant loss of output power. In addition, conventional generators nearby are more affected when there is a high fluctuation in the relative rotor angle, the reactive energy emitted when it occurs, and after clearing faults.

Therefore, to minimize the influence of LSPVs, support services must be considered to help the system operate in the long run, primarily when an incident occurs. For example, modern technologies can be used as a SVC application that supports reactive power: backup diesel generators; currently, a very concerned solution is using storage systems with dispatching solutions.

In the future, this research may be integrated with storage systems and optimization algorithms to enable methodology and optimal operating costs considering the hosting capacity of LSPVs. At the same time, it still has to meet the requirements to overcome incidents in the grid.

References

Abdullah, A. N., & Ibrahim, M. S. (2015). December *Improvement of power and voltage quality in power system distribution using facts (SVC & DSTATCOM)* (pp. 266–327). Second Engineering Scientific Conference College of Engineering, University of Diyala.

Achilles, S., Schramm, S., & Bebic, J. (2008). Transmission system performance analysis for high penetration photovoltaics. *NREL*. Rep. No. SR 581-42300. [Online]. Available online: http://www.nrel.gov/docs/fy08osti/42300.pdf.

Ahmadi, M., Lotfy, M. E., Shigenobu, R., Yona, A., & Senjyu, T. (2018). Optimal sizing and placement of rooftop solar photovoltaic at Kabul city real distribution network. *IET Generation, Transmission & Distribution, 12*(2), 303–309.

Barbose, G.L. (2019). U.S. Renewables Portfolio Standards: 2019 Annual Status Update, Lawrence. Berkeley National Laboratory, 2019. Available website: https://eta-publications.lbl.gov/sites/default/files/rps_annual_status_update-2019_edition.pdf.

Barragán-Escandón, E., & Zalamea-León, J. (2019). Terrados-Cepeda "Incidence of photovoltaics in cities based on indicators of occupancy and urban sustainability". *Energies, 12,* 810.

Bawazir, R. O., & Cetin, N. S. (2020). Comprehensive overview of optimizing PV-DG allocation in power system and solar energy resource potential assessments. *Energy Report*, *6*, 173−208.

Bianchi, N. (2015). Electrical Machine Analysis Using Finite Elements, Boca Raton, 1st Edition, https://doi.org/10.1201/9781315219295.

Castro, L. M., Rodríguez-Rodríguez, J. R., & Martin-del-Campo, C. (2020). Modelling of PV systems as distributed energy resources for steady-state power flow studies. *International Journal of Electrical Power & Energy Systems*, *115*105505.

Chapter 25, (2020). ETAP 20.60 User Guide, Operation Technology INC.

Circular No. 39/2015/TT-BCT of the Government of Vietnam. (n.d.) [Online], Available: http://www.moit.gov.vn.

Duong, M., Tran, N., Hossain, C. (2019). The Impact of Photovoltaic Penetration with Real Case: ThuaThienHue−Vietnamese Grid, International Conference on Robotics, Electrical and Signal Processing Techniques (ICREST).

Duong, M.Q., Tran, N.T.N., Sava, G.N., Scripcariu, M. (2017a). "The Impacts of Distributed Generation Penetration into the Power System," 2017 International Conference on Electromechanical and Power Systems (SIELMEN), pp. 295−301.

Duong, M.Q., Tran, N.T.N., Sava, G.N., Scripcariu, M. (2017b). The Impacts of Distributed Generation Penetration into the Power System, International Conference on Electromechanical and Power Systems (SIELMEN), pp. 295−301.

D'Adamo, I., Falcone, P. M., Gastaldi, M., & Morone, P. (2018). The economic viability of photovoltaic systems in public buildings: Evidence from Italy. *Energy*, *81*, 2460−2471.

Eftekharnejad, S., Vittal, V., Heydt, G. T., Keel, B., & Loehr, J. (2013). Impact of increased penetration of photovoltaic generation on power systems. *IEEE Transactions on Power Systems*, *28*(2), 893−901.

Gagnon, P., Stoll, B., Ehlen, A., Mai, T., Barbose, G., & Mills, A. (2018). Estimating the value of improved distributed photovoltaic adoption forecasts for utility resource planning. *National Renewable Energy Laboratory*. NREL/PR-6A20−71505, https://www.nrel.gov/docs/fy18osti/71505.pdf.

Glover, J. D., Overbye, T. J., & Sarma, M. S. (2015). ISBN: 978-1-305−63213- 4 *Power system analysis & design* (sixth edition). Cengage Learning.

Gope, S., Dawn, S., Mitra, R., Goswami, A. K., & Tiwari, P. K. (2019). Transmission congestion relief with integration of photovoltaic power using lion optimization algorithm. In J. Bansal, K. Das, A. Nagar, K. Deep, & A. Ojha (Eds.), *Soft computing for problem solving. Advances in intelligent systems and computing* (816). Singapore: Springer.

Guerrero, J., Gebbran, D., Mhanna, S., Chapman, A. C., & Verbič, G. (2020). Towards a transactive energy system for integration of distributed energy resources: Home energy management, distributed optimal power flow, and peer-to-peer energy trading. *Renewable and Sustainable Energy Reviews*, *132*110000.

Haifeng, L., Licheng, J., Le, D., & Chowdhury, A. A. (2010). Impact of high penetration of solar photovoltaic generation on power system small signal stability. *Proceedings of International Conference on Power System Technology*, 1−7.

Huang, Y., & Wang, D. (2018). Effect of control-loops interactions on power stability limits of VSC integrated to AC system. *IEEE Transactions on Power Delivery*, *33*(1), 301−310. Available from https://doi.org/10.1109/TPWRD.2017.2740440.

IEEE. (2003). 1547 Standard for Interconnecting Distributed Resources with Electric Power Systems, Oct. 2003. [Online]. Available: http://grouper.ieee.org/groups/scc21/1547/1547/index.html.

Jamil, M., & Anees, A. S. (2016). Optimal sizing and location of SPV (solar photovoltaic) based MLDG (multiple location distributed generator) in distribution system for loss reduction, voltage profile improvement with economical benefits. *Energy, 103,* 231−239.

Joshi, A. S., Dincer, I., & Reddy, B. V. (2009). Performance analysis of photovoltaic systems: A review. *Renewable and Sustainable Energy Reviews, 13*(8), 1884−1897.

Kassakian, J. G., & Schmalensee, R. (2011). *The future of the electric grid* (p. 2011) Massachusetts: Massachusetts Institute of Technology.

Kemp, L. (2018). A systems critique of the 2015 Paris Agreement on climate. In M. Hossain, R. Hales, & T. Sarker (Eds.), *Pathways to a sustainable economy* (pp. 25−41). Cham: Springer.

Lacal Arantegui, R., & Jäger-Waldau, A. (2018). Photovoltaics and wind status in the European union after the paris agreement. *Renewable and Sustainable Energy Reviews, 81,* 2460−2471.

Liu, Y., Bebic, J., Kroposki, B., de Bedout, J., & Ren, W. (2008). Distribution system voltage performance analysis for high-penetration PV, in *Proc. IEEE Energy 2030 Conf.,* Atlanta, GA, Nov. 2008, pp. 1−8.

Nguyen, B., Nguyen, V., Duong, M., Le, K., Nguyen, H., & Doan, A. (2020). Propose a mppt algorithm based on thevenin equivalent circuit for improving photovoltaic system operation. *Energy Res, 8,* 14.

Onshore Grid Code. (2015). Gridcode for High and Extra High Voltage [Online], Available online: https://www.tennet.eu/electricity-market/germancustomers/grid-customers/grid-connection-regulations.

Ostertagová, E. (2012). Modelling using polynomial regression. *Procedia Engineering, 48,* 500−506.

Paul, K. (2022). Modified grey wolf optimization approach for power system transmission line congestion management based on the influence of solar photovoltaic system. *International Journal of Energy and Environmental Engineering, 13,* 751−767.

Prakash, P., & Khatod, D. K. (2016). Optimal sizing and siting techniques for distributed generation in distribution systems: A review. *Renewable and Sustainable Energy Reviews, 57,* 111−130.

Refaat, S. S., Haitham, A. R., Sanfilippo, A. P., & Amira, M. (2018). Impact of grid-tied large-scale photovoltaic system on dynamic voltage stability of electric power grids. *IET Renewable Power Generation, 12*(2), 157−164.

Savić, Ž., & Đurišić. (2014). Optimal sizing and location of SVC devices for improvement of voltage profile in distribution network with dispersed photovoltaic and wind power plants. *Applied Energy, 134*(Supplement C), 114−124.

Smith, J.W., Sunderman, W., Dugan, R., & Seal, B. (2011). Smart inverter volt/var control functions for high penetration of PV on distribution systems, in *Proc. IEEE Power Systems Conf. Expo.,* Phoenix, AZ, Mar. 2011, pp. 1−6.

The Vietnamese Prime Minister. (2016). Approval of the Revised National Power Development Master Plan for the 2011−2020 Period with the Vision to 2030, Decision No. 428/QD-TTg, March 18, 2016.

Tran, T., & Duong, Q. (2019). Design and performance assessment of hybrid-maximum power point tracking algorithm. *Journal Of Science And Technology: Issue On Information And Communications Technology, 17*(12.2), 28−34. Available from https://doi.org/10.31130/ict-ud.2019.96.

Tran, N. T. N., Yang, H. T., & Duong, M. Q. (2020). Integrated transient stability analysis with multi-large-scale solar photovoltaic in distribution network. *Journal of Science & Technology*, *142*(C), 40−45.

Turitsyn, K., Sulc, P., Backhaus, S., & Chertkov, M. (2010). "Distributed control of reactive power flow in a radial distribution circuit with high photovoltaic penetration," in *Proc. IEEE Power and Energy Society General Meeting*, Minneapolis, MN, Jul. 2010, pp. 1−6.

UL. (n.d.) 1741 Standard for Inverters, Converters, Controllers and Interconnection System Equipment for Use with Distributed Energy Resources. [Online].

Ullah, H., Kamal, I., Ali, A., & Arshad, N. (2018). Investor focused placement and sizing of photovoltaic grid-connected systems in Pakistan. *Renewable Energy*, *121*, 460−473.

Van Tan, N., Nam, N. B., Hieu, N. H., Hung, L. K., Duong, M. Q., & Lam, L. H. (2020). A proposal for an MPPT algorithm based on the fluctuations of the PV output power, output voltage, and control duty cycle for improving the performance of PV systems in microgrid. *Energies*, *13*(17), 4326.

Xu, L. (2013). *Modeling, analysis and control of voltage-source converter in microgrids and HVDC. Graduate Theses and Dissertations* (pp. 17−47). University of South Florida.

Yahia Baghzouz, P. E. (2011). *EE 340 Synchronous Generators I*. Las Vegas: University of Nevada.

Zhu, J., Booth, C. D., Adam, G. P., Roscoe, A. J., & Bright, C. G. (2013). Inertia emulation control strategy for VSC-HVDC transmission systems. *IEEE Transactions on Power Systems*, *28*(2), 1277−1287. Available from https://doi.org/10.1109/TPWRS.2012.2213101.

Power management and control of an all-electric ship powered by solar photovoltaic and hydrogen energy system

Muhammad Maaruf[1] and Muhammad Khalid[2,3]
[1]Control and Instrumentation Engineering Department & Center for Smart Mobility and Logistics, King Fahd University for Petroleum and Minerals, Dhahran, Saudi Arabia, [2]Electrical Engineering Department, King Fahd University of Petroleum and Minerals, Dhahran, Saudi Arabia, [3]Interdisciplinary Research Center for Sustainable Energy Systems, King Fahd University of Petroleum and Minerals, Dhahran, Saudi Arabia

14.1 Introduction

In recent years, due to the depletion of fossil fuel sources and escalating environmental pollution, the promotion of green energy has become the focal center for researchers. Due to this fact, the maritime industries are moving away from using diesel engines and toward the all-electric architecture to provide power for both propulsion systems and ship services. The advancement of power electronics technology allows the integration of various clean energy sources such as fuel cells, battery energy systems, solar photovoltaic (PV) cells, ultracapacitors, flywheels, and supermagnetic energy sources to power the all-electric ship. This architecture has several merits compared to the combustion engine, such as higher power availability, reduced noise, carbon emissions, etc.

The operation of the all-electric ship is monitored by the integrated power system (IPS). The IPS supplies electric power to auxiliary loads in the ship and propulsion load and at the same time coordinates the balance between the load and supply (Zhao et al., 2022). A power management technique is essentially needed to maintain power quality and ensure the continuous supply of power to meet the load requirement. Moreover, an effective power management method ensures effective and appropriate power sharing, which enhances the system's reliability, robust operation, control performance, and overall efficiency (Wang et al., 2022).

Various control strategies have been proposed for the power management of an all-electric ship with multiple energy sources. A feasible energy management was proposed to enhance the reliability of a ferry ship supplied by battery and fuel cells (Banaei et al., 2021; Rafiei et al., 2021). In Ma et al. (2021), a supervisory control approach was proposed to coordinate the pulsed-power load of an all-electric ship.

Performance Enhancement and Control of Photovoltaic Systems. DOI: https://doi.org/10.1016/B978-0-443-13392-3.00015-3

In Sun et al. (2022), the authors presented a coordinated optimal control to suppress voltage fluctuations and maintain the power balance. In Terriche et al. (2020), an optimal control approach was employed to mitigate harmonic distortions and increase the power factor of a zero-emission ship. In M. Gan et al. (2022), a model predictive control (MPC) technique was applied to an all-electric ship to obtain optimal power management. In Hou et al. (2018), a multiobjective MPC was developed to enhance the overall efficiency of the propulsion system of an electric ship. In Steinsland et al. (2020), a feedback linearization control is utilized to coordinate the multiple energy sources powering an all-electric ship. In Khooban et al. (2020), an intelligent control strategy is implemented to achieve effective energy management of an electric ferry ship. In Khan et al. (2017), an intelligent power storage management was designed for an all-electric ship to enhance the health of the energy storage unit. The authors in Alafnan et al. (2018) considered a hybrid energy system comprising supermagnetic energy storage and battery to stabilize the DC-bus voltage and mitigate the negative effects of propulsion-load fluctuations. In Peng Wu, Partridge, and Bucknall (2020), a reinforcement learning-based power management system is employed to coordinate the propulsion system and the power supply unit consisting of a battery storage system and fuel cell.

The literature mentioned earlier has successfully used different control and optimization strategies for power management of all-electric ships with multiple energy sources and loads. However, the authors ignored the dynamics of the power converters and/or the propulsion motor. As a result, their analyses cannot be accurate enough due to the simplification of the system model. In this chapter, a comprehensive model of the all-electric ship is derived without considering model simplification. The propulsion system and the ship's onboard loads are powered by solar PV panels, fuel cells, and electrolyzers, which produce and store the hydrogen fuel. The propulsion system comprises a propeller and permanent magnet synchronous motor (PMSM). The microgrid architecture of the all-electric ship is depicted in Fig. 14.1.

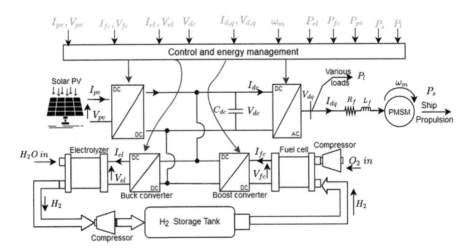

Figure 14.1 Microgrid topology of the all-electric ship.

In addition, this chapter employed nonlinear control and power management strategies to coordinate between the energy system and various loads. The control and power management strategies guarantee the continuous supply of power to the propulsion system and ship services, the stability of the ship microgrid, and accurate tracking performance.

14.2 System modeling

14.2.1 PV array model

The PV array generates electricity through the photoelectric effect. The magnitude of the PV output depends on the intensity of the radiation. The PV array transfers the maximum power to the DC bus based on the incremental conductance method. The PV output current is given as (Nyanya et al., 2021) follows:

$$I_{pv} = N_p \left(I_{pho} - I_0 \left[\exp \left(\frac{q(V_{pv} + I_{pv}R_s)}{N_s A \Lambda_b T} \right) - 1 \right] - \frac{V_{pv} + I_{pv}R_s}{R_p} \right) \tag{14.1}$$

where V_{pv} and I_{pv} are the PV output current and voltage, respectively; I_0 and I_{pho} are the saturation current and photogenerated current, respectively; R_s and R_p are the series and parallel resistances, respectively; and T is the PV operating temperature.

The dynamic equation of the PV boost converter is expressed as follows:

$$L_{pv} \frac{dI_{pv}}{dt} = V_{pv} - \left(1 - \mu_{pv} \right) V_{dc} - I_{pv}R_{pv} \tag{14.2}$$

where L_{pv}, R_{pv}, and μ_{pv} are the boost converter inductance, resistance, and duty-cycle, respectively.

The electric power generated by the PV is given as follows:

$$P_{pv} = I_{pv} V_{pv} \tag{14.3}$$

The dynamics of the PV power is calculated as follows:

$$\frac{dP_{pv}}{dt} = V_{pv} I_{pv} + V_{pv} I_{pv} = F_{pv} + G_{pv} u_{pv} \tag{14.4}$$

where

$$F_{pv} = \dot{V}_{pv} I_{pv} + L_{pv}^{-1} V_{pv} [V_{pv} - V_{dc} - I_{pv} R_{pv}]; G_{pv} = L_{pv}^{-1} V_{pv} V_{dc}$$

14.2.2 Water electrolyzer model

The electrolyzer splits water molecules into hydrogen and oxygen through a chemical reaction. The generated hydrogen is stored. The reaction is facilitated by the passage of DC current into the electrolyzer. The output voltage of the electrolyzer is given as follows (Kong et al., 2019):

$$V_{el} = V_0 + \frac{r_1 + r_2 T_{el}}{A_{el}} + \left(a_1 + a_2 T_{el} + a_3 T_{el}^2\right)\log\left(\frac{b_1 + b_2/T_{el} + b_3/T_{el}^2}{A_{el}}I_{el} + 1\right)$$

(14.5)

where V_{el}, I_{el}, and T_{el} are the electrolyzer voltage, current, and temperature, respectively; r_i ($i = 1,2$) are ohmic resistance parameters of the electrolyte; and a_i, b_i ($i = 1,2,3$) are coefficients of overvoltage at the electrodes.

The dynamic equation of the electrolyzer buck converter is expressed as follows:

$$L_{el}\frac{dI_{el}}{dt} = \mu_{el}V_{dc} - V_{el} - I_{el}R_{el}$$

(14.6)

where L_{el}, R_{el}, and μ_{el} are the buck converter inductance, resistance, and duty cycle, respectively.

The electric power transferred to the electrolyzer is given as follows:

$$P_{el} = I_{el}V_{el}$$

(14.7)

The dynamics of the electrolyzer power is calculated as follows:

$$\frac{dP_{el}}{dt} = V_{el}I_{el} + V_{el}I_{el} = F_{el} + G_{el}u_{el}$$

(14.8)

where $F_{el} = V_{el}I_{el} + L_{el}^{-1}V_{el}[-I_{el}R_{el} - V_{el}]$; $G_{el} = L_{el}^{-1}V_{el}V_{dc}$

14.2.3 PEM fuel cell model

The PEM fuel cell generates electrical energy via a chemical reaction of hydrogen and oxygen. The by-products of this reaction are water and heat. The output voltage of the fuel cell is given by P. Wu and Bucknall (2020):

$$V_{fc} = V_{oc} - V_{act} - I_{fc}R_{in}$$
$$V_{oc} = 4.39 \times 10^{-5} ln\left(P_{H_2}P_{O_2}^{0.5}\right)T_{fc} + 1.04 \times 10^{-4}\left(T_{fc} - 298.2\right)$$
$$V_{act} = \frac{NA_{no}}{\tau s + 1}ln\left(I_{fc}/I_0 s\right)$$

(14.9)

where V_{fc}, I_{fc}, and T_{fc} are the fuel cell voltage, current, and temperature, respectively; V_{act} and V_{oc} are the activation voltage drop and open circuit voltage, respectively; R_{in} is the internal resistance of the fuel cell; and T_{fc} is the operating temperature of the fuel cell.

The dynamics equation of the fuel cell boost converter is expressed as follows:

$$L_{fc}\frac{dI_{fc}}{dt} = V_{fc} - (1 - \mu_{fc})V_{dc} - I_{fc}R_{fc} \tag{14.10}$$

where L_{fc}, R_{fc}, and μ_{fc} are the boost converter inductance, resistance, and duty cycle, respectively.

The electric power generated by the fuel cell is given as follows:

$$P_{fc} = I_{fc}V_{fc} \tag{14.11}$$

The dynamics of the fuel cell power is calculated as follows:

$$\frac{dP_{fc}}{dt} = V_{fc}I_{fc} + V_{fc}I_{fc} = F_{fc} + G_{fc}u_{fc} \tag{14.12}$$

where $F_{fc} = V_{fc}I_{fc} + L_{fc}^{-1}V_{fc}[V_{fc} - V_{dc} - I_{fc}R_{fc}]$; $G_{fc} = L_{fc}^{-1}V_{fc}V_{dc}$

14.2.4 PMSM model

The PMSM converts electrical power into mechanical power, which propels the ship. The mathematical model of the PMSM in the d-q reference frame is given as follows (B. Gan et al., 2021):

$$J\frac{d\omega_s}{dt} = p(L_d - L_d)I_dI_q + p\phi_fI_q - B\omega_s - T_s$$
$$L_d\frac{dI_d}{dt} = V_d + L_qp\omega_sI_q - R_sI_d \tag{14.13}$$
$$L_q\frac{dI_q}{dt} = V_q - p\phi_f\omega_s - L_dp\omega_sI_d - R_sI_q$$

where ω_s is the angular speed of the propeller; I_d and I_q are d- and q-axis currents of the AC motor, respectively; L_d, and L_q are d- and q-axis stator inductances of the AC motor, respectively; R_s is the stator resistance; p is the number of poles; T_s is the load torque; and J is the moment of inertia.

14.2.5 DC-bus model

The DC bus transfers real power to the load and reduces voltage fluctuations as the AC load demand sporadically increases. The DC-bus dynamic model is derived using Kirchoff's current law as follows:

$$C_{dc}\frac{dV_{dc}}{dt} = \left(1 - \mu_{fc}\right)I_{fc} + \left(1 - \mu_{pv}\right)I_{pv} - \mu_{el}I_{el} - \frac{V_dI_d}{V_{dc}} \tag{14.14}$$

where V_{dc} and I_{dc} are the DC-bus voltage and capacitance, respectively.

14.2.6 Ship model

The dynamic equation of the ship comprises the ship's speed, propulsion force, hydrodynamic resistance, wind, and wave excitations (Hou et al., 2018).

$$M\frac{dV_s}{dt} = F_s - \frac{\rho V_s^2}{2}\left[C_{fric}A_s + C_{drag}A_s + C_{wind}A_T\right] \tag{14.15}$$

where V_s is the velocity of the ship; F_s is the ship's thrust force; C_{fric}, C_{drag} and C_{wind} are drag coefficients; A_s is the submerged area of the ship; and A_T is the area of the ship in the air.

The thrust force F_s should be large enough to overcome the opposing forces and rotate the propeller that propels the ship. The corresponding electrical power P_s that the PMSM needs to propel the ship is calculated as follows:

$$P_s = F_s V_s = \left(M\frac{dV_s}{dt} + \frac{\rho V_s^2}{2}\left[C_{fric}A_s + C_{drag}A_s + C_{wind}A_T\right]\right)V_s \tag{14.16}$$

The relationship between the angular speed of the propeller ω_s and the ship velocity V_s is

$$\omega_s^r = \frac{V_s}{D_{pr}/2} \tag{14.17}$$

where ω_s is the reference angular speed and D_{pr} is the diameter of the propeller.

14.3 Control design and power management

Fig. 14.2 depicts the power management flowchart of the all-electric ship.
The total power in the system P_T is given by the following:

- When $P_T > 0$, the fuel cells transfer power to meet the power deficit and maintain the load-generation balance. The reference fuel cell power is set as ($P_{el}^r = P_T$).
- When $P_T < 0$, the surplus power generated by the PV unit is transferred to the electrolyzer where hydrogen is produced and stored. The reference electrolyzer power is set as ($P_{el}^r = P_T$).
- When ($P_T = 0$), the load generation is balanced.

Define the following tracking errors:

$$Z_1 = P_{pv} - P_{pv}^r \quad Z_2 = P_{el} - P_{el}^r \quad Z_3 = P_{fc} - P_{fc}^r \quad Z_4 = \omega_s - \omega_s^r$$
$$Z_5 = V_{dc} - V_{dc}^r \quad Z_6 = I_d - I_d^r \quad Z_7 = I_q - I_q^r \tag{14.18}$$

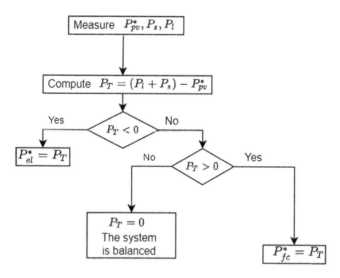

Figure 14.2 Power management flowchart for coordinating multiple energy sources and loads.

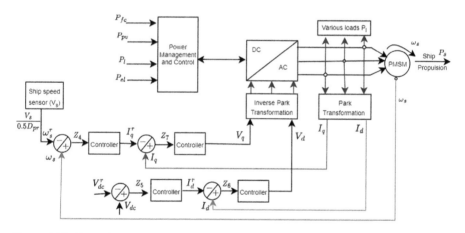

Figure 14.3 Control block diagram of the all-electric ship.

The control block diagram of the closed-loop system is depicted in Fig. 14.3. The control signals that will ensure coordinated, reliable, and accurate performance of the power sources to maintain load-generation balance are designed as follows:

$$\mu_{pv} = [\dot{P}^r_{pv} - F_{pv} - \Lambda_{pv}Z_1]/G_{pv} \quad \mu_{el} = [\dot{P}^r_{el} - F_{el} - \Lambda_{el}Z_2]/G_{el}$$
$$\mu_{fc} = [\dot{P}^r_{fc} - F_{fc} - \Lambda_{fc}Z_3]/G_{fc} \tag{14.19}$$

where Λ_{pv}, Λ_{el}, and Λ_{fc} are positive constants.

The controller that will ensure that the propeller is rotating at the required angular speed is designed as follows:

$$I_q^r = [\dot{\omega}_r - J^{-1}(-T_s - B\omega_s) - \Lambda_\omega Z_4]/G_\omega \tag{14.20}$$

where $G_\omega = [p(L_d - L_d)I_d + p\phi_f]/J$ and Λ_ω is a positive constant.

The DC-bus voltage fluctuates due to sporadic variations in PV power and propulsion power. The DC-bus voltage regulator that will minimize the fluctuations is designed as follows:

$$I_d^r = (\dot{V}_{dc}^r - [(1 - \mu_{fc})I_{fc} + (1 - \mu_{pv})I_{pv} - \mu_{el}I_{el}]/C_{dc} - \Lambda_{dc}Z_5)/G_{dc} \tag{14.21}$$

where $G_{dc} = -V_dI_d/(C_{dc}V_{dc})$ and Λ_{dc} is a positive constant.

The I_d^r from Eq. (14.21) is used as the reference signal for I_d. The controller for I_d is designed as follows:

$$V_d = L_d(\dot{I}_d^r - [L_qp\omega_sI_q - R_sI_d]/L_d - \Lambda_dZ_6) \tag{14.22}$$

The I_q^r from Eq. (14.21) is used as the reference signal for I_q. The controller for I_q is designed as follows:

$$V_q = L_q(\dot{I}_q^r - [-p\phi_f\omega_s - L_dp\omega_sI_d - R_sI_q]/L_q - \Lambda_qZ_7) \tag{14.23}$$

where Λ_q is a positive constant.

14.4   Stability of the system

In this section, the asymptotic stability of the closed-loop system is highlighted. Consider the following Lyapunov function:

$$W = \frac{1}{2}Z_1^2 + \frac{1}{2}Z_2^2 + \frac{1}{2}Z_3^2 + \frac{1}{2}Z_4^2 + \frac{1}{2}Z_5^2 + \frac{1}{2}Z_6^2 + \frac{1}{2}Z_7^2 \tag{14.24}$$

The derivative of Eq. (14.24) with respect to time yields:

$$\begin{aligned}\dot{W} = &Z_1(\dot{P}_{pv} - \dot{P}_{pv}^r) + Z_2(\dot{P}_{el} - \dot{P}_{el}^r) + Z_3(\dot{P}_{fc} - \dot{P}_{fc}^r) + Z_4(\dot{\omega}_s - \dot{\omega}_s^r) \\ &+ Z_5(\dot{V}_{dc} - \dot{V}_{dc}^r) + Z_6(\dot{I}_d - \dot{I}_d^r) + Z_7(\dot{I}_q - \dot{I}_q^r)\end{aligned} \tag{14.25}$$

Considering the dynamic Eqs. (14.4, 14.8, 14.12, 14.13, and 14.14) and the control laws (14.19, 14.20, 14.21, 14.22, and 14.23), Eq. (14.25) becomes

$$\dot{W} = -\Lambda_{pv}Z_1^2 - \Lambda_{el}Z_2^2 - \Lambda_{fc}Z_3^2 - \Lambda_\omega Z_4^2 - \Lambda_{dc}Z_5^2 - \Lambda_dZ_6^2 - \Lambda_qZ_7^2 \tag{14.26}$$

From Eq. (14.26), it can be concluded that the closed-loop system is asymptotically stable.

14.5 Simulation results

This section presents the results of simulating the comprehensive model of the all-electric ship along with the designed control signals. The parameters of the controller were tuned by trial and error and are given as follows: $\Lambda_{pv} = 20$, $\Lambda_{el} = 25$, $\Lambda_{fc} = 25$, $\Lambda_{dc} = 18$, $\Lambda_d = 20$, $\Lambda_q = 20$, and $\Lambda_\omega = 8$.

The speed profile of the all-electric ship is depicted in Fig. 14.4. The power distribution among the energy sources is shown in Fig. 14.5. It is clear that the power

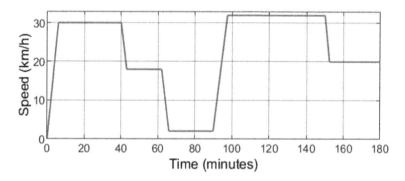

Figure 14.4 The all-electric ship speed profile.

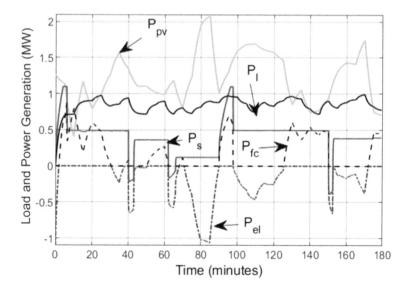

Figure 14.5 Power distribution based on the power management.

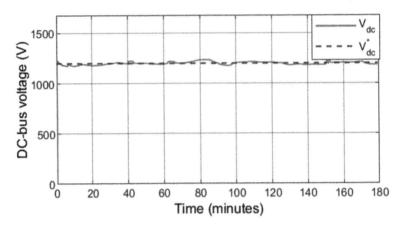

Figure 14.6 DC-bus voltage regulation.

management and control schemes are able to balance the loads and the power supplied by the energy sources. The DC-bus voltage regulation is depicted in Fig. 14.6. It can be observed that the controller is able to maintain the DC-bus voltage very close to the reference value.

14.6   Conclusion

This chapter presented power management and nonlinear control of an all-electric ship powered by solar PV and hydrogen energy systems. The comprehensive mathematical model of the all-electric ship has been driven. Then, an energy management algorithm was developed to obtain power distribution among the energy sources. In order to ensure that the system is operating according to the power management with accurate regulation of the DC-bus voltage, a nonlinear control technique was developed. The asymptotic stability of the closed-loop system has been highlighted using the Lyapunov function.

References

Alafnan, H., Zhang, M., Yuan, W., Zhu, J., Li, J., Elshiekh, M., & Li, X. (2018). Stability improvement of DC power systems in an all-electric ship using hybrid SMES/battery. *IEEE Transactions on Applied Superconductivity*, *28*(3), 1–6. Available from https://doi.org/10.1109/tasc.2018.2794472.

Banaei, M., Boudjadar, J., Ebrahimy, R., Madsen, H. (2021). IECON Proceedings (Industrial Electronics Conference) Available from https://doi.org/10.1109/IECON48115.2021.9589512, 9781665435543, IEEE Computer Society Denmark Optimal Control Strategies of Fuel cell/ Battery Based Zero-Emission Ships: A Survey 2021.

Gan, B., Zhang, B., & Feng, G. (2021). Design and analysis of modular permanent magnet fault-tolerant motor for ship direct-drive propulsion. *IEEJ Transactions on Electrical and Electronic Engineering, 16*(9), 1260−1278. Available from https://doi.org/10.1002/tee.23424, http://onlinelibrary.wiley.com/journal/10.1002/(ISSN)1931-4981.

Gan, M., Hou, H., Wu, X., Liu, B., Yang, Y., & Xie, C. (2022). Machine learning algorithm selection for real-time energy management of hybrid energy ship. *Energy Reports, 8,* 1096−1102. Available from https://doi.org/10.1016/j.egyr.2022.02.200, http://www.journals.elsevier.com/energy-reports/.

Hou, J., Sun, J., & Hofmann, H. F. (2018). Mitigating power fluctuations in electric ship propulsion with hybrid energy storage system: Design and analysis. *IEEE Journal of Oceanic Engineering, 43*(1), 93−107. Available from https://doi.org/10.1109/JOE.2017.2674878.

Khan, M. M. S., Faruque, M. O., & Newaz, A. (2017). Fuzzy logic based energy storage management system for MVDC power system of all electric ship. *IEEE Transactions on Energy Conversion, 32*(2), 798−809. Available from https://doi.org/10.1109/TEC.2017.2657327.

Khooban, M. H., Gheisarnejad, M., Farsizadeh, H., Masoudian, A., & Boudjadar, J. (2020). A new intelligent hybrid control approach for DC-DC converters in zero-emission ferry ships. *IEEE Transactions on Power Electronics, 35*(6), 5832−5841. Available from https://doi.org/10.1109/TPEL.2019.2951183, https://ieeexplore.ieee.org/xpl/mostRecentIssue.jsp?punumber = 63.

Kong, L., Yu, J., & Cai, G. (2019). Modeling, control and simulation of a photovoltaic /hydrogen/ supercapacitor hybrid power generation system for grid-connected applications. *International Journal of Hydrogen Energy, 44*(46), 25129−25144. Available from https://doi.org/10.1016/j.ijhydene.2019.05.097, http://www.journals.elsevier.com/international-journal-of-hydrogen-energy/.

Ma, Y., Oslebo, D., Maqsood, A., & Corzine, K. (2021). Pulsed-power load monitoring for an all-electric ship: Utilizing the fourier transform data-driven deep learning approach. *IEEE Electrification Magazine, 9*(1), 25−35. Available from https://doi.org/10.1109/mele.2020.3047164.

Nyanya, M. N., Vu, H. B., Schönborn, A., & Ölçer, A. I. (2021). Wind and solar assisted ship propulsion optimisation and its application to a bulk carrier. *Sustainable Energy Technologies and Assessments, 47.* Available from https://doi.org/10.1016/j.seta.2021.101397, http://www.journals.elsevier.com/sustainable-energy-technologies-and-assessments.

Rafiei, M., Boudjadar, J., & Khooban, M. H. (2021). Energy management of a zero-emission ferry boat with a fuel-cell-based hybrid energy system: feasibility assessment. *IEEE Transactions on Industrial Electronics, 68*(2), 1739−1748. Available from https://doi.org/10.1109/TIE.2020.2992005, http://ieeexplore.ieee.org/xpl/tocresult.jsp?isnumber = 5410131.

Steinsland, V., Kristensen, L.M., Arghandeh, R. (2020). Design of Modular Multilevel Converters for the Shipnet in Medium Voltage DC All-Electric Ships. 2020 IEEE 21st Workshop on Control and Modeling for Power Electronics (COMPEL). IEEE; pp. 1−8.

Sun, X., Qiu, J., Tao, Y., Yi, Y., & Zhao, J. (2022). Distributed optimal voltage control and berth allocation of all-electric ships in seaport microgrids. *IEEE Transactions on Smart Grid, 13*(4), 2664−2674. Available from https://doi.org/10.1109/TSG.2022.3161647, https://ieeexplore.ieee.org/servlet/opac?punumber = 5165411.

Terriche, Y., Mutarraf, M. U., Golestan, S., Su, C. L., Guerrero, J. M., Vasquez, J. C., & Kerdoun, D. (2020). A hybrid compensator configuration for VAR control and harmonic suppression in all-electric shipboard power systems. *IEEE Transactions on Power Delivery, 35*(3), 1379−1389. Available from https://doi.org/10.1109/TPWRD.2019.2943523, https://ieeexplore.ieee.org/servlet/opac?punumber = 61.

Wang, B., Peng, X., Zhang, L., & Su, P. (2022). Real time power management strategy for an all-electric ship using a predictive control model IET generation. *Transmission and Distribution*, *16*(9), 1808−1821. Available from https://doi.org/10.1049/gtd2.12419, https://ietresearch.onlinelibrary.wiley.com/journal/17518695.

Wu, P., Partridge, J., & Bucknall, R. (2020). Cost-effective reinforcement learning energy management for plug-in hybrid fuel cell and battery ships. *Applied Energy*, *275*115258. Available from https://doi.org/10.1016/j.apenergy.2020.115258.

Wu, P., & Bucknall, R. (2020). Hybrid fuel cell and battery propulsion system modelling and multi-objective optimisation for a coastal ferry. *International Journal of Hydrogen Energy*, *45*(4), 3193−3208. Available from https://doi.org/10.1016/j.ijhydene.2019.11.152, http://www.journals.elsevier.com/international-journal-of-hydrogen-energy/.

Zhao, T., Qiu, J., Wen, S., & Zhu, M. (2022). *Efficient onboard energy storage system sizing for all-electric ship microgrids via optimized navigation routing under onshore uncertainties. IEEE Transactions on Industry Applications* (2, pp. 1664−1674). China: Institute of Electrical and Electronics Engineers Inc. Available from https://doi.org/10.1109/TIA.2022.3145775, https://ieeexplore.ieee.org/servlet/opac?punumber = 2858, 19399367.

Accurate and fast MPPT procedure for metaheuristic algorithm under partial shading effect

15

Hicham Oufettoul[1,2], Najwa Lamdihine[1], Saad Motahhir[3],
Ibtihal Ait Abdelmoula[2], Nassim Lamrini[2], Hicham Karmouni[4] and
Ghassane Aniba[1]

[1]Mohammadia School of Engineers, Mohammed V University in Rabat, Rabat, Morocco,
[2]Green Energy Park research platform (GEP, IRESEN/UM6P), Ben Guerir, Morocco, [3]ENSA,
SMBA University, Fez, Morocco, [4]National School of Applied Cadi Ayyad University,
Marrakech, Fez, Morocco

15.1 Introduction

Today, society faces many challenges in achieving a sustainable civilization. Climate change is one of the most pressing issues; accelerating energy transition and achieving a circular economy are priorities for many countries and stakeholders (Brockway et al., 2019). Thus, transitioning to a renewable energy system promotes independence from fossil fuels, but can create new supply uncertainties (Moriarty & Honnery, 2016). Therefore, research in the solar energy sector, particularly in the PV field, is of considerable interest. Photovoltaic technologies allow for immediate energy production from sunlight, which is a free and sustainable source. Moreover, PV systems can be operated as grid-connected or standalone structures (Lu et al., 2015). Despite recent advances in the exploitation of PV systems, their low energy-conversion efficiency is a significant barrier to power generation, and their ability to achieve maximum power point tracking (MPPT) is critically important (Karmouni et al., 2022; Oufettoul et al., 2021; Oufettoul et al., 2022). The essential component of a PV system is an array of PV modules connected in series or parallel to fulfill the desired electrical parameters. The voltage and current exhibited a non-linear correlation. Hence, maximum power is produced at a single operating voltage (Rasheed et al., 2021). Hence, the main objective of the current study is to extract the MPP under partially shaded conditions to improve the PV power generation efficiency since the PV system must operate permanently at its maximum power point (Karmouni et al., 2022). Currently, perturbation and observation (P&O) and incremental conductance are the most widely used methodologies in industry. The primary advantages of these approaches are their simple structure and rapid convergence to MPP (Motahhir et al., 2020). Furthermore, they are applied only under a uniform irradiation distribution and deliver a reliable duty-cycle signal since the

Performance Enhancement and Control of Photovoltaic Systems. DOI: https://doi.org/10.1016/B978-0-443-13392-3.00016-5

P—V curve of the PV array exhibits only one global peak. However, the PV array P—V curve comprises multiple peaks due to changing conditions, including partial shadowing that may originate from dust deposition, passing clouds, and nearby objects (Oufettoul, Motahhir, Aniba, Masud, et al., 2022; Pachauri et al., 2022; Yadav et al., 2022). Traditional MPPT methods fail to distinguish the global maximum power point from the local MPP within a PV system, including more than one PV module subjected to nonuniform irradiation. In this context, Coppola et al. recently presented an MPPT distribution strategy, that is considered an alternative to the conventional approach for overcoming the shortcomings of MPPT regarding partial shading scenarios on the one hand and mitigating hot spot failure on the other hand. Thus, the DMPPT concept is achieved by including an MPPT optimizer along with a microinverter to extract the maximum power point for each PV unit independently (Balato et al., 2018; Balato et al., 2020). Nevertheless, the major drawback of such a configuration is the intensive use of sensors, particularly when the PV array size is large. Despite recent trade, the literature favors machine learning approaches. However, priority is given to the metaheuristic algorithms for MPPT applications. They are labeled "soft computing techniques" and are the most robust MPPT algorithms recommended for industrialization (Dhimish, 2019). Several efficient nature-inspired algorithms have recently been proposed, including Particle Swarm Optimization (PSO) (Alshareef et al., 2019), differential evolution (DE) (Tey et al., 2014), the Flower Pollination Algorithm (FPA) (Prasanth Ram & Rajasekar, 2017), and Teaching and Learning-Based Optimization (TLBO) (Chao & Wu, 2016). These methods are utilized for global search assignments and are effective under nonuniform solar irradiance and temperature conditions, partial shading, and rapidly changing environmental conditions. However, these approaches have certain limitations (Jantsch et al., n.d.). For instance, Sarvi et al. proposed a PSO-based MPPT for PV systems operating under the PS effect. However, this approach exhibits oscillations around the steady state (Sarvi et al., 2015). Accordingly, multiple investigations have proven that PSO has certain constraints, including slow convergence, poor local search capabilities, additional settings to modify, and oscillations around the GMPP (Eltamaly, 2021). Furthermore, most MPPT metaheuristic algorithms are imperfect because typical CSA algorithms require a lengthy conversion time, a higher failure rate, and substantial oscillations in a steady state (Eltamaly, 2021). Additionally, the standard GWO can track the GMPP with multiple iterations and considerable power loss. It may converge to the GMPP with fewer power fluctuations than the PSO (Jiang et al., 2013). The search speed of the conventional ABC algorithm is slower and exhibits premature convergence, which limits its application to PV systems. It is efficient in terms of exploration but insufficient in terms of exploitation (Benyoucef et al., 2015). Therefore, this approach is unsuitable for the practical identification of the GMPP. Hybrid approaches have also piqued the curiosity of researchers by combining two or more MPPT methods to explore their benefits (Ma et al., n.d.). Consequently, several researchers sought to improve the MPPT application to reduce oscillations, response time, and complexity (Bendib et al., 2015). Based on previous literature, Fig. 15.1 provides an overview of the MPPT technique. Ultimately, the present framework seeks to achieve fast and accurate maximum power points by exploring novel procedures for conventional metaheuristic algorithms, thereby

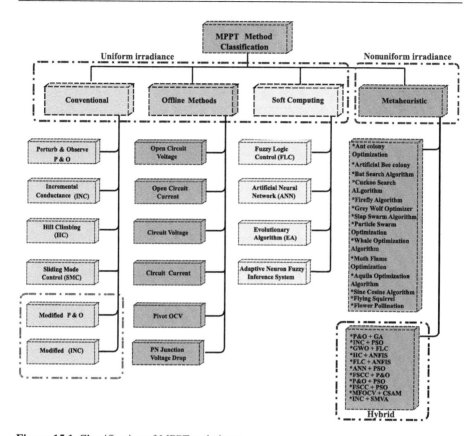

Figure 15.1 Classification of MPPT techniques.

increasing energy production under partial shade conditions. Hence, the main contribution of this work is to shed new light on the quick MPPT response of traditional metaheuristic approaches with well-defined starting points via a series of practical experiments performed on the PV array curve. Finally, the robustness and effectiveness of several metaheuristic methods are validated using MATLAB software and a Raspberry Pi 4 board.

15.1.1 Paper organization

The remaining paper is arranged as follows:

- The second section covers solar cell modeling currently available in the literature, as well as the partial shading effect on PV systems.
- The third section describes the MPPT metaheuristic techniques investigated in this research.
- The fourth section is devoted to the simulation running on MATLAB software and subsequently on the Raspberry embedded board under a nonuniform radiation distribution. Finally, concise findings, comprehensive discussion, and a clear conclusion outlining the merits and drawbacks of the proposed MPPT technique are presented, and potential avenues for further research are highlighted (see Fig. 15.2; Ali et al., 2020).

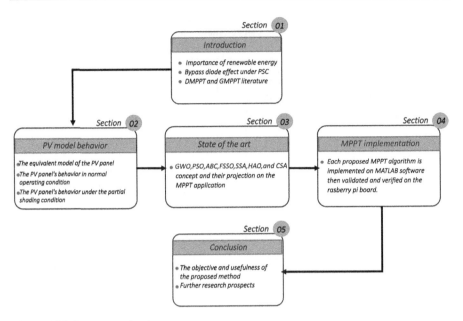

Figure 15.2 Paper organization.

15.2   Equivalent circuit of a photovoltaic cell

The predominant technology used to convert solar energy into electricity is solar cells, the assembly of which forms a PV panel (Oufettoul et al., 2022). Consequently, modeling PV cells is a crucial step in the simulation of PV units, enabling an in-depth evaluation of solar system management and assessing the effects of temperature and irradiance on the panel's behavior. Various PV models, including implicit and explicit models, have been proposed and utilized to quantify PV curves (Boutana et al., 2017). However, explicit alternatives require more computational effort than implicit options, which only require a few mathematical computations (Yaqoob et al., 2021). The following are the most widely cited PV cell equivalent circuits in the literature (1D1R, 1D2R, 2D4R, 2D2R, 3D2R, 3D5R, and xD2R; see Fig. 15.3) (Araújo et al., 2020; 2020; Bellia et al., 2019; 2019). Although the 1D1R model reflects the behavior of the actual PV cells better than the ideal model, it may be inaccurate in certain instances, particularly when the PV cell undergoes several failures and severe temperature changes. The issue discussed in this article is power improvement by utilizing the MPPT application.

Consequently, a single-diode model has been selected owing to its simplicity, ease of design, and limited number of components (Alqahtani et al., 2016; Motahhir et al., 2017). The physical model's five parameters are as follows: The current source I_{PV} is linked in parallel, and the ideality coefficient of the diode is connected in reverse to reflect the output voltage. Besides the diode saturation current, the series resistance

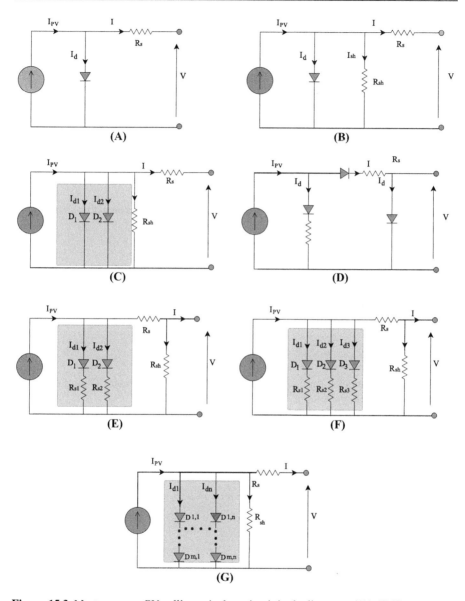

Figure 15.3 Most common PV cell's equivalent circuit in the literature ((A) 1D1R, (B) 1D2R, (C) 2D4R, (D) 2D2R, (E) 3D2R, (F) 3D5R, and (G) xD5R).

R_s and shunt resistance R_{sh} have small and high values, respectively, based on the electrical circuit counterpart of the single-diode model, shown in Fig. 15.3B.

Kirchhoff's law can be applied to determine the output current of the PV cell as follows:

$$I_{PV} = I + I_d + I_{sh} \qquad (15.1)$$

Where I_d indicates the forward diode current and I_{sh} represents the shunt resistor current. Thus, using Shockley's diode law, I_d can be computed as follows:

$$I_d = I_o \times \left[exp^{\frac{V+I \times R_s}{V_t}} - 1 \right] \tag{15.2}$$

Where V is the output voltage and V_t is the diode's thermal voltage; it may be expressed by the following formula (15.3):

$$V_t = \frac{d \times K \times T_c}{q} \tag{15.3}$$

T_c denotes the cell temperature, K corresponds to Boltzmann constant, q represents the electron's charge, and the Eq. (15.4) describes I_{sh} current.

$$I_{sh} = \frac{V + I \times R_s}{R_{sh}} \tag{15.4}$$

The I consequently is written as as shown in Equation (15.5):

$$I = I_{PV} - I_o \left[exp^{\frac{V+I \times R_s}{V_t}} - 1 \right] - \frac{V + I \times R_s}{R_{sh}} \tag{15.5}$$

Thus, the following formula (15.5) provides the current output of the dual diode model depicted in Fig. 15.3C.

$$I = I_{PV} - I_o \left[exp^{\frac{V+I \times R_s}{V_{t1}}} - 1 \right] - I_o \left[exp^{\frac{V+I \times R_s}{V_{t2}}} - 1 \right] - \frac{V + I \times R_s}{R_{sh}} \tag{15.6}$$

Where I_{o1} and I_{o2} are the reverse saturation currents in diodes D1 and D2, respectively. While V_{t1} and V_{t2} present the thermal voltages of the diodes; the mathematical formula is written as follows:

$$V_{t1} = \frac{d_1 \times K \times T_c}{q} \tag{15.7}$$

$$V_{t2} = \frac{d_2 \times K \times T_c}{q} \tag{15.8}$$

15.3   Effect of partial shading condition on PV system

Photovoltaic modules are typically mounted outdoors under harsh environmental conditions. Therefore, it might occasionally be vulnerable to uneven radiation dispersion over the PV array owing to the shadow of the nearby building, dust

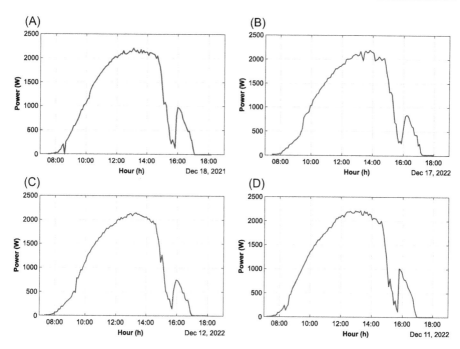

Figure 15.4 Daily energy of a PV array subject to shading effect in portrait mode with conventional MPP extraction.

accumulation, or bird droppings, which may partially cover the PV panel or array. Fig. 15.4 depicts the daily energy curve produced by a mounter solar system in portrait mode on the Moroccan Green Energy Park platform subjected to dynamically changing irradiance using the conventional maximum power point tracking method (Hicham Oufettoul et al., 2023).

The PV array subjected to partial-shading conditions experienced an imbalance in sunlight distribution. Consequently, this may provoke the appearance of a hot spot on the PV module receiving comparatively less irradiance (Wu et al., 2020). Therefore, the module acts as a load rather than a generator that is susceptible to damage owing to a hot spot abnormality. Thus, the system's power output may decrease, resulting in diminished efficiency. Thus, bypass diodes are inserted antiparallel to the PV cell group to prevent the module from self-heating and improve their efficiency under the partial-shading effect (Vieira et al., 2020). Furthermore, the literature argues that bypass diodes improve PV quality by providing an alternating current path. The subparts of each module that underwent partial shading exhibited distinct currents. Consequently, the P−V curve exhibited several peaks, wherein the overall peak reflected the MPP of the global PV system. However, the remaining peaks are presented as local peaks.

Figs. 15.5−15.7 show the outcomes of three experiments performed under an inhomogeneous irradiation distribution on a PV array located at the Green Energy Park platform. The adopted PV array consists of four units connected in series, each comprising

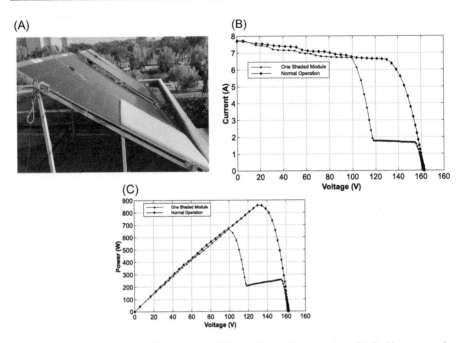

Figure 15.5 First partial shading scenario, (A) actual shading situation, (B) I−V curve, and (C) P−V curve using the PVPMCX1000 tracer.

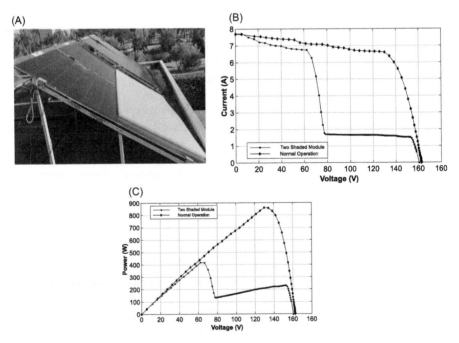

Figure 15.6 Second partial shading scenario, (A) actual shading situation, (B) curve, and (C) P−V curve using the PVPMCX1000 tracer.

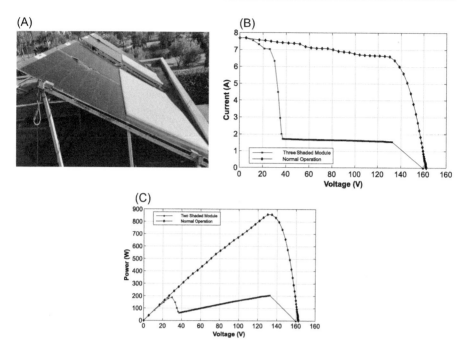

Figure 15.7 Third partial shading scenario, (A) actual shading situation, (B) I−V curve, and (C) P−V curve using the PVPMCX1000 tracer.

three bypass diodes. The first shadowing scenario involved shading a single PV unit, whereas the remaining modules received radiation as high as $812W/m^2$. The second test involved shading the two PV panels with an irradiance of $180W/m^2$. Finally, the third trial darkened the three shaded PV modules with irradiation of $180W/m^2$.

As demonstrated in Fig. 15.8, the shading models applied in the previous tests (see Figs. 15.5−15.7) are incorporated into the MATLAB software, whereby the panels are subjected to different irradiation levels to obtain the voltages corresponding to the global and local maximum points.

Fig. 15.9 demonstrates the voltage values corresponding to the maximum power output based on the previously acquired P−V curves and the physical model of the MPPT given in Eqs. (15.5)−(15.7). Therefore, the initial duty-cycle values chosen for implementation are 75%, 63%, 52%,and 5%.

$$V_{mpp2} = N_s \times V_{mpp,STC} \tag{15.9}$$

$$V_{mpp1} = (N_s - N_{sh}) \times V_{mpp,STC} - N_{sh} \times 0.7 \tag{15.10}$$

Where V_{mpp1} and V_{mpp2} represent the voltage corresponding to the first and second peak, respectively. While N_s is the total number of PV panels, N_{sh} is the number of shaded panels, and $V_{mpp,STC}$ represents the voltage at the MPP once the PV array is exposed to STC (see Fig. 15.10).

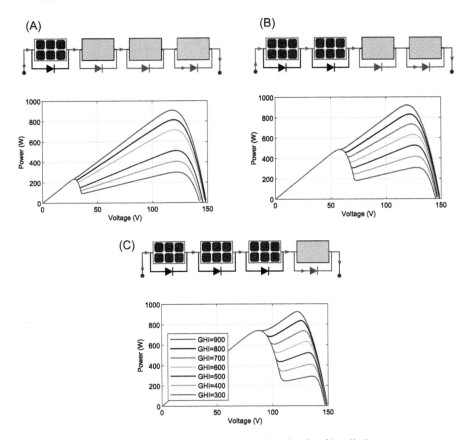

Figure 15.8 Different applied shading modes at varying levels of irradiation.

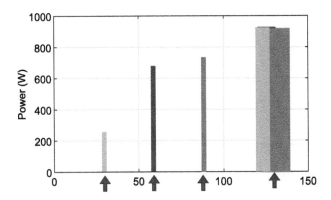

Figure 15.9 Voltage position range corresponds to the potential maximum points.

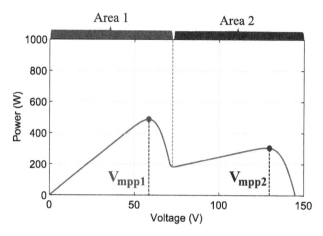

Figure 15.10 First and second peak's voltage of PV array.

15.4 MPPT technique under partially shaded conditions

Various MPPT optimization methods are addressed in this section to verify the validity of the proposed procedure over a wide range of algorithms and discover the strengths and weaknesses of each MPPT strategy.

15.4.1 Grey wolf optimizer

15.4.1.1 Theory

The grey wolf optimizer is a novel metaheuristic algorithm designed to reflect the leadership hierarchy and hunting mechanism of the grey wolf, which has been proposed by Mirjalili et al. (Mirjalili et al., 2014). GWO consists of four subspecies of grey wolves: Alpha, Beta, Delta, and Omega, as illustrated in Fig. 15.11. Alpha wolves are the pack's leader, providing an optimizing lifestyle solution, whereas the beta member class is subordinate and supports the alpha's decision-making. Meanwhile, the delta belongs to the third class, commanding the lowest GW level (Omega), which serves as a scapegoat. The GWO methodology has been applied to solve optimization issues based on the population distribution of potential solutions. The GWO approach involves three main processes: encirclement, hunting, and prey attack. Therefore, the following mathematical formulae describe the GWO process (Nadimi-Shahraki et al., 2021).

15.4.1.1.1 Encircling
After identifying and locating the prey position, each search agent surrounds the target. The following mathematical model represents circular behavior:

$$D_{wolf} = \left| C \times X_{wolf}(t) - X(t) \right| \tag{15.11}$$

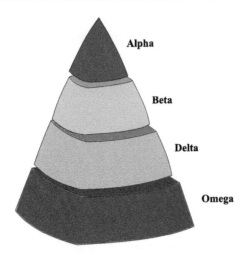

Figure 15.11 Grey wolf social hierarchy.

$$X_{wolf}(t + 1) = X_{wolf}(t) - A \times D_{wolf} \tag{15.12}$$

$$\vec{A} = 2 \times \vec{a} \times \vec{r}_1 - \vec{a} \tag{15.13}$$

$$\vec{C} = 2 \times \vec{r}_2 \tag{15.14}$$

Where t represents the iteration number and $X(t)$ is the current Grey wolf position. Hence, $X(t + 1)$ is the position it will eventually reach, while $X_{wolf}(t)$ explicitly refers to α, β, δ wolves. The A and C are coefficient vectors that maintain a suitable balance between exploration and exploitation, r_1 and r_2 are random integers ranging between $[0, 1]$, and a's components are progressively lowered from 2 to 0 throughout the iterations.

15.4.1.1.2 Hunting
Based on the placements of all agents α, β, and δ, the grey wolves chased the prey after encircling it. The following equations reflect the updated concept of this procedure.

$$D_\alpha = \left| C \times X_\alpha(t) - X(t) \right|, D_\beta = \left| C \times X_\beta(t) - X(t) \right|, D_\delta = \left| C \times X_\delta(t) - X(t) \right| \tag{15.15}$$

$$\vec{d}_\alpha = \vec{X}_\alpha - \vec{D}_\alpha \times \vec{A}_1, \ \vec{d}_\beta = \vec{X}_\beta - \vec{D}_\beta \times \vec{A}_1, \ \vec{d}_\alpha = \vec{X}_\alpha - \vec{D}_\alpha \times \vec{A}_1 \tag{15.16}$$

$$\overrightarrow{d_{out}} = \frac{\overrightarrow{d_1} + d_2 + \overrightarrow{d_3}}{3} \tag{15.17}$$

15.4.1.1.3 Attacking

Grey wolves surrounded the prey and began to prepare for capture due to A
$[-2a, 2a]$. When A is larger than one, the grey wolves stay away from the prey to
perform global scanning; however, if the A value is smaller than 1, the grey wolf
pack approaches the prey and eventually completes it.

15.4.1.2   GWO application for maximum power point tracking technique

During the operation of the PV system under partial shading conditions, the P−V curve
is distorted into several peaks, reflecting the applied shading profile. Nevertheless, a
single peak indicated the maximum actual power of the predicted target. Consequently,
population-based search algorithms, such as the GWO, are essential for achieving such
a target. Figs. 15.12 and 15.13 exemplify the MPPT strategy, wherein the controller

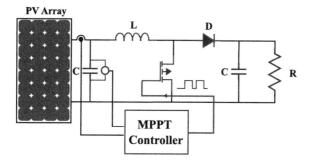

Figure 15.12 Block diagram of MPPT stand-alone PV system.

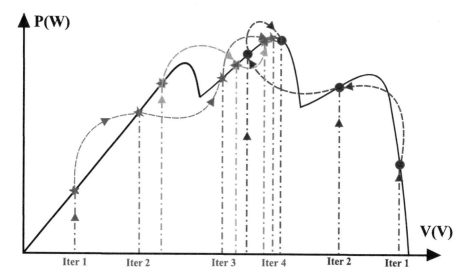

Figure 15.13 Search under partial shade.

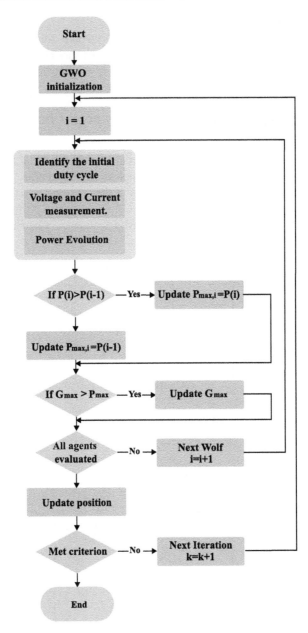

Figure 15.14 Flowchart of the GWO algorithm for the MPPT application.

acquires the output voltage and current to determine the power output (Chtita et al., 2022; Mohanty et al., 2016, 2017). The grey wolf agent indicates the duty cycle values. Therefore, Eq. (15.11) can be modified into Eq. (15.15), whereas the GWO fitness function is deduced from Eq. (15.16). Fig. 15.14 describes a flowchart of the MPPT strategy based on the GWO algorithm.

15.4.2 Particle swarm optimization

15.4.2.1 Theory

Particle swarm optimization (PSO) is a stochastic optimization methodology based on the intelligence of swarm behavior. It uses the concept of social interaction between birds in flight or schools of fish, wherein all particles triumph by simulating the success of their neighbors (Gad, 2022; Kennedy & Eberhart, 1995). Therefore, many agents navigate through the space to update their respective generations. The optimal location of a neighboring particle influences its position, thus revealing the best solution. The PSO mathematical algorithm is described as follows:

$$x_i^{k+1} = x_i^k + v_i^{k+1} \tag{15.18}$$

$$v_i^{k+1} = \left(w \times v_i^k\right) + c_1 \times r_1\left(P_{best,i} - x_i^k\right) + c_2 \times r_2\left(G_{best} - x_i^k\right) \tag{15.19}$$

Where x and v denote the particle's location and velocity, respectively. k is the number of iterations, while r_1 and r_2 are random variables with a uniform distribution centered in [0, 1]. The acceleration coefficients are denoted as c_1 and c_2; the inertia weight is expressed as w.

15.4.2.2 PSO application for maximum power point tracking technique

The PSO strategy outlined in the preceding section is now applied in MPPT applications utilizing the direct control technique for power stations running under partial shading conditions, where the P−V curve includes numerous peaks. Owing to the singularity of such a situation, PSO will satisfies the practical considerations of the power generation system under the partial shading effect.

The position is defined as the actual duty cycle delivering four duty cycles (retrieved from the experimental test, as shown in Fig. 15.9) to the DC/DC converter. In contrast, the resulting power is defined as the fitness function. Fig. 15.15 illustrates the MPPT application flowchart based on the PSO algorithm (Abdulkadir et al., 2014; Koad et al., 2017).

15.4.3 Artificial bee colony

15.4.3.1 Theory

The artificial bee colony (ABC) algorithm is a straightforward bio-inspired strategy that utilizes limited controllable parameters and is independent of the initial circumstances. This innovative metaheuristic handles multidimensional and multimodal optimization challenges. The artificial bee colonies consist of workers, observers, and scout bees (Hancer, 2020; Kumar et al., 2017). A worker bee is

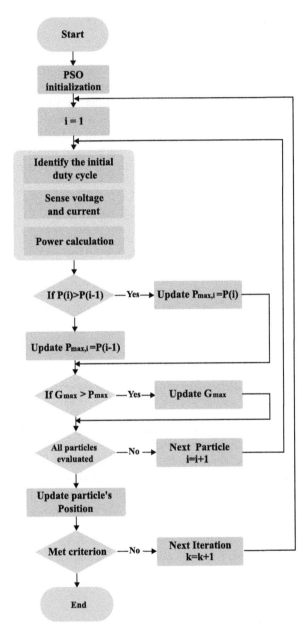

Figure 15.15 Flowchart of the PSO algorithm for the MPPT application.

involved in searching for food as long as the observer waits in the hive to make a decision, whereas a scout bee randomly explores a new potential food source. The ABC algorithm is divided into four phases: initialization, worker bees, observers, and scout bees (Crawford et al., 2014).

- The ABC initialization step involves creating an initial population using a random distribution according to the following equation:

$$x_i^j = x_{min}^j + r \text{ and } [0,1](x_{max}^j - x_{min}^j) \tag{15.20}$$

- Where $(i = 1, 2, 3, \ldots n)$ and $(j = 1, 2, 3, \ldots m)$. In addition, x_{min} and x_{max} are the minimum and maximum parameters of the search agents, respectively, and n presents the number of parameters to be optimized, while m indicates the number of employed or spectator bees.
- During the employed bee process, after each cycle, each employed bee updates its position with respect to its neighborhood, adopting Eq. (15.21):

$$x_i^{(k+1)} = x_i(k) + B \times (x_i(k) - x_j(k)) \tag{15.21}$$

Parameter B is a random number chosen from $[0, 1]$. Hence, this variable significantly influences all bees to redirect themselves to the optimal or nearby solution, while k represents iteration number.

- Observer bees patiently wait in the dancing area to move closer to the employed bee position where the nectar quantity is high. This pattern can be expressed as follows:

$$x_i(k+1) = x_s(k) + \frac{B \times (x_i(k) - x_j(k))}{\frac{N_p}{2} - 1} \tag{15.22}$$

Where N_p represents the bee number and x_s denotes the food source position with the highest amount of nectar.

- During the scout bee phase, the abandoned solutions are determined by investigating the least fitness results with a randomly selected new fitness.

15.4.3.2 ABC application for maximum power point tracking technique

This section adopts the ABC algorithm in the MPPT context by applying Eqs. (15.20)–(15.22). Thus, most ABC parameters are fixed. Initially, we assigned several candidate solutions corresponding to the number of duty cycles in an MPPT application. Furthermore, the duty cycle is evaluated as follows: The ABC algorithm is adopted in the MPPT context by applying Eqs. (15.20)–(15.22). Thus, most ABC parameters are fixed. Initially, we assigned several candidate solutions corresponding to the number of duty cycles in the MPPT application. Furthermore, the duty cycle is evaluated as follows:

$$d_{curr} = d_{min} + r \times (d_{max} - d_{min}) \tag{15.23}$$

$$d_{new} = d_{curr} + cts \times (d_{curr} - d_{old}) \tag{15.24}$$

The current duty cycle is denoted by d_{curr}. However, d_{old} represents the prior value of the duty cycle. Similarly, $d_{\max}$ and $d_{\min}$ reflect the maximum and lowest anticipated duty cycle values, respectively. *cts* represents a constant value bounded by $[-1, 1]$. Fig. 15.16 shows the MPPT application flowchart based on the ABC algorithm (Gonzalez-Castano et al., 2021; Soufyane Benyoucef et al., 2015).

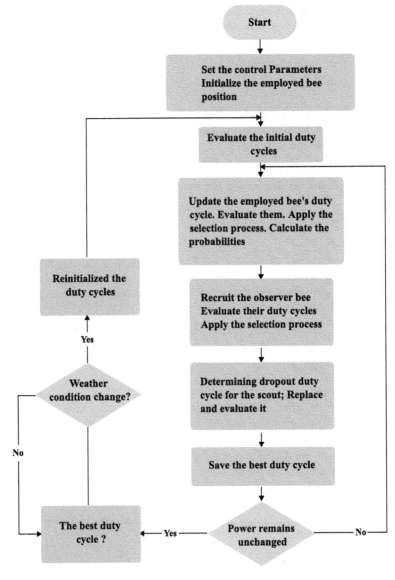

Figure 15.16 Flowchart of the ABC algorithm for the MPPT application.

15.4.4   Cuckoo search algorithm

15.4.4.1   Theory

In 2009, Yang et al. developed a cuckoo algorithm inspired by cuckoo breeding behavior. Indeed, the latter parasitically lay eggs in other bird nests instead of building their own. It randomly searches for nestmates. The best nest is then selected. In addition, the cuckoo takes significant steps to promote hatching by meticulously placing its eggs in a convenient location and, exceptionally, by depositing the eggs of the host bird outside the nest to decrease the chance of discovery (Mohanty et al., 2017). Furthermore, certain cuckoo species can lay eggs that mimic those of other species. Three types of brood parasitism have been reported in the literature: intraspecific, cooperative, and nest takeover. Hence, a Lévy flight is a defining feature of the cuckoo search process (Yang & Deb, 2009; Yang, 2010). A Lévy flight is mathematically defined as a random walk, wherein the step sizes are extracted from the Lévy distribution according to the power law (see Eq. 15.25).

$$y = l^{-\lambda} \tag{15.25}$$

Where l represents the flight length and y indicates the variance, emphasizing that $1 < \lambda < 3$ has an infinite variance. This framework implies three idealized criteria for the search behavior of cuckoos. First, each cuckoo lays one egg at a time and deposits it into a randomly selected nest. Second, the best nest, with the highest quality eggs, is inherited by the next generation. Third, accessible nests are evaluated, and a conceptual probability, P is assigned to the number of eggs detected by the host bird. Indeed, when generating a new solution $x(k + 1)$ for a cuckoo bird, a Levy flight is performed as dictated by the following mathematical formula:

$$x_i^{(k+1)} = x_i^k + \alpha \oplus levy(\lambda) \tag{15.26}$$

Where x and i represent the sample and sample number, respectively, as well as k denotes the iteration number. Considering the constraints imposed by the optimization process, adopting the desired step size is relevant, which is performed by applying Eq. (15.27).

$$\alpha = \alpha_0 \left(x_j^k - x_i^k \right) \tag{15.27}$$

Where α and α_0 signify the desired step size and the initial step, respectively.

15.4.4.2   CSA application for maximum power point tracking

The CSA structure is slightly changed in the scope of the MPPT algorithm design, incorporating a few parameters (Abo-Elyousr et al., 2020) considering a specific nest number, initial population, and initial step.

Initialization involves appealing the resulting samples to the PV generator and setting power as the initial fitness value. The most significant power of the corresponding voltage is considered to be the current best sample (Ding et al., 2019). After applying Levy's flight, new voltage samples are produced according to Eq. (15.30).

$$S_u = \frac{T \times (1 + B) \times \sin\left(\pi \times \frac{\beta}{2}\right)}{T \times \left(\pi \times \frac{\beta}{2}\right) \times \beta \times 2^{\frac{\beta-1}{2}}} \ and \ S_v = 1 \tag{15.28}$$

$$d_{i+1}^k = d_i^k + \alpha \frac{|u|}{v^{1/\beta}} \times \left(d_{best} - d_i^k\right) \tag{15.29}$$

$$u = \text{Normal } r \text{ and}\left(0, S_u{}^2\right) \text{ and } v = \text{Normal } r \text{ and } \left(0, S_v{}^2\right) \tag{15.30}$$

i and k represent the numbers of particles and the search iteration, respectively. Thus α is the step size chosen according to the addressed issue. Although $\alpha = 0.7$ and $\beta = 1.4$ are constant, they are suggested in several studies. At the same time, u and v are uniformly distributed matrices.

The flowchart (see Fig. 15.17) shows the different steps of using the CS algorithm to build the MPPT application.

15.4.5   Horse herd optimization

15.4.5.1   Theory

The horse herd optimization approach (HOA) is a novel approach introduced in 2021 (MiarNaeimi et al., 2021). Inspired by the gregarious behavior of horses, it is a metaheuristic algorithm devoted to high-dimensional optimization challenges. Furthermore, this approach mimics the social performance of horses of different ages using six crucial attributes: grazing, hierarchy, sociability, imitation, defense mechanisms, and roaming. Therefore, this approach exhibits particularly interesting efficiencies in the literature compared to recently popular optimization algorithms (Sarwar et al., 2022). Horses are placed at each stage according to Eq. (15.31).

$$x_m^{k,age} = V_m^{k,age} + x_m^{(k-1),age} \tag{15.31}$$

Where the age can be either $(\alpha, \beta, \lambda, \gamma)$, namely, the age value is determined in such a manner that $\alpha, \beta, \lambda,$ and γ represent horses aged $[0, 5]$, $[5, 10]$, $[10, 15]$, and over 15 years, respectively.

$x_m^{k,age}$: represents the nth horse position, age: indicates each horse's range, $v_m^{k,age}$: denotes the horse's velocity vector, k: presents the current iteration.

Eqs. (15.32)–(15.35) may be expressed as horse movement vectors in each process cycle.

$$V_m^{k,\alpha} = Gr_m^{k,\alpha} + Dm_m^{k,\alpha} \tag{15.32}$$

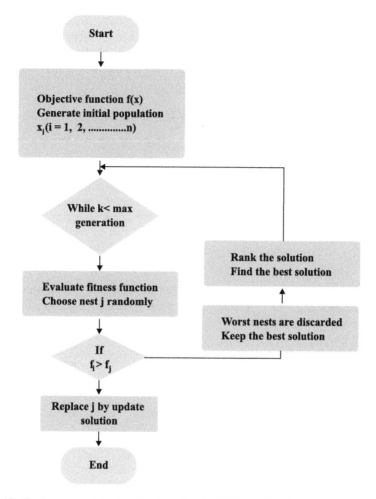

Figure 15.17 Flowchart of the CS algorithm for the MPPT application.

$$V_m^{k,\beta} = Gr_m^{k,\beta} + Dm_m^{k,\beta} + Hi_m^{k,\beta} + So_m^{k,\beta} \tag{15.33}$$

$$V_m^{k,\gamma} = Gr_m^{k,\gamma} + Dm_m^{k,\gamma} + Im_m^{k,\gamma} + Hi_m^{k,\gamma} + So_m^{k,\gamma} \tag{15.34}$$

$$V_m^{k,\delta} = Gr_m^{k,\delta} + Im_m^{k,\delta} + Ro_m^{k,\delta} \tag{15.35}$$

These are critical phases of a horse's individual and social intelligence. The aforementioned formula is completed only once at each stage (grazing, hierarchy, sociability, imitation, defense mechanism, and roam) and has been explained in detail.

15.4.5.1.1 Grazing (Gr)

Horses eat grasses, bushes, and other forage. They graze for more than $16h$ per day, with infrequent breaks. Grazing is uninterrupted feeding. Model the grazing regions for each horse using HOA. Thus, grazing can be described using Eqs. (15.36) and (15.37).

$$Gr_m^{k,age} = g_k \times (l + pu) \times \left[x_m^{k-1}\right], age = (\alpha, \beta, \lambda, \text{ and } \gamma) \tag{15.36}$$

$$g_m^{k,age} = g_m^{(k-1),age} \times w_g \tag{15.37}$$

Where $Gr_m^{k,age}$ represents the movement and grazing capability of the horses; thus, grazing was reduced by weight per cycle. The parameter r is a random parameter that ranges from 0 to 1. l and u are the lower and upper limits of grazing space, respectively.

15.4.5.1.2 Hierarchy (Hi)

Horses follow the leader, which is usually a human. According to the rule of dominance, a mature stallion or filly leads a group. Hi represents a group's tendency to follow the strongest horse. Eqs. (15.38) and (15.39) are suitable for defining the hierarchical law.

$$Hi_m^{k,age} = h_m^{k,age} \times \left(x_*^{k-1} - x_m^{k-1}\right), age = (\alpha, \beta, \text{ and } \gamma) \tag{15.38}$$

$$h_m^{k,age} = h_m^{(k-1),age} \times w_h \tag{15.39}$$

Where $Hi_m^{k,age}$ denotes the best horse location effect, with respect to the velocity parameter, as well as x_* denotes the best horse location.

15.4.5.1.3 Sociability (So)

Horses are natural social animals that often cohabit with other animals. The herd is preferable for ensuring sustainability and the horses' security. The social behavior, which signifies the movement towards a middle position relative to other horses, is designated by the So-factor. Eqs. (15.40) and (15.41) are provided to explain the sociability law.

$$So_m^{k,age} = s_m^{k,age} \left(\left(\frac{1}{N} \sum_{n=1}^{N} x_n^{k-1} \right) - x_m^{k-1} \right), age = (\beta \text{ and } \gamma) \tag{15.40}$$

$$s_m^{k,age} = s_m^{(k-1),age} \times w_s \tag{15.41}$$

Where $So_m^{k,age}$ and $s_m^{k,age}$ represent the social motion vector and herd orientation, respectively, in a particular iteration k, and N represents the total number of horses.

15.4.5.1.4 Imitation (Im)

Horses imitate each other and adopt positive and negative patterns of other horses, including the location of the best grazing area. Horse imitation behavior is also addressed as a factor i. The imitation process is described by Eqs. (15.42) and (15.43).

$$Im_m^{k,age} = i_m^{k,age}\left(\left(\frac{1}{PN}\sum_{n=1}^{PN}x_n^{k-1}\right) - x_m^{k-1}\right), age = \gamma \qquad (15.42)$$

$$i_m^{k,age} = i_m^{(k-1),age} \times w_i \qquad (15.43)$$

Where PN is the number of horses with the best positions, w_i denotes the cycle-by-cycle reduction factor, and $Im_m^{k,age}$ represent the movement vector from the ith horse to the best horse on average at location x_m.

15.4.5.1.5 Defense mechanism (DM)

Horses defend themselves or fight to keep themselves safe. In the event of a trap, they attempt to escape before falling. Further, they fight each other for food and water or against an unsafe or unfriendly atmosphere. The horse herd algorithm's defense mechanism is to flee horses that behave unacceptably. Eqs. (15.44) and (15.45) provide explanations for the horse's defense mechanisms.

$$Dm_m^{k,age} = d_m^{k,age}\left(\left(\frac{1}{dN}\sum_{n=1}^{dN}x_n^{k-1}\right) - x_m^{k-1}\right), age = \alpha, \gamma \text{ and } \beta \qquad (15.44)$$

$$d_m^{k,age} = d_m^{(k-1),age} \times w_d \qquad (15.45)$$

Where $Dm_m^{k,age}$ refers to the escape vector of the ith horse from the average of some horses with the worst locations, which are represented by the vector x_m, dN is the number of horses with the worst location, and d_m is the mitigation factor per cycle.

15.4.5.1.6 Roam (Ro)

It is obvious that horses forage outdoors for food. A horse may graze at an unexpected location. Horses are often quite inquisitive and explore new areas to find fresh pastures or become familiar with their communities. The horse's tendency to wander is often seen during youth and eventually fades as the animal matures. The following equations describe this stage.

$$Ro_m^{k,age} = r_m^{k,age}\partial x_m^{k-1}, age = \gamma \text{ and } \delta \qquad (15.46)$$

$$r_m^{k,age} = r_m^{(k-1),age} \times w_r \qquad (15.47)$$

Where $Ro_m^{k,age}$ represents the random velocity vector of the ith horse for a local area search and escape from local minima, while w_r denotes a cycle-by-cycle damping ratio.

15.4.5.2 HOA application for maximum power point tracking

The horse herd optimization approach is invoked in this framework to perform the MPPT assignment; consequently, four horses are utilized to obtain a strong performance. Thus, the objective function is computed for each horse based on the power output magnitude corresponding to each horse's location, which is subsequently applied to determine the horse's motion, and the expressions are incorporated to update each horse's movement. Fig. 15.18 depicts a flowchart of the HOA for the MPPT application (Hosseinalipour & Ghanbarzadeh, 2022; Sarwar et al., 2022).

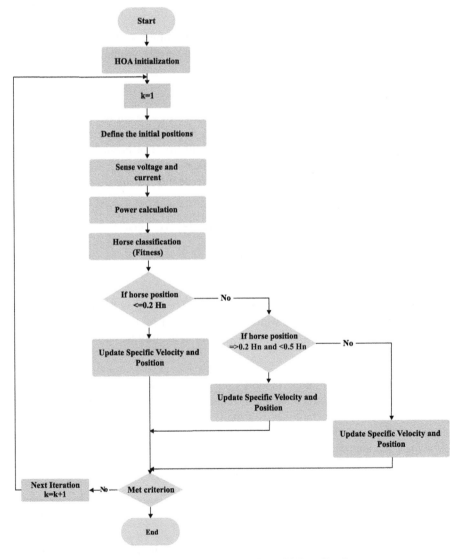

Figure 15.18 Flowchart of the HOA algorithm for the MPPT application.

15.4.6 Flying squirrel search optimization

The flying squirrel search optimization is a new swarm-based metaheuristic optimization algorithm developed by Singh and his collaborators (Singh et al., 2021). This method is established to address the mathematical and technological challenges. These behaviors propose two main theories of motion to simulate flying in trees and wandering on the ground or a tree branch in search of food. The posture traits of the flying squirrel and food source correspond to the expected outcome vector and equivalent welfare, respectively. Therefore, the posture is divided into three districts based on the welfare value, with each district handling the optimal (hickory tree), near-optimal (acorn tree), and unplanned (normal tree) solution sets. In MPPT implementation, the power output is determined as an objective function, and the duty cycle is defined as a decision variable (Miarnaeimi et al., 2019). The flowchart in Fig. 15.19 shows the different steps of the FSSO algorithm for MPPT applications. First, initialize the FSSO parameters and duty-cycle values specified in Section 15.3. Subsequently, the DC/DC converter is commanded successively by each duty cycle, allowing the corresponding power output to be investigated for each process. This refers to the objective fitness function, as shown in Eq. (15.48).

$$f(d) = \max(P_{PV}(d)) \tag{15.48}$$

The hickory tree is declared as the best position after evaluating the power output. The tree acorn represents the next-best flying squirrel (FS) position. In anticipation of this, other normal flying squirrels are located on normal trees. Once the seasonal monitoring status is verified, the duty cycle is updated using Eq. (15.53) or (15.56), depending on the following condition $S_C^k < S_{min}$.

15.4.6.1 Occasional observations

This criterion for occasional observations prevents the algorithm from being trapped in local maxima. Estimate the periodic coherence S_C and its base value S_{min} for a space with a single dimension using Eq. (15.49).

$$S_c^k = \left| duty_{at}^k - duty_{ht}^k \right| \tag{15.49}$$

Where $duty_{at}$ and $duty_{ht}$ denote the squirrel's position at an acorn and hickory tree, respectively. Hence, k denotes the number of iterations. The Levy distribution s is applied to explore the hunting space more efficiently. Consequently, by modifying the duty cycle,

$$duty_{ht}^{k+1} = duty_{ht}^k + s \tag{15.50}$$

$$S_u = \frac{T \times (1 + B) \times \sin\left(\pi \times \frac{\beta}{2}\right)}{T \times \left(\pi \times \frac{\beta}{2}\right) \times \beta \times 2^{\frac{\beta-1}{2}}} \quad and \quad S_v = 1 \tag{15.51}$$

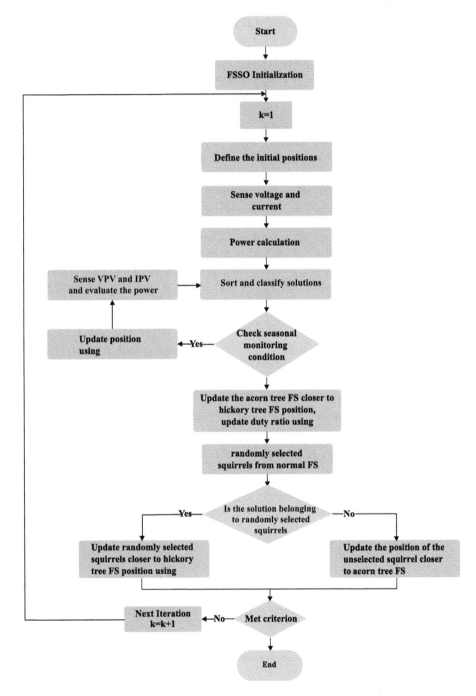

Figure 15.19 Flowchart of the FSSO algorithm for the MPPT application.

$$d_{i+1}^k = d_i^k + \alpha \frac{|u|}{v^{1/\beta}} \times \left(d_{best} - d_i^k\right) \tag{15.52}$$

$$u = \text{Normal } r \text{ and } \left(0, S_u^2\right) \text{ and } v = \text{Normal } r \text{ and } \left(0, S_v^2\right) \tag{15.53}$$

Where α indicates the step size chosen according to the addressed issue. Although $\alpha = 0.7$ and $\beta = 1.4$ are constant, they are suggested in several studies. At the same time, u and v are uniformly distributed matrices.

15.4.6.1.1 Regular updates

The squirrels on hickory trees are immobile. The squirrel on the acorn tree advances toward the hickory tree. Some randomly selected squirrels from the normal trees moved toward the hickory tree, whereas the remaining squirrels migrated toward the acorn tree. The corresponding duty cycles are updated according to Eqs. (15.54)−(15.56):

$$duty_{at}^{k+1} = duty_{at}^k + g_d \times G_c \left(duty_{ht}^k - duty_{at}^k\right) \tag{15.54}$$

$$duty_{nt}^{k+1} = duty_{nt}^k + g_d \times G_c \left(duty_{ht}^k - duty_{nt}^k\right) \tag{15.55}$$

$$duty_{nt}^{k+1} = duty_{nt}^k + g_d \times G_c \left(duty_{at}^k - duty_{nt}^k\right) \tag{15.56}$$

Where G_c and g_d represent the sliding distance and the constant, respectively. The expression for the sliding distance g_d is given as follows:

$$g_d^{k+1} = \frac{h_g}{S_f \times \tan\left(\frac{C_D}{C_L}\right)} \tag{15.57}$$

Where the height loss magnitude h_g after flying is assumed to be $8m$; the scaling factor led to limit the perturbations in (23), (24), and (25). The drag coefficient C_D is fixed at 0.6, whereas the coefficient of lift C_L is arbitrarily determined between $[0.5, 1.8]$.

Once the maximum number of iterations is reached, the improved algorithm obtains the cyclic ratio corresponding to the GMPP.

15.4.7 Salp swarm algorithm

15.4.7.1 Theory

Mirjalili et al. are the first researchers to introduce the Salp Swarm algorithm. It is a recently developed metaheuristic algorithm inspired by the swarming behavior of salps in the ocean (Abdulkadir et al., 2014; Mirjalili et al., 2017). According to this concept, the algorithm has proven its efficiency in various applications. The salp chain comprises of a single leader and several followers. The leader's salp is at the top of the chain, whereas the remaining salps are followers (Faris et al., 2020). The location of the salps is determined in an n-dimensional search space, where n denotes the number of variables in the problem. Therefore, a two-dimensional matrix can represent the location data of the salps. The position of the leader salp is updated as follows:

$$x_1^k = F + C_1(ub - lb) \times C_2 + lb \times C_3 \geq 0.5 \tag{15.58}$$

$$x_1^k = F + C_1(ub - lb) \times C_2 + lb \times C_3 < 0.5 \tag{15.59}$$

$$C_1 = 2 \times \exp\left(-\frac{4k}{K}\right)^2 \tag{15.60}$$

Where x and F represent the leader's position and food supply, respectively; thus, ub represents the upper bound and lb represents the lower bound. C_1, C_2, and C_3 are random values. Coefficient C_1 is a crucial metric for balancing the exploitation and exploration of the SSA, as defined in Eq. (15.60). K and k denote the maximum and current number of iterations, respectively. The formula for updating the position is expressed by Eq. (15.61).

$$x_i^k = \frac{1}{2} a \times t^2 + v_0 \times t \tag{15.61}$$

Where v_0 is the initial velocity at time $t = 0$. Given that $v_0 = 0$, the formula changes as follows:

$$x_i^k = \frac{1}{2}\left(x_{i-1}^k - x_i^k\right) \tag{15.62}$$

15.4.7.2 SSA application for maximum power point tracking

The MPPT controller delivers a duty cycle as an output. Therefore, the duty cycle is considered the salp swarm position, and the PV panels' power output is the fitness function (Jamaludin et al., 2021). Fig. 15.20 illustrates the complete flowchart of the MPPT controller based on the SSA under partial shading conditions. The procedures are as follows: Step one: Initialize the salp population x_i^k with $i = 1, 2, 3...M$ and $k = 1, 2, 3, ...N$. Step two: Assess the entire population. Step three: Identification of the most appropriate salp. Step four: Update the duty cycle using Eqs. (15.58) and (15.59). Step five: Update the potential solutions for leaders. Step six: Update candidate solutions for followers. Step seven: Modify solution candidates that exceed the maximum and minimum values. Step eight: Met the criteria (Jamaludin, Tajuddin, Ahmed, Azmi, et al., 2021; Mao et al., 2020).

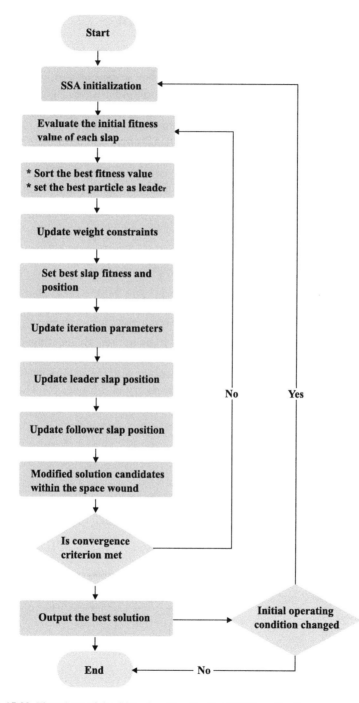

Figure 15.20 Flowchart of the SSA algorithm for the MPPT application.

15.5 Simulation results and discussion

15.5.1 MPPT validation based on MATLAB software

This section describes the simulation results of the aforementioned approaches to assess the performance of the MPPT control algorithm based on the proposed process. The PV model adopted for this investigation consists of four panels connected in series and a DC/DC boost converter. The irradiance levels distributed across the PV panels are uniform to emphasize the mismatch effect and induce several peaks on the P-V curve. Each PV module terminal has been mounted in parallel to the bypass diode to ensure safety and allow high currents to flow. Each MPPT method is carefully executed according to the flowchart instructions and the initial duty cycle values presented in Section 15.3. Fig. 15.21 illustrates the shading model circuit in MATLAB/Simulink.

15.5.2 MPPT validation based on Raspberry Pi 4

The MPPT controller validation proposed in this paper is implemented on "Raspberry Pi 4 Model B" to exploit the results obtained by following the V-cycle development process (see Fig. 15.22), applying "Model In Loop Approves", "Software In Loop" and "Process In Loop". Furthermore, all the simulations are performed using MATLAB/Simulink software on a core $i5$, $1.9GHz$ computer with $16GBRAM$ to verify the compliance of this software with the requirements for integration on an embedded board such as the "Raspberry Pi 4 Model B." Thus, it provides a low-cost solution for possible hardware implementations.

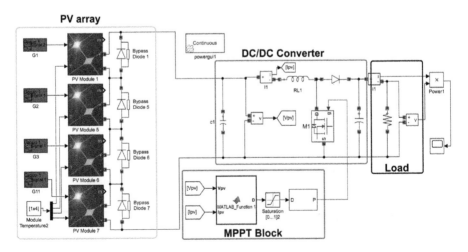

Figure 15.21 Photovoltaic system under MATLAB software.

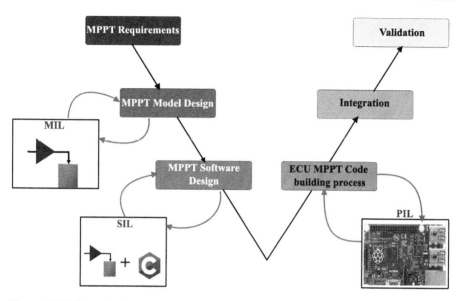

Figure 15.22 V-cycle development process to verify and validate MPPT application.

15.5.3 Analysis and discussion

This section seeks to discover the optimal solution for MPPT applications by analyzing the performance of each strategy. Seven adopted MPPT algorithms are based on a stochastic research interface boost converter to retrieve the PV system's maximum power point. Each approach initiates tracking based on the various duty cycles provided in the privilege section to optimize the performance of the PV system. Consequently, the analysis focused on several aspects, including the speed of reaching the MPP, precision, and accuracy. Figs 15.23−15.29 show seven output power curves corresponding to each algorithm tested in the MIL, SIL, and PIL. Indeed, all the results obtained during the validation tests (MIL, SIL, and PIL) are perfectly similar.

The GWO outputs shown in Fig. 15.23 demonstrates slight oscillations at the beginning of MPP restitution, which may be expected owing to random components. However, the adopted initial duty cycle improved the speed of reaching the maximum power compared to the literature. Thus, an ultimate power of 640 W is achieved at 0.158 s.

Fig. 15.24 demonstrates the power output produced by implementing the proposed PSO flowchart. The output power of the PV generator is maintained at 640 W after 0.08 s. Therefore, this approach is accurate; moreover, it is quicker than the GWO algorithm. As far as oscillations are concerned, they are almost nonexistent compared to GWO algorithm.

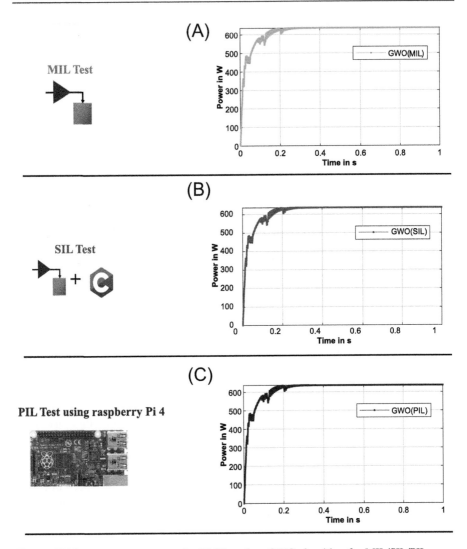

Figure 15.23 PV system power under PSC based on GWO algorithm for MIL/SIL/PIL testing.

Fig. 15.25 demonstrates the power output of the PV system handled by the CSA algorithm; obviously, it is relatively quick to reach the MPP compared to the prede-termined procedures; in fact, this algorithm is recognized as the fastest metaheuris-tic technique, as proven in the literature, besides the appearance of various oscillations immediately once the tracking begins. In addition, this approach has inaccuracy inconvenience; it is settled on a power production of 630 W instead of 640 W.

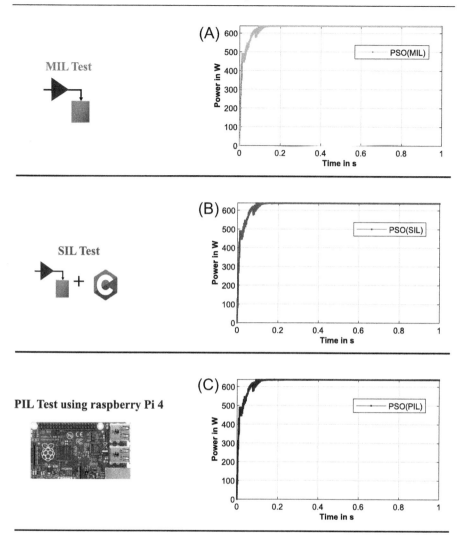

Figure 15.24 PV system power under PSC based on PSO algorithm for MIL/SIL/PIL testing.

Fig. 15.26 demonstrates the ABC algorithm results, wherein the output power rapidly reaches the maximum power the PV system could produce after approximately 0.05 s. Nevertheless, the oscillations are perceptible and appear during around 0.2 s. Hence, it has extracted the maximum power of 640 watt in terms of accuracy.

The HAO technique adheres to the same philosophy as the ABC algorithm, in terms of speed. The power output results are shown in Fig. 15.27 demonstrate clearly that the HAO approach is highly accurate as the algorithm rapidly achieves

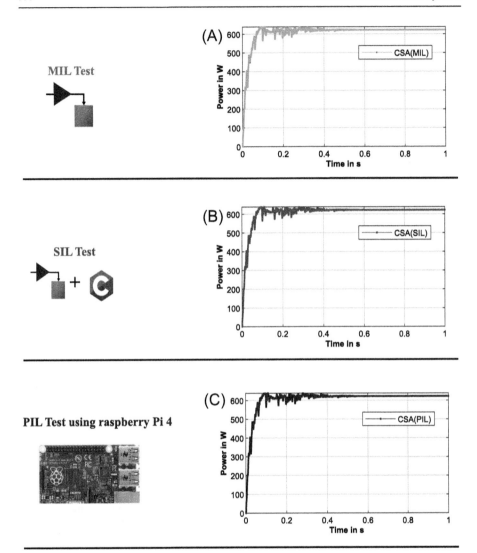

Figure 15.25 PV system power under PSC based on CSA for MIL/SIL/PIL testing.

the actual MPP value. However, this technique is unsuitable regarding the oscillation challenge. Indeed, it shows frequent oscillation within the first 0.5 s.

The results achieved using the FSSO algorithm correctly tracked the 640 W maximum power point after a duration of 0.18 s. This particular algorithm exhibits fast MPP holding compared to FSSO in the literature. Nevertheless, this technique undergoes multiple oscillations, as seen in the 0.1 s and 0.175 s instants, as illustrated in Fig. 15.28.

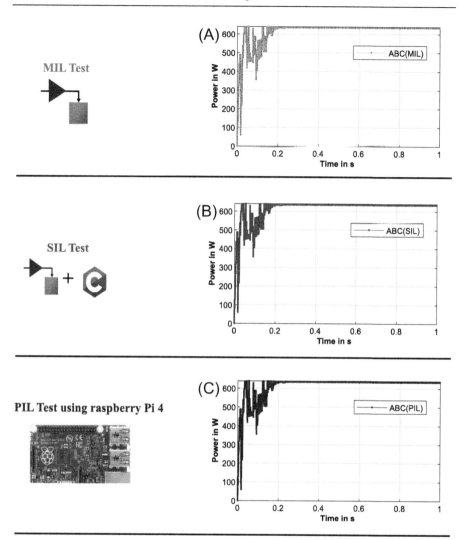

Figure 15.26 PV system power under PSC based on ABC algorithm for MIL/SIL/PIL testing.

Fig. 15.29 displays the results of the SSA algorithm deployment in an MPPT application, where the maximum output power of the PV system has been achieved after only 0.17 s. Furthermore, this method is highly accurate insofar as the algorithm can extract a power of 640 W. Besides, the acquired results do not exhibit any meaningful oscillation.

In light of the thorough deliberation elaborated through the acquired results, PSO and SSA are preferable to all other strategies suggested in this investigation by

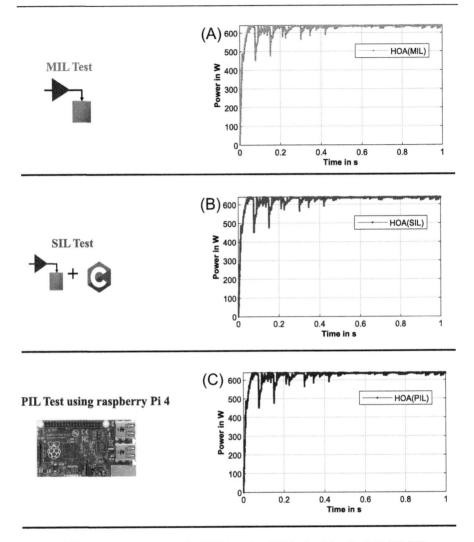

Figure 15.27 PV system power under PSC based on HOA algorithm for MIL/SIL/PIL testing.

assessing through a mixture of optimization factors. Depending on the outcome, we may infer that certain methods are speedy, but have a greater proportion of oscillations, while other approaches are promising in terms of oscillations, but take longer to achieve MPP. Despite the fact that the SSA and PSO approaches have random elements, which could lead to a failure to seek the maximum power point, the adoption of a particular initial operating cycle makes them perform faster and speed up

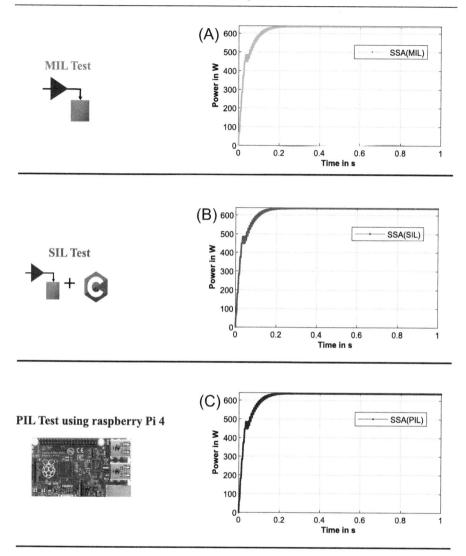

Figure 15.28 PV system power under PSC based on SSA algorithm for MIL/SIL/PIL testing.

the actual MPP tracking. In contrast, HAO and CSA are faster but suffer losses due to substantial oscillations at the beginning of the process.

The ABC strategy may become the preferred alternative if oscillations are anticipated. However, it is also critical to note that the rapidity with which ABC achieves MPP is acceptable. After extensive discussion and simulation, the PSO, SSA, and GWO are congruent with the initial duty cycle distribution and selections.

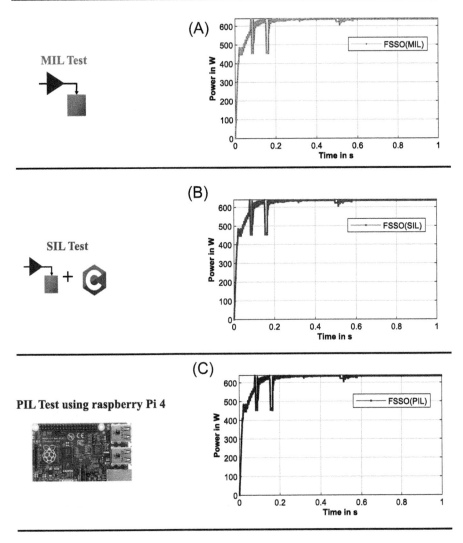

Figure 15.29 PV system power under PSC based on FSSO algorithm for MIL/SIL/PIL testing.

15.6 Conclusion

This paper provides both theoretical and simulation analyses of seven recent and well-recognized techniques. The purpose of this study is to identify the perfect techniques to be incorporated into the list of upcoming research projects. After ascertaining that PSO, SSA, and GWO surpass other approaches in terms of speed, oscillations, and accuracy, these algorithms enable the full potential of solar panels to be exploited. The main conclusions drawn from this investigation are as follows:

- The algorithms, including PSO, SSA, and GWO, are fast and accurate.
- ABC may require the sitting of constants to ensure a perfect algorithm.
- HAO and CSA are extremely fast algorithms, although they are unstable

Consequently, the PSO, SSA, and GWO techniques are more appropriate. Hence, they are worth exploring in future studies, particularly in areas where the influence of dust and urban premises is very significant, as well as in static and dynamic reconfiguration techniques. As a perspective for further research, the partial shading effect on panel efficiency is also enhanced by reconfiguration techniques and intelligent topologies; in this regard, the upcoming work will be based on dynamic reconfiguration using GWO and SSA techniques.

Acknowledgment

This work was supported by the PV INTELLIGENT project through the Institute for Solar Energy Research and New Energies (IRESEN). The authors would like to convey their appreciation to the Green Energy Park research platform for the exceptional working conditions and the seamless access to the required measuring and testing equipment.

References

Abdulkadir, M., Yatim, A. H. M., & Yusuf, S. T. (2014). An improved PSO-based MPPT control strategy for photovoltaic systems. *International Journal of Photoenergy* (2014). Available from https://doi.org/10.1155/2014/818232, http://www.hindawi.com/journals/ijp/contents/.

Abo-Elyousr, F. K., Abdelshafy, A. M., & Abdelaziz, A. Y. (2020). *MPPT-based particle swarm and cuckoo search algorithms for PV systems. Green Energy and Technology* (pp. 379—400). Egypt: Springer Verlag. Available from https://doi.org/10.1007/978-3-030-05578-3_14, http://www.springer.com/series/8059.

Ali, A., Almutairi, K., Padmanaban, S., Tirth, V., Algarni, S., Irshad, K., Islam, S., Zahir, M. H., Shafiullah, M., & Malik, M. Z. (2020). Investigation of MPPT techniques under uniform and non-uniform solar irradiation condition—A retrospection. *IEEE Access, 8*, 127368—127392. Available from https://doi.org/10.1109/ACCESS.2020.3007710, http://ieeexplore.ieee.org/xpl/RecentIssue.jsp?punumber = 6287639.

Alqahtani, A., Alsaffar, M., El-Sayed, M., & Alajmi, B. (2016). Data-driven photovoltaic system modeling based on nonlinear system identification. *International Journal of Photoenergy* (2016). Available from https://doi.org/10.1155/2016/2923731, http://www.hindawi.com/journals/ijp/contents/.

Alshareef, M., Lin, Z., Ma, M., & Cao, W. (2019). Accelerated particle swarm optimization for photovoltaic maximum power point tracking under partial shading conditions. *Energies, 12*(4), 623. Available from https://doi.org/10.3390/en12040623.

Araújo, N. M. F. T. S., Sousa, F. J. P., & Costa, F. B. (2020). Equivalent models for photovoltaic cell—A review. *Revista de Engenharia Térmica, 19*(2), 77. Available from https://doi.org/10.5380/reterm.v19i2.78625.

Crawford, B., Soto, R., Cuesta, R., & Paredes, F. (2014). Application of the artificial bee colony algorithm for solving the set covering problem. *The Scientific World Journal, 2014,* 1−8. Available from https://doi.org/10.1155/2014/189164.

Balato, M., Costanzo, L., & Vitelli, M. (2018). *DMPPT PV System: Modeling and Control Techniques Advances in Renewable Energies and Power Technologies* (pp. 163−205). Italy: Elsevier. Available from https://doi.org/10.1016/B978-0-12-812959-3.00005-8, http://www.sciencedirect.com/science/book/9780128129593.

Balato, M., Liccardo, A., & Petrarca, C. (2020). Dynamic boost based DMPPT emulator. *Energies, 13*(11), 2921. Available from https://doi.org/10.3390/en13112921.

Bellia, H., Youcef, R., & Fatima, M. (2019). A detailed modeling of photovoltaic module using MATLAB. *NRIAG Journal of Astronomy and Geophysics, 3*(1), 53−61. Available from https://doi.org/10.1016/j.nrjag.2014.04.001.

Bendib, B., Belmili, H., & Krim, F. (2015). A survey of the most used MPPT methods: Conventional and advanced algorithms applied for photovoltaic systems. *Renewable and Sustainable Energy Reviews, 45,* 637−648. Available from https://doi.org/10.1016/j.rser.2015.02.009.

Boutana, N., Mellit, A., Lughi, V., & Massi Pavan, A. (2017). Assessment of implicit and explicit models for different photovoltaic modules technologies. *Energy, 122,* 128−143. Available from https://doi.org/10.1016/j.energy.2017.01.073, http://www.elsevier.com/inca/publications/store/4/8/3/.

Brockway, P. E., Owen, A., Brand-Correa, L. I., & Hardt, L. (2019). Estimation of global final-stage energy-return-on-investment for fossil fuels with comparison to renewable energy sources. *Nature Energy, 4*(7), 612−621. Available from https://doi.org/10.1038/s41560-019-0425-z, http://www.nature.com/nenergy/.

Chao, K. H., & Wu, M. C. (2016). Global maximum power point tracking (MPPT) of a photovoltaic module array constructed through improved teaching-learning-based optimization. *Energies, 9*(12). Available from https://doi.org/10.3390/en9120986, http://www.mdpi.com/journal/energies/.

Chtita, S., Motahhir, S., El Hammoumi, A., Chouder, A., Benyoucef, A. S., El Ghzizal, A., Derouich, A., Abouhawwash, M., & Askar, S. S. (2022). A novel hybrid GWO−PSO-based maximum power point tracking for photovoltaic systems operating under partial shading conditions. *Scientific Reports, 12*(1). Available from https://doi.org/10.1038/s41598-022-14733-6, http://www.nature.com/srep/index.html.

Dhimish, M. (2019). Assessing MPPT techniques on hot-spotted and partially shaded photovoltaic modules: Comprehensive review based on experimental data. *IEEE Transactions on Electron Devices, 66*(3), 1132−1144. Available from https://doi.org/10.1109/TED.2019.2894009, https://ieeexplore.ieee.org/xpl/mostRecentIssue.jsp?punumber = 16.

Ding, M., Lv, D., Yang, C., Li, S., Fang, Q., Yang, B., & Zhang, X. (2019). Global maximum power point tracking of PV systems under partial shading condition: A transfer reinforcement learning approach. *Applied Sciences, 9*(13), 2769. Available from https://doi.org/10.3390/app9132769.

Eltamaly, A. M. (2021). An improved Cuckoo search algorithm for maximum power point tracking of photovoltaic systems under partial shading conditions. *Energies, 14*(4), 953. Available from https://doi.org/10.3390/en14040953.

Faris, H., Mirjalili, S., Aljarah, I., Mafarja, M., & Heidari, A. A. (2020). *Salp swarm algorithm: Theory, literature review, and application in extreme learning machines Studies in Computational Intelligence* (pp. 185−199). Jordan: Springer Verlag. Available from http://www.springer.com/series/7092, https://doi.org/10.1007/978-3-030-12127-3_11.

Gad, A. G. (2022). Particle swarm optimization algorithm and its applications: A systematic review. *Archives of Computational Methods in Engineering, 29*(5), 2531−2561. Available from https://doi.org/10.1007/s11831-021-09694-4, http://www.springerlink.com/content/1134-3060.

Gonzalez-Castano, C., Restrepo, C., Kouro, S., & Rodriguez, J. (2021). MPPT algorithm based on artificial bee colony for pv system. *IEEE Access, 9*, 43121−43133. Available from https://doi.org/10.1109/ACCESS.2021.3066281, http://ieeexplore.ieee.org/xpl/RecentIssue.jsp?punumber = 6287639.

Hancer, E. (2020). *Artificial Bee Colony: Theory, Literature Review, and Application in Image Segmentation Studies in Computational Intelligence* (pp. 47−67). Turkey: Springer. Available from https://doi.org/10.1007/978-981-15-1362-6_3, http://www.springer.com/series/7092.

Hosseinalipour, A., & Ghanbarzadeh, R. (2022). A novel approach for spam detection using horse herd optimization algorithm. *Neural Computing and Applications, 34*(15), 13091−13105. Available from https://doi.org/10.1007/s00521-022-07148-x, http://link.springer.com/journal/521.

Jamaludin, M. N. I., Tajuddin, M. F. N., Ahmed, J., Azmi, A., Azmi, S. A., Ghazali, N. H., Babu, T. S., & Alhelou, H. H. (2021). An effective salp swarm based MPPT for photovoltaic systems under dynamic and partial shading conditions. *IEEE Access, 9*, 34570−34589. Available from https://doi.org/10.1109/ACCESS.2021.3060431, http://ieeexplore.ieee.org/xpl/RecentIssue.jsp?punumber = 6287639.

Jamaludin, M.N.I., Tajuddin, M.F.N.B., Ahmed, J., Sengodan, T. (2021). 4th International Conference on Electrical, Computer and Communication Technologies, ICECCT 2021, https://doi.org/10.1109/ICECCT52121.2021.9616622, 9781665414807, Institute of Electrical and Electronics Engineers Inc. Malaysia Hybrid Bio-Intelligence Salp Swarm Algorithm for Maximum Power Point Tracking (MPPT) of Photovoltaic Systems under Gradual Change in Irradiance Conditions, http://ieeexplore.ieee.org/xpl/mostRecentIssue.jsp?punumber = 9616617.

Jantsch, M., Real, M., Häberlin, H., Whitaker, C., Kurokawa, K., Blässer, G., Kremer, P., Verhoeve, C. W. (n.d.). Measurement, pv, power, tracking performance.

Jiang, L. L., Maskell, D. L., & Patra, J. C. (2013). A novel ant colony optimization-based maximum power point tracking for photovoltaic systems under partially shaded conditions. *Energy and Buildings, 58*, 227−236. Available from https://doi.org/10.1016/j.enbuild.2012.12.001.

Karmouni, H., Chouiekh, M., Motahhir, S., Dagal, I., Oufettoul, H., Qjidaa, H., Sayyouri, M. (2022). 1 2022/01 11th IEEE International Conference on Renewable Energy Research and Applications, ICRERA 2022, https://doi.org/10.1109/ICRERA55966.2022.9922834, 9781665471404, 446−451, Institute of Electrical and Electronics Engineers Inc. Morocco A Novel MPPT Algorithm based on Aquila Optimizer under PSC and Implementation using Raspberry http://ieeexplore.ieee.org/xpl/mostRecent Issue.jsp?punumber = 9921418.

Karmouni, H., Chouiekh, M., Motahhir, S., Qjidaa, H., Ouazzani Jamil, M., & Sayyouri, M. (2022). Optimization and implementation of a photovoltaic pumping system using the sine−cosine algorithm. *Engineering Applications of Artificial Intelligence, 114*, 105104. Available from https://doi.org/10.1016/j.engappai.2022.105104.

Kennedy, J., Eberhart, R. (1995). 12 1995/12 IEEE International Conference on Neural Networks - Conference Proceedings 1942−1948 IEEE undefined Particle swarm optimization 4.

Koad, R. B. A., Zobaa, A. F., & El-Shahat, A. (2017). A novel MPPT algorithm based on particle swarm optimization for photovoltaic systems. *IEEE Transactions on Sustainable Energy*, *8*(2), 468−476. Available from https://doi.org/10.1109/TSTE.2016.2606421.

Kumar, A., Kumar, D., & Jarial, S. K. (2017). A review on artificial bee colony algorithms and their applications to data clustering. *Cybernetics and Information Technologies*, *17*(3), 3−28. Available from https://doi.org/10.1515/cait-2017-0027, http://www.cit.iit.bas.bg/CIT_2017/v-17-3/01_paper.pdf.

Lu, Y., Wang, S., & Shan, K. (2015). Design optimization and optimal control of grid-connected and standalone nearly/net zero energy buildings. *Applied Energy*, *155*, 463−477. Available from https://doi.org/10.1016/j.apenergy.2015.06.007, https://www.journals.elsevier.com/applied-energy.

Mao, M., Zhang, L., Yang, L., Chong, B., Huang, H., & Zhou, L. (2020). MPPT using modified salp swarm algorithm for multiple bidirectional PV-Ćuk converter system under partial shading and module mismatching. *Solar Energy*, *209*, 334−349. Available from https://doi.org/10.1016/j.solener.2020.08.078, http://www.elsevier.com/inca/publications/store/3/2/9/index.htt.

Ma, W., Ma, M., Wang, H., Zhang, Z., Zhang, R., & Wang, J. (2021). Shading fault detection method for household photovoltaic power stations based on inherent characteristics of monthly string current data mapping. *CSEE Journal of Power and Energy Systems*. Available from http://doi.org/10.17775/CSEEJPES.2021.09520.

MiarNaeimi, F., Azizyan, G., & Rashki, M. (2021). Horse herd optimization algorithm: A nature-inspired algorithm for high-dimensional optimization problems. *Knowledge-Based Systems* (213). Available from https://doi.org/10.1016/j.knosys.2020.106711, https://www.journals.elsevier.com/knowledge-based-systems.

Miarnaeimi, F., Azizyan, G., Shabakhty, N., Rashki, M. (2019). Flying Squirrel Optimizer (FSO): A novel SI-based optimization algorithm for engineering problems. 11.

Mirjalili, S., Gandomi, A. H., Mirjalili, S. Z., Saremi, S., Faris, H., & Mirjalili, S. M. (2017). Salp swarm algorithm: A bio-inspired optimizer for engineering design problems. *Advances in Engineering Software*, *114*, 163−191. Available from https://doi.org/10.1016/j.advengsoft.2017.07.002, http://www.journals.elsevier.com/advances-in-engineering-software/.

Mirjalili, S., Mirjalili, S. M., & Lewis, A. (2014). Grey Wolf optimizer. *Advances in Engineering Software*, *69*, 46−61. Available from https://doi.org/10.1016/j.advengsoft.2013.12.007, http://www.journals.elsevier.com/advances-in-engineering-software/.

Mohanty, S., Subudhi, B., & Ray, P. K. (2016). A new MPPT design using grey Wolf optimization technique for photovoltaic system under partial shading conditions. *IEEE Transactions on Sustainable Energy*, *7*(1), 181−188. Available from https://doi.org/10.1109/TSTE.2015.2482120.

Mohanty, S., Subudhi, B., Ray, P. K., & Grey Wolf-Assisted, A. (2017). Perturb & observe MPPT algorithm for a PV system. *IEEE Transactions on Energy Conversion*, *32*(1), 340−347. Available from https://doi.org/10.1109/TEC.2016.2633722.

Moriarty, P., & Honnery, D. (2016). Can renewable energy power the future. *Energy Policy*, *93*, 3−7. Available from https://doi.org/10.1016/j.enpol.2016.02.051, http://www.journals.elsevier.com/energy-policy/.

Motahhir, S., Chalh, A., Ghzizal, A. E., Sebti, El. G., & Derouich, A. (2017). Modeling of photovoltaic panel by using proteus. *Journal of Engineering Science and Technology Review*, *10*(2), 8−13. Available from https://doi.org/10.25103/jestr.102.02.

Motahhir, S., El Hammoumi, A., & Ghzizal, A. E. (2020). The most used MPPT algorithms: Review and the suitable low-cost embedded board for each algorithm. *Journal of Cleaner Production*, *246*, 118983. Available from https://doi.org/10.1016/j.jclepro.2019.118983.

Nadimi-Shahraki, M. H., Taghian, S., & Mirjalili, S. (2021). An improved grey wolf optimizer for solving engineering problems. *Expert Systems with Applications*, *166*, 113917. Available from https://doi.org/10.1016/j.eswa.2020.113917.

Oufettoul, H., Aniba, G., Motahhir, S. (2021). Proceedings of 2021 9th International Renewable and Sustainable Energy Conference, IRSEC 2021, https://doi.org/10.1109/IRSEC53969.2021.9741122, 9781665413190, Institute of Electrical and Electronics Engineers Inc. Morocco MPPT Techniques Investigation in Photovoltaic System http://ieeexplore.ieee.org/xpl/mostRecentIssue.jsp?punumber = 9740722.

Oufettoul, H., Lamdihine, N., Motahhir, S., Lamrini, N., Ait Abdelmoula, I., & Aniba, G. (2023). Comparative performance analysis of PV module positions in a solar PV array under partial shading conditions. *IEEE Access*, *11*, 12176−12194. Available from https://doi.org/10.1109/access.2023.3237250.

Oufettoul, H., Motahhir, S., Aniba, G., Abdelmoula, I. A. (2022). 11th IEEE International Conference on Renewable Energy Research and Applications, ICRERA 2022, https://doi.org/10.1109/ICRERA55966.2022.9935687, 9781665471404, 352−359, Institute of Electrical and Electronics Engineers Inc. Morocco Comprehensive Analysis of MPPT Control Approaches under Partial Shading Condition http://ieeexplore.ieee.org/xpl/mostRecentIssue.jsp?punumber = 9921418 2022-.

Oufettoul, H., Motahhir, S., Aniba, G., Ait Abdelmoula, I. (2022). 11th IEEE International Conference on Renewable Energy Research and Applications, ICRERA 2022, https://doi.org/10.1109/ICRERA55966.2022.9935686, 9781665471404, 360−368, Institute of Electrical and Electronics Engineers Inc. Morocco Sensor Placement Strategy for Locating Photovoltaic Array Failures. http://ieeexplore.ieee.org/xpl/mostRecentIssue.jsp?punumber = 9921418.

Oufettoul, H., Motahhir, S., Aniba, G., Masud, M., & AlZain, M. A. (2022). Improved TCT topology for shaded photovoltaic arrays. *Energy Reports*, *8*, 5943−5956. Available from https://doi.org/10.1016/j.egyr.2022.04.042, http://www.journals.elsevier.com/energy-reports/.

Pachauri, R. K., Motahhir, S., Gupta, A. K., Sharma, M., Minai, A. F., Hossain, M. S., & Yassine, A. (2022). Game theory based strategy to reconfigure PV module arrangements for achieving higher GMPP under PSCs: Experimental feasibility. *Energy Reports*, *8*, 10088−10112. Available from https://doi.org/10.1016/j.egyr.2022.08.006, http://www.journals.elsevier.com/energy-reports/.

Prasanth Ram, J., & Rajasekar, N. (2017). A novel flower pollination based global maximum power point method for solar maximum power point tracking. *IEEE Transactions on Power Electronics*, *32*(11), 8486−8499. Available from https://doi.org/10.1109/TPEL.2016.2645449, http://ieeexplore.ieee.org/xpl/tocresult.jsp?isnumber = 4712525.

Rasheed, M., Shihab, S., Mohammed, O. Y., & Al-Adili, Aqeel (2021). Parameters estimation of photovoltaic model using nonlinear algorithms. *Journal of Physics: Conference Series*, *1795*(1). Available from https://doi.org/10.1088/1742-6596/1795/1/012058, 012058.

Sarvi, M., Ahmadi, S., & Abdi, S. (2015). A PSO-based maximum power point tracking for photovoltaic systems under environmental and partially shaded conditions. *Progress in Photovoltaics: Research and Applications*, *23*(2), 201−214. Available from https://doi.org/10.1002/pip.2416, http://onlinelibrary.wiley.com/journal/10.1002/(ISSN)1099-159X.

Sarwar, S., Hafeez, M. A., Javed, M. Y., Asghar, A. B., & Ejsmont, K. (2022). A horse herd optimization algorithm (HOA)-based MPPT technique under partial and complex partial shading conditions. *Energies*, *15*(5). Available from https://doi.org/10.3390/en15051880, https://www.mdpi.com/1996-1073/15/5/1880/pdf.

Singh, N., Gupta, K. K., Jain, S. K., Dewangan, N. K., & Bhatnagar, P. (2021). A flying squirrel search optimization for MPPT under partial shaded photovoltaic system. *IEEE Journal of Emerging and Selected Topics in Power Electronics*, *9*(4), 4963−4978. Available from https://doi.org/10.1109/JESTPE.2020.3024719, http://ieeexplore.ieee. org/xpl/RecentIssue.jsp?punumber = 6245517.

Soufyane Benyoucef, A., Chouder, A., Kara, K., Silvestre, S., & Ait Sahed, O. (2015). Artificial bee colony based algorithm for maximum power point tracking (MPPT) for PV systems operating under partial shaded conditions. *Applied Soft Computing*, *32*, 38−48. Available from https://doi.org/10.1016/j.asoc.2015.03.047.

Tey, K. S., Mekhilef, S., Yang, H. T., & Chuang, M. K. (2014). A differential evolution based MPPT method for photovoltaic modules under partial shading conditions. *International Journal of Photoenergy* (2014). Available from https://doi.org/10.1155/ 2014/945906, http://www.hindawi.com/journals/ijp/contents/.

Vieira, Romênia, Araújo, F. ábio de, Dhimish, Mahmoud, & Guerra, Maria (2020). A comprehensive review on bypass diode application on photovoltaic modules. *Energies*, *13*(10), 2472. Available from https://doi.org/10.3390/en13102472.

Wu, Z., Hu, Y., Wen, J. X., Zhou, F., & Ye, X. (2020). A review for solar panel fire accident prevention in large-scale PV applications. *IEEE Access*, *8*, 132466−132480. Available from https://doi.org/10.1109/ACCESS.2020.3010212, http://ieeexplore.ieee.org/xpl/Recent Issue.jsp?punumber = 6287639.

Yadav, V. K., Yadav, A., Yadav, R., Mittal, A., Wazir, N. H., Gupta, S., Pachauri, R. K., & Ghosh, S. (2022). A novel reconfiguration technique for improvement of PV reliability. *Renewable Energy*, *182*, 508−520. Available from https://doi.org/10.1016/j.renene.2021. 10.043, http://www.journals.elsevier.com/renewable-and-sustainable-energy-reviews/.

Yang, X.S., Deb, S. (2009). 12 2009/12 2009 World Congress on Nature and Biologically Inspired Computing, NABIC 2009—Proceedings. https://doi.org/10.1109/NABIC.2009. 5393690. 210−214. United Kingdom Cuckoo search via Lévy flights.

Yang, S. (2010). Nature-inspired Metaheuristic Algorithms.

Yaqoob, S. J., Saleh, A. L., Motahhir, S., Agyekum, E. B., Nayyar, A., & Qureshi, B. (2021). Comparative study with practical validation of photovoltaic monocrystalline module for single and double diode models. *Scientific Reports*, *11*(1). Available from https://doi. org/10.1038/s41598-021-98593-6, http://www.nature.com/srep/index.html.

Implementation of variable-step P&O MPPT control for PV systems based on dSPACE: processor-in-the-loop test

16

Mohamed Said Adouairi[1], Saad Motahhir[2] and Badre Bossoufi[1]
[1]Laboratoire LIMAS, Faculté des Sciences Dhar El Mahraz, Université Sidi Mohammed Ben Abdellah, Fès, Morocco, [2]ENSA, SMBA University, Fez, Morocco

16.1 Introduction

The world's energy consumption continues to grow, raising crucial questions about the problem of global warming due to greenhouse gases on the one hand and the depletion of fossil fuels on the other. As a result of this awareness, environmentally friendly economic development is absolutely necessary. In order to make the supply of electricity more environmentally friendly, PV solar energy is derived from the direct conversion of part of the sun's radiation into electrical energy. The voltage generated can vary depending on the material used to manufacture the cell.

The internal characteristics of PV cells are not very linear due to rapid variations in sunlight and weather conditions. Therefore, MPPT algorithms are used to optimize the PV production efficiency and extract the maximum power from the PV panels regardless of the weather conditions. There are several research works on these MPPT techniques (Motahhir et al., 2020; Adouairi et al., 2023). In the perturbation and observation (P&O) MPPT algorithm based on a fixed step, the disadvantages are that the power extracted from the PV array with a larger step contributes to faster dynamics but excessive steady-state oscillations when a smaller step is required and the steady-state output cannot be achieved (Bendib et al., 2015). The advantages of the method (P&O) are ease of implementation and simplicity of the control scheme compared to other techniques (Bendib et al., 2015).

In order to extract the maximum possible power from the PV module, we used a variable step P&O algorithm to overcome the sensitivity limits of MPP tracking and to control the buck converter. The simulation of our system is done by the processor-in-the-loop (PIL) test, which is then performed, The purpose of this test is to verify the efficiency of our command before using it in the real system to avoid redundant testing activities after each improvement (Motahhir et al., 2017). This step is performed after the TargetLink is configured on MATLAB®/Simulink, as described in detail in Section 16.6. After that, the C code is generated and executed on the target hardware (dSPACE ds1104), which acts as a controller, while the installation is still running in the MATLAB/Simulink environment (on the computer), as shown in Fig. 16.1.

Performance Enhancement and Control of Photovoltaic Systems. DOI: https://doi.org/10.1016/B978-0-443-13392-3.00017-7

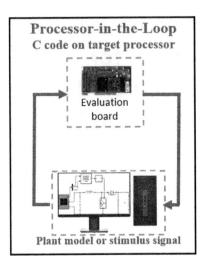

Figure 16.1 Processor-in-the-loop (PIL) closed loop.

16.2   Mathematical model of a PV module

The performance of the solar cell is normally evaluated under standard test conditions, where there is an average solar spectrum, illuminance is normalized to 1000 W/m^2, and the cell temperature is defined as 25°C. In the literature, several models have been identified (Townsend, 1989); for our work, we have chosen the five-parameter model (I_{ph}, I_0, n, R_s, and R_p), represented in Fig. 16.2.

This equivalent diagram is constituted by the following parameters:

I_{ph}: The photonic current
R_s: The series resistance
R_p: The shunt resistance
D: a diode characterizing the junction

According to Fig. 16.1, the mathematical equation (current−voltage) of the equivalent electric circuit of the PV cell is written as follows (Motahhir et al., 2018):

$$I = I_{ph} - I_d - \frac{V + R_s I}{R_p} \tag{16.1}$$

$$I = I_{ph} - I_d - \left(\left[\exp \frac{q}{mkT_c} (V + IR_s) \right] - 1 \right) - \frac{V + R_s I}{R_p} \tag{16.2}$$

The panel used in this work is KCP 12060 panel, and as presented in Table 16.1, the datasheet of the PV panel provides only some characteristics of the PV panel.

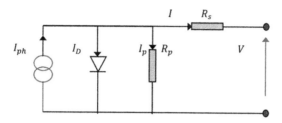

Figure 16.2 Electrical equivalent diagram of a photovoltaic cell.

Table 16.1 Characteristics of KCP 12060 panel.

PV panel parameters	Values
Maximum power, P_{mp}	59.5 W
Maximum power voltage, V_{mp}	17 V
Maximum power current, I_{mp}	3.50 A
Short circuit current, I_{sc}	4.03 A
Open circuit voltage, V_{oc}	21.20 V

16.3 Buck converter

A DC−DC converter is a power electronics device that performs DC−DC conversion. Generally interposed between a generator and a receiver, both with direct current, it makes it possible to regulate the voltage applied to the receiver or the current that circulates in it. Within the framework of our work, we used a buck converter as shown in Fig. 16.3.

The buck converter controls the flow:

- a voltage generator whose test is always positive.
- a current receiver whose current cannot become negative.

This allows to replace the switch K with a transistor because the current is always positive, and moreover, the switches must be controlled at the blocking and at the ignition.

Three operating phases are possible:

T on and D off.
T off and D on
T and D blocked.

16.3.1 Transfer function in continuous operation mode

The transfer function of the output voltage as a function of the duty cycle is as follows:

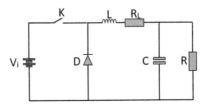

Figure 16.3 Buck converter.

$$\frac{v_s}{\alpha} = v_i \frac{1}{1 + \frac{L}{R}P + LCP^2} \tag{16.3}$$

The transfer function of the output voltage as a function of the input voltage is as follows:

$$\frac{v_s}{v_i} = \alpha \frac{1}{1 + \frac{L}{R}P + LCP^2} \tag{16.4}$$

The transfer function of the output voltage as a function of the inductance current is as follows:

$$\frac{v_s}{i_L} = R \frac{1}{1 + RCP} \tag{16.5}$$

16.3.2   Buck converter state model

Based on the application of Kirchhoff's laws to the operating regime (switch condition) and taking into account the resistance of the inductance, we obtain the mathematical model of the Buck converter as follows:

$$L\frac{di_L}{dt} = v_i - v_s - ri_L \Leftrightarrow \frac{di_L}{dt} = -\frac{r}{L}i_L - \frac{v_s}{L} + \frac{v_i}{L} \tag{16.6}$$

$$C\frac{dv_s}{dt} = i_c \Leftrightarrow C\frac{dv_s}{dt} = i_L - i_s \Leftrightarrow \frac{dv_s}{dt} = \frac{i_L}{C} - \frac{v_s}{RC} \tag{16.7}$$

$$\begin{cases} \dfrac{di_L}{dt} = -\dfrac{r}{L}i_L - \dfrac{v_s}{L} + \dfrac{v_i}{L} \\[2mm] \dfrac{dv_s}{dt} = \dfrac{i_L}{C} - \dfrac{v_s}{RC} \end{cases}$$

With $X_1 = i_L$ and $X_2 = v_s$, we have $\dot{X}_1 = \frac{di_L}{dt}$ and $\dot{X}_2 = \frac{dv_s}{dt}$

Hence the state model:

$$\begin{bmatrix} \dot{X}_1 \\ \dot{X}_2 \end{bmatrix} = \begin{bmatrix} -\dfrac{r}{L} & -\dfrac{1}{L} \\ \dfrac{1}{C} & -\dfrac{1}{RC} \end{bmatrix} \begin{bmatrix} X_1 \\ X_2 \end{bmatrix} + \begin{bmatrix} \dfrac{1}{L} \\ 0 \end{bmatrix} v_i \qquad (16.8)$$

with $A = \begin{bmatrix} -\dfrac{r}{L} & -\dfrac{1}{L} \\ \dfrac{1}{C} & -\dfrac{1}{RC} \end{bmatrix}$ and $B = \begin{bmatrix} \dfrac{1}{L} \\ 0 \end{bmatrix}$ The buck converter parameter values are

as shown in Table 16.2.

16.4 Maximum power point tracking (MPPT)

In variable weather conditions, the operating point of the PV array and load, which is the intersection between the current–voltage characteristic and the load line, may deviate from the maximum power point. This maximum power depends largely on the insulation, the load impedance, and the cell temperature. The use of DC–DC converters with tracking algorithms is a solution used to allow systems to operate at the maximum power point (MPP) despite weather-related disturbances; in the literature, this control is called maximum power point tracking (MPPT) (Reza Reisi et al., 2013).

The MPPT control is connected to a DC–DC converter that allows the adaptation between the PV panels and the load so that the generated power corresponds to its maximum value (Zhang et al., 2016). This control technique acts on the duty cycle to bring the operating point of the generator to its optimum value, regardless of whether there are variations or sudden variations in the load. Fig. 16.4 shows an elementary PV conversion chain associated with an MPPT control.

16.4.1 Principle of "Perturb and Observe" (P&O) commands

It is a method of maximum power tracking in an iterative way, i.e., under the effect of the disturbance, voltage and current sampling are carried out via sensors to determine the power. The variation of this last one is evaluated compared to that of the voltage with the aim of an eventual decision in order to be around the point of maximum power and that each time the point of operation moves to the right or to the left of the maximum power (Bendib et al., 2015).

Table 16.2 Data specification of buck converter.

V_i	V_s	f_s	L	R_L	C	R
38 V	15 V	100 kHz	0.04 H	0.001 Ω	6.25 μF	10 Ω

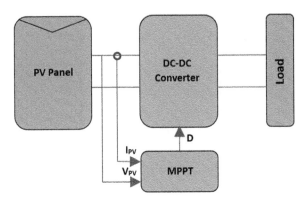

Figure 16.4 Photovoltaic conversion chain with a static converter controlled by MPPT.

As we can see in Fig. 16.5, when Dp and Dv are positive, we move away from the point of maximum power to the left, so the control must bring the voltage back to its operating value by making increments.

16.4.2 P&O-based variable-step MPPT

The perturbation and observation algorithm is simple and easy to implement compared to other algorithms. In this work, the converter control realized by the variable-step P&O method is shown in Fig. 16.6. This algorithm is based on increasing or decreasing the step ΔD of duty cycle. In the flowchart in Fig. 16.6, if $dP/dV < 0$, the duty cycle is increased, and if $dP/dV > 0$, the duty cycle is decreased. Thus, the increase and decrease in duty cycle are based on the step size ΔD. If the fixed step size of the chosen perturbation ΔD is small, the convergence rate of the system to MPP is slow and thus the desired level of duty cycle is not achieved. If ΔD is large, the system oscillates around the MPP.

The incremental step of the duty cycle is determined by Sivaraman and Nirmalkumar (2015) and is represented by the following equation:

$$Stp = \frac{N}{I} \left| \frac{dP}{dV} \right| \tag{16.9}$$

For the increment of the duty cycle in the variable-pitch P&O, we used two different increments of the duty cycle to adopt by measuring the variation of current.

Thus, the current increases when the irradiation level increases; therefore the step increases. In Eqs. (16.10 and 16.11), which present these two cases of proposed step, two coefficients C1 and C2 are used, and Fig. 16.5 shows the flowchart of the P&O with variable step.

$$Stp_1 = \frac{N}{I} \left| \frac{dP}{dV} \right| C2 \tag{16.10}$$

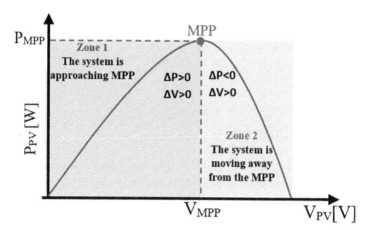

Figure 16.5 Curve of variation of power in relation to voltage.

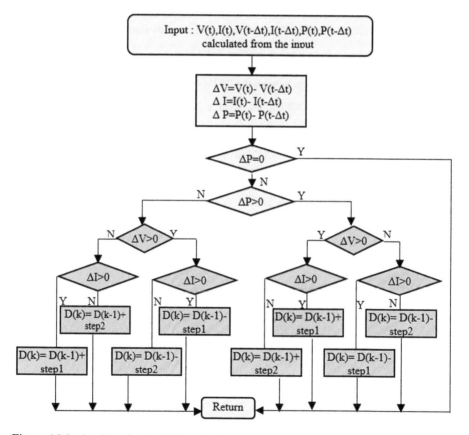

Figure 16.6 Algorithm for the P&O method.

$$Stp_2 = \frac{N}{I} \left| \frac{dP}{dV} \right| C1 \qquad\qquad (16.11)$$

16.5   dSPACE platform

The dSPACE DS1104 board is a real-time processor board used for the development and simulation of control systems in automotive, aerospace, power electronics, and other industries.

Real-time control of continuous systems is done using a PC connected to the dSPACE DS1104 board. Programming is done using the Simulink modeling tool, which helps to pose the problem graphically using interconnected blocks. In fact, many DSP-based real-time development systems now come with an interface to Simulink by which they can convert the Simulink blocks into machine code that can be run on a DSP-based system. This greatly reduces the development and prototyping time for system control. The prototyping then goes through three stages:

- Construction of the control system using Simulink blocks.
- Simulation of the system to see the results in different scenarios.
- Running the model in real time through the DS1104 board.

The main processor is an MPC8240, with a 250-MHz clock. The characteristics of the board are given in Table 16.3.

The DS1104 (Master PPC) board has eight ADCs with input voltage between -10 V and $+10$ V and eight DACs with output voltage between -10 V and $+10$ V. It also has several interfaces including digital inputs and outputs, incremental encoders, etc. (see Fig. 16.7). The DS1104 board also has a slave DSP, the TMS320F240 DSP, which will be used to generate the PWM signals.

The PPC master controls two types of analog to digital converters:

- A multiplexed A/D converter (ADC1) for signals (ADCH1 to ADCH4):
  - 16-bit resolution
  - Voltage range ± 10 V
  - Offset error margin ± 5 mV
  - Margin of error on gain $\pm 0.25\%$ signal-to-noise ratio
  - Signal-to-noise ratio SNR >80 dB (at 10 kHz)
- Four A/D converters (ADC2 to ADC5) for the signals (ADCH5 to ADCH8). The characteristics of these converters are as follows:
  - 12-bit resolution
  - Input voltage range ± 10 V
  - Offset error margin ± 5 mV
  - Margin of error on gain $\pm 0.5\%$
  - Signal to noise ratio >70 dB

Table 16.3 Main features of DS1104.

Parameter		Specification
Processor		• MPC8240 processor with PPC 603e core and on-chip peripherals • 64-bit floating-point processor • CPU clock: 250 MHz • 2 × 16 kB cache, on-chip • On-chip PCI bridge (33 MHz)
Memory Timer	Global memoryFlash memory Four general-purpose timers One sampling rate timer (decremental) One time base counter	• 32 MB SDRAM • 8 MB • 32-bit down counter • Reload by hardware • 80-ns resolution • 32-bit down counter • Reload by software • 40-ns resolution • 64-bit up counter • 40-ns resolution

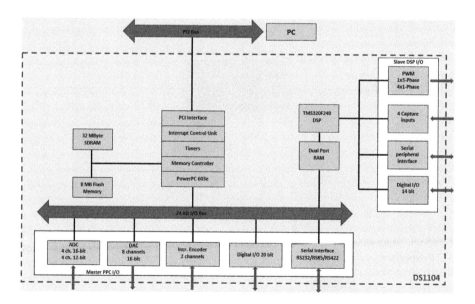

Figure 16.7 Architecture of the DS1104.

The card can be installed in almost any PC with a free 5-V PCI slot. Thanks to the TargetLink function, you can easily run several tests on your Simulink model and the DS1104 controller board at the same time. Among these tests is the PIL test, which is more easily configured in the Simulink model window thanks to this function.

16.6   The processor-in-the-loop test

The processor-in-the-loop (PIL) test can be applied in many applications such as wind turbines, chemical reactors, robots, drones, and many others. This testing technique is primarily used to verify the system's ability to handle potential problems. It consists of predicting that the simulated controllers behave in the same way as in a real installation. In addition, this test can be repeated many times to make potential corrections until the problems are resolved, which reduces the risk, time, and cost of testing. The test is shown in Fig. 16.8; the PIL block is generated and connected to the installation model in order to acquire the PV voltage and current, after which the PIL block will calculate the duty cycle in order to control the buck in the MATLAB/Simulink environment; the connection is made through the PCI bus.

PIL testing is one of the most powerful tools used to test and design embedded systems with a low-cost option. The configuration steps are presented as follows:

- Prepare a subsystem preparation process:

  To prepare a Simulink system for automatic code generation with TargetLink, we first **select the subsystem for which we want to generate the product code, as shown in** Fig. 16.9. This subsystem must be initializable and must contain only Simulink blocks that can be improved and/or supported.

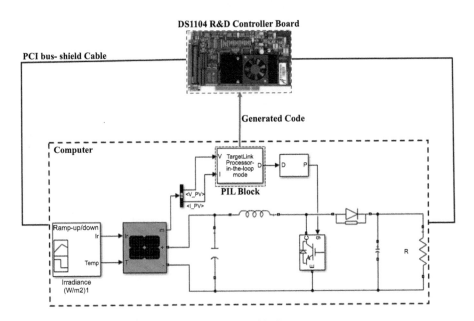

Figure 16.8 The PV generation system using PIL block.

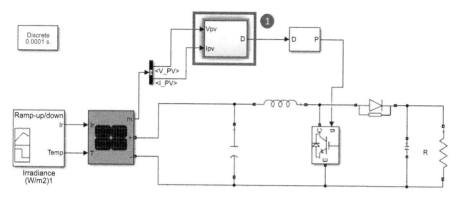

Figure 16.9 The system in the Simulink environment.

By typing **"tl_prepare_system" in the MATLAB command window, the system preparation dialog box opens to allow you to set other preparation options. Leave all the options unchanged, and for the selected "subsystem," click on prepare system as shown in** Fig. 16.10.

The system preparation starts and messages about the preparation process are displayed in the MATLAB command window. When the preparation is finished, the TargetLink message browser opens.

The MPPT subsystem is now a TargetLink subsystem (Fig. 16.11).

In the final step, to make the TargetLink model valid, add the flowing blocks from the **TargetLink block library to the root level of the model: a target link main dialog block, a MIL handler block, and a PIL block** (Fig. 16.12).

- Simulation configuration

In PIL simulation, the generated code of the controller model is compiled with the target compiler and simulated on an evaluation board equipped with the same target processor as the final ECU.

After configuring the TargetLink model, it is very easy to perform the PIL simulation now: First, make sure that the evaluation board is connected to the PC and that in the TargetLink main dialog the appropriate combination of target compiler and evaluation board is selected (Fig. 16.13).

If you are also interested in information on the execution time and/or stack consumption of the generated code, select the corresponding checkbox (Fig. 16.14).

- PIL simulation

Double-click the PIL mode button in the model to generate the production code and compile it for the PIL simulation (Fig. 16.15).

TargetLink automatically handles the download and communication process between the PC and the evaluation board, so no further user interaction is required. The TargetLink subsystem is now in a PIL simulation mode (Fig. 16.16).

This means that the generated code of the controller model is simulated on the target processor.

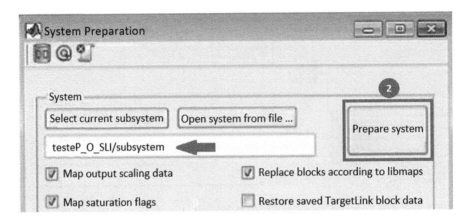

Figure 16.10 The system preparation dialog box.

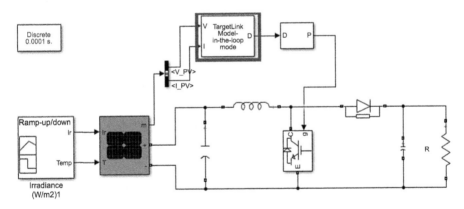

Figure 16.11 MPPT TargetLink subsystem.

As a result, on the next simulation, the TargetLink plot overview window is displayed (Fig. 16.17), showing the simulation results of our PIL block. The two plots above the simulation result of the additional subpoints show the execution time and stack consumption if you have previously selected the corresponding checkboxes.

- Code size information

TargetLink also provides a code summary, listing the size of each generated C code file as well as the RAM and ROM consumption of the generated code (Fig. 16.18).

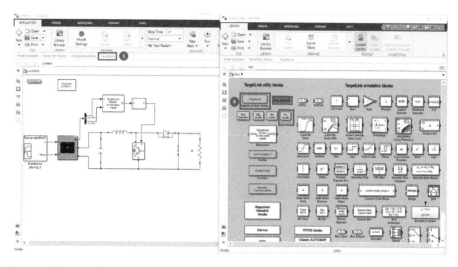

Figure 16.12 The TargetLink block library.

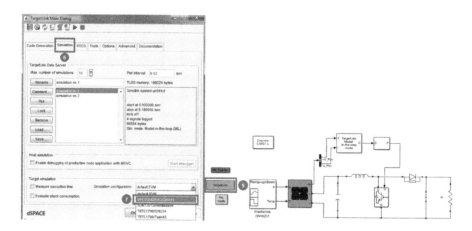

Figure 16.13 TargetLink main dialog.

16.7 Simulation and experimental results

We simulated our system that consists of a buck converter driven by two MPPT controllers, a PV panel and a 10 Ω load in MATLAB/Simulink software by the PIL test on the use of the dSPACE DS1104 board. Fig. 16.19 shows the different values

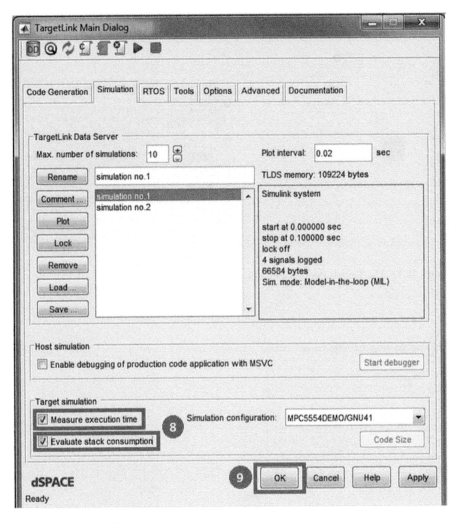

Figure 16.14 Target simulation setup.

of solar irradiation, current, and power output of the PV module KCP 12060 used in our work.

We presented in Figs. 16.20 and 16.21 the system output power extracted by the fixed pitch P&O control and the power extracted by the variable pitch P&O control compared to the PV power. We used two types of pitch: a large pitch $\Delta D_{\max} = 0.08$ and the small pitch $\Delta D_{\min} = 0.002$; the fixed-pitch P&O shows high oscillations; on the other hand, the variable-pitch P&O's oscillations around the MPP are low.

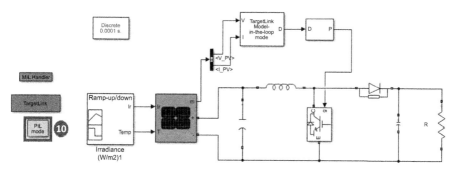

Figure 16.15 PIL mode button.

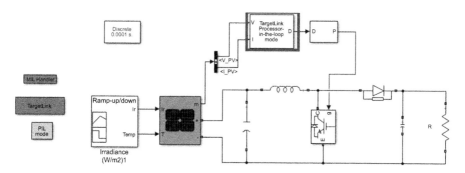

Figure 16.16 The TargetLink subsystem in mode processor in the loop.

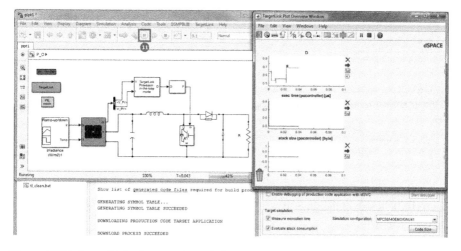

Figure 16.17 TargetLink plot overview window.

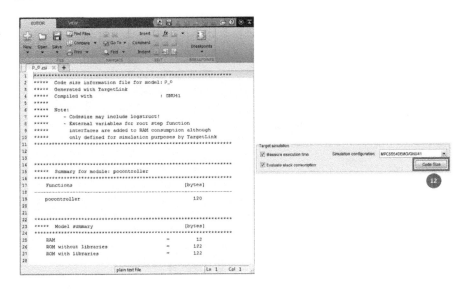

Figure 16.18 Code size information.

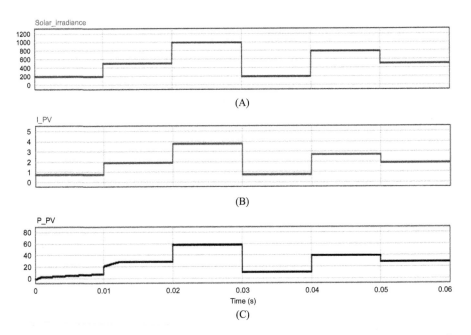

Figure 16.19 Result of the simulation: (A) different irradiation values, (B) output current of PV. (C) PV power.

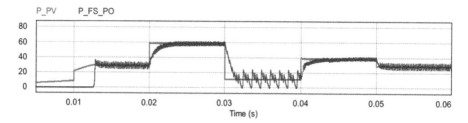

Figure 16.20 PV output power compared to the power extracted by the fixed-step P&O control

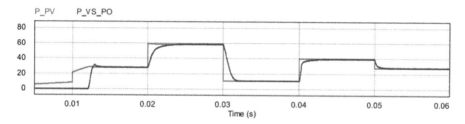

Figure 16.21 PV output power compared to the power extracted by the variable-step P&O control.

16.8 Conclusion

This paper presents the performance of the variable-step P&O algorithm for the PV system. In general, the fixed-step P&O algorithm is widely used, but we notice that the control performed the maximum power point tracking by varying the duty cycle. However, we observe significant oscillations during the creation of the maximum power point. On the other hand, the oscillations are very low during the creation of the maximum power point due to the variable-step P&O control. The validation of the variable-step P&O control is done by the PIL test based on the dSPACE DS1104, and we have presented the different steps to perform this test.

References

Adouairi, M. S., Bossoufi, B., Motahhir, S., & Saady, I. (2023). Application of fuzzy sliding mode control on a single-stage grid-connected PV system based on the voltage-oriented control strategy. *Results in Engineering, 17*. Available from https://doi.org/10.1016/j.rineng.2022.100822.

Bendib, B., Belmili, H., & Krim, F. (2015). A survey of the most used MPPT methods: Conventional and advanced algorithms applied for photovoltaic systems. *Renewable and Sustainable Energy Reviews, 45*, 637–648. Available from https://doi.org/10.1016/j.rser.2015.02.009.

Motahhir, S., El Ghzizal, A., Sebti, S., Derouich, A., & Meng, W. (2017). MIL and SIL and PIL tests for MPPT algorithm. *Cogent Engineering*, *4*(1). Available from https://doi.org/ 10.1080/23311916.2017.1378475.

Motahhir, S., El Hammoumi, A., & El Ghzizal, A. (2018). Photovoltaic system with quantitative comparative between an improved MPPT and existing INC and P&O methods under fast varying of solar irradiation. *Energy Reports*, *4*, 341−350. Available from https://doi.org/10.1016/j.egyr.2018.04.003.

Motahhir, S., El Hammoumi, A., & El Ghzizal, A. (2020). The most used MPPT algorithms: Review and the suitable low-cost embedded board for each algorithm. *Journal of Cleaner Production*, *246*. Available from https://doi.org/10.1016/j.jclepro.2019.118983.

Reza Reisi, A., Hassan Moradi, M., & Jamasb, S. (2013). Classification and comparison of maximum power point tracking techniques for photovoltaic system: A review. *Renewable and Sustainable Energy Reviews*, *19*, 433−443. Available from https://doi. org/10.1016/j.rser.2012.11.052.

Sivaraman, P., & Nirmalkumar, A. (2015). A new method of maximum power point tracking for maximizing the power generation from an SPV plant. *Journal of Scientific and Industrial Research*, *74*(7), 411−415. Available from: http://nopr.niscair.res.in/bit- stream/123456789/31771/1/JSIR%2074%287%29%20411-415.pdf.

Townsend, T.U. (1989). A method for estimating the long-term performance of direct- coupled photovoltaic systems.

Zhang, N., Sutanto, D., & Muttaqi, K. M. (2016). A review of topologies of three-port DC−DC converters for the integration of renewable energy and energy storage system. *Renewable and Sustainable Energy Reviews*, *56*, 388−401. Available from https://doi. org/10.1016/j.rser.2015.11.079.

Design and development of a low-cost single-axis solar tracking system

17

Mohamed Boujoudar[1,2], Ibtissam Bouarfa[2,3], Massaab El Ydrissi[1,2], Mounir Abraim[1,2], Omaima El Alani[1,2], El Ghali Bennouna[2] and Hicham Ghennioui[1]

[1]Laboratory of Signals, Systems, and Components, Sidi Mohamed Ben Abdellah University, Fez, Morocco, [2]Green Energy Park research platform (IRESEN/UM6P), Ben Guerir, Morocco, [3]Laboratory of Innovative Technologies, Sidi Mohamed Ben Abdellah University, Fez, Morocco

17.1   Introduction

Solar trackers are used in solar energy conversion into electricity or thermal energy using either PV or CSP technologies. The system output depends strongly on the amount of solar energy collected using direct or indirect methods. In recent years, new technologies were developed to increase this latter. Control systems were proposed to optimize the tilt angle in order to collect the most available solar radiation (Masters, 2013; Shi & Chew, 2012). Ideal systems must accurately point the sun's position and compensate for the changes in daily altitude angle, seasonal latitude offset, and azimuth angle. However, a number of limitations are still presented that reduce the amount of energy collected in the tracker's systems (Bentaher et al., 2014; De Castro et al., 2013). For this reason, more focus should be addressed on the tracking axes configuration, moving fixtures, and control systems optimization (Sallaberry et al., 2015; Skouri et al., 2016). Solar trackers could be classified into two groups based on their movement: trackers that rotate around one axis called single-axis trackers (Chang, 2009; Li et al., 2011), and double-axis trackers that rotate around two axes (Arbab et al., 2009; Sun et al., 2017). In addition, other studies have proposed other types of trackers with complicated structural designs, where they are not commercially used as single- or double-axis trackers (Abu-Khader et al., 2008; Sungur, 2009).

Solar trackers are indispensable for CSP technology as they use Direct Normal Irradiation (DNI), while they are quite necessary for PV technology since they use Global Horizontal Irradiation (GHI), and solar panels could be installed at a fixed tilt angle. Single-axis trackers could increase PV production by 12%−20% compared to fixed flat panels (Lazaroiu et al., 2015). For trackers with a double axis, a 20.4% yearly energy gain was achieved compared to inclined solar panels (Ismail et al., 2013), while a yearly energy gain of 43.9% was collected in a similar study

Performance Enhancement and Control of Photovoltaic Systems. DOI: https://doi.org/10.1016/B978-0-443-13392-3.00018-9

(Abdallah, 2004). Several studies were proposed that compare the performance when using single- or double-axis solar trackers. They found an energy gain of 3%−5% for double axis compared to single axis (Koussa et al., 2011). From the literature, several researchers come up with the fact that double-axis trackers are more efficient and cost-effective (Huang et al., 2011; Nsengiyumva et al., 2018). As such, further design and cost optimization, as well as maintenance costs, should be more addressed to increase their maturity.

17.2 Materials and methods

17.2.1 Geographical location

The research has been conducted in the Green Energy Park research platform in Ben Guerir. The map coordinates of the location are $32°14'9.3696''$ north latitude and $−7°57'13.8168''$ west longitude.

17.2.2 Sun tracking algorithm

17.2.2.1 Hour angle

To describe the earth's rotation around its polar axis, the hour angle concept could be used as shown in Fig. 17.1 illustrating the angular distance between the observer's meridian and the meridian whose plane contains the sun. This measurement is made from the south, turning toward the west. It is often expressed in hours, minutes, and seconds, but sometimes in degrees (from 0 to 360 degrees, in the retrograde direction).

$$\omega = 15(TSV - 12) \tag{17.1}$$

where TSV is the local solar time.

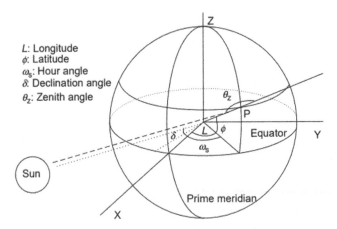

Figure 17.1 Hour angle, declination angle, and latitude angle during the earth's rotation about its polar axis.

17.2.2.2 Declination angle

The declination is the angle between the equatorial plane and the line to the sun. This angle corresponds to the Earth's axial tilt relative to its orbital plane, which is approximately 23.45 degrees. It oscillates with the period of 1 year. The declination angle varies from 23.45 to −23.45 degrees on June 21st and December 22nd, respectively, and 0 degree represents either fall or spring. The declination angle can be mathematically expressed as follows:

$$\delta = 23.5' \sin\left(\frac{360}{365}(D - 80)\right) \tag{17.2}$$

where D is the number of days.

17.2.2.3 Latitude angle (∅)

It is an angular measurement. It characterizes the angle that the plane of the equator makes with the direction that links the earth's center to the point considered. The attitude of the earth's equator is therefore equal to 0 degree.

17.2.2.4 Sunrise (SRT), sunset time (SST), and solar noon (SN)

Predicting the time and location of sunrise and sunset during the day is necessary for tracking algorithms. Sunrise (SRT) and sunset (SST) are defined as the times when the upper limb of the sun is on the horizon where the hour angles of sunrise and sunset could be then calculated:

$$\omega_s = \cos(-\tan(\delta)\tan(\phi))^{-1} \tag{17.3}$$

The daylight can therefore be estimated as follows:

$$T_{\text{hours of daylight}} = \frac{\omega_s}{15} \tag{17.4}$$

Solar noon can be calculated from SST and SRT as follows:

$$SN = SRT + \left(\frac{SST - SRT}{2}\right) \tag{17.5}$$

17.2.2.5 Equation of time (EOT)

The equation of time is the difference between local mean time and solar time, and it is crucial for the determination of the tracking algorithm. It can be calculated by the following equation by Lamm (Lamm, 1981)

$$EOT = 60 \sum_{k=0}^{5} a_k \cos\left(\frac{360kN}{360.25}\right) + b_k \sin\left(\frac{360kN}{360.25}\right) (\text{min}) \tag{17.6}$$

where N is the number of days and a_k and b_k are the coefficients (Table 17.1).

Table 17.1 Coefficients of the EOT.

K	a_k	b_k
0	2.087×10^{-4}	0
1	9.2869×10^{-3}	1.2229×10^{-1}
2	5.2252×10^{-2}	1.5698×10^{-1}
3	1.3077×10^{-3}	5.1602×10^{-3}
4	2.1867×10^{-3}	2.9823×10^{-3}
5	1.51×10^{-4}	2.3463×10^{-4}

17.2.2.6 Local clock time (LCT)/longitude correction (LC)

Since local time is not essential in a sun tracking system, it is very crucial in time conversion analysis. The knowledge of the local standard, the day of the year, and the location are needed for conversion between solar time and clock time, which can be calculated as follows:

$$LCT = TSV - \frac{EOT}{60} + LC + d \qquad (17.7)$$

where

$$LC = \frac{\text{local longitude} - \text{longitude of standart time zone meridian}}{15} \text{(hours)}$$

17.2.2.7 Zenith angle and altitude angle

The altitude angle is the angle between the horizontal line and the sun line. The zenith angle indicates the angle between the vertical (pointing above the observer's position) and the line to the sun. The relationship between zenith and altitude angle is described as follows:

$$\theta_z = 90 - \alpha \qquad (17.8)$$

where

$$\alpha = \sin^{-1}(\sin(\delta)\sin(\phi) + \cos(\delta)\cos(\omega)\cos(\phi))$$

17.2.2.8 Solar azimuth angle (A)

Solar azimuth angle is the angular displacement from the south of the projection of beam radiation on the horizontal plane. Fig. 17.2 illustrates the relationships between the three angles (altitude, zenith, and azimuth angles). Solar azimuth angle can be calculated as follows:

$$\sin(A) = \frac{\cos(\delta)\sin(\omega)}{\cos(\alpha)} \qquad (17.9)$$

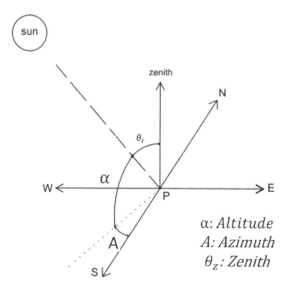

Figure 17.2 The altitude angle, zenith angle, and azimuth angle of the earth's surface coordinate system.

17.2.2.9 Incidence angle and tracking angle

The incidence angle is the angle between the direct solar ray and the normal of the plane surface. It can be expressed by the following equation for horizontal signal axis north tracking:

$$\theta = \cos^{-1}\left(\sqrt{1 - \cos(\alpha)\cos^2(A)} \right) \tag{17.10}$$

In commercial applications, the north-south horizontal orientation is used as this configuration (south = 0 degree and west = 90 degree) ensures high energy and lower daily fluctuations.

The tracking angle can then be calculated as follows:

$$s = \frac{\pi}{2} - \tan^{-1}\left(\frac{\sin(\gamma)}{\sin(\alpha)} \right) \tag{17.11}$$

17.2.2.10 Sun tracking system

A solar tracker ensures that the solar panels track in real time the sun variation during the day where the maximum amount of sunlight will be collected by the solar panels. The first solar tracking system was introduced by McFee in 1975 (McFee, 1975). These solar trackers allow maintaining the orientation of the collectors toward the sun. Thus, the angle of incidence remains constant, and, therefore, the maximum energy is collected. There are different types of sun trackers that could be classified according to the tracking method and according to the number of axes.

17.2.2.11   Solar tracker according to the number of axes

A solar tracker can have one or two degrees of freedom of rotation: the horizontal to adjust the azimuth and the vertical for the inclination. There are different types, mainly the single-axis tracker; this type allows following the sun according to one degree of freedom in relation to a single axis. This is generally the north-south axis as shown in Fig. 17.3A. In order to optimize solar irradiation on the solar collector, the axis can be in an inclined position for a certain angle of optimum inclination; this angle depends on several parameters such as longitude and season. For the northern hemisphere, the solar collector is tilted toward the south, and for the southern hemisphere, the solar panel is tilted toward the north for an improved irradiation rate. The second type is the two-axis tracker (shown in Fig. 17.3B) with two axis of rotation. Generally, one is perpendicular to the other, which allows following the sun with two degrees of freedom, as represented in the figure below. This configuration requires two motors (transmission chain) that introduce extra costs compared to a single axis characterized by a less complex configuration. However, a single-axis tracker cannot take into account seasonal changes and therefore lower energy compared to a dual-axis tracker.

17.2.2.12   Solar tracker according to the tracking technique

The main purpose of a solar tracker is to continuously follow the position of the sun throughout the months and the day. This ensures the maximum efficiency of the collector. Two different techniques are used to track the sun's position: closed loop system or active system; this type of tracker is equipped with optical sensors where the optimal position of the sun is instantaneously determined. Following is the open loop system, or astronomical system, where the sun's position is calculated in function of a preprogrammed astronomical algorithm. The active systems are very sensitive to the sky variation compared to astronomical calculation as they are based on photoresistors and voltage amplifiers, yet they are low cost and can be easily implemented.

In this work, we have opted to realize a solar tracking system that will be based on an astronomical algorithm, in particular, the Solar Position Algorithm (SPA). Therefore, the SPA algorithm (Reda & Andreas, 2004) calculates the sun's position with high accuracy based on the time, date, and location (longitude and latitude). The elevation and the azimuth angles are the main output of this algorithm.

The global synoptic of the prototype takes into account the required sensors and a microelectronic circuit, which allows the user to receive, analyze, and transmit information while performing the necessary calculations. The synoptic diagram is shown in Fig. 17.4.

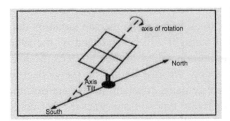 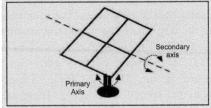

Figure 17.3 (A) Single-axis tracker and (B) two-axis tracker.

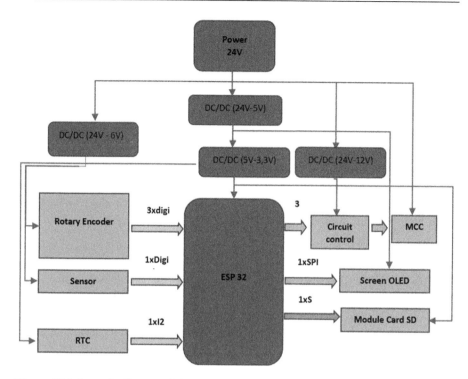

Figure 17.4 Synoptic diagram of the prototype.

17.2.2.13 Drive mechanism/transmission

In order to select the optimal drive mechanism adequate to our case, a comparative study between different types of electric motors is presented according to various aspects, flexibility, and simplicity with respect to assembly, control, as well as sensitivity (Table 17.2).

Considering the different aspects of this comparison, we opted for a DC motor. It has a higher starting torque, fast start and stop, reversal, and variable speed depending on the input voltage as well as simplicity of control compared to an AC motor. The rotational movement of the solar tracker is normally faster than the rotational speed of the sun (0.000694 rpm). For this reason, on/off solar tracking control systems synchronize the actuators' angular rotation with the sun's motion. Based on the work of Prinsloo and Dobson (2015), a worm wheel gearbox was chosen with crown gear to ensure the right reduction ratio for the tracking system.

To size the suitable gear motor for our tracking system, many steps were followed as follows:

- Determination of the mechanical characteristics of the load
- Calculation of the output torque of the reducer (load torque)
- Calculation of the reduction ratio
- Calculation of the torque at the motor output
- Mechanical characteristics of the load
- Weight: 1000 kg
- Radii: 416 mm

- Manufacturing materials: galvanized steel
- Calculation of the various torques required to drive the collector (resistant, friction, and motor torque)

The specifications of the selected motor are summarized in Table 17.3:

17.2.2.14 Control system and sensing device

Our study is based on several criteria, namely tracking methodology, transmission chain, communication interface, CPU, type of sensors, protection mode, etc. Regarding these studies and analyses, we were able to proceed with the design of our tracking system. However, before initiating the large-scale implementation, it is very important to design and validate first a prototype. The electronic materials and sensor used in this prototype are described in Table 17.4.

Table 17.2 Comparison of electric motors.

Motor type	Asynchronous	Synchronous	Direct current	Step by step
Cost	low	high	high	low
Benefits	Robust and low maintenance	Fixed speed	Simplicity of control/ variable speed	High precision/ good response for start/stop
Limitations	• Dependence between speed and load • High current peak	• Maintenance of the rings • Direct start is not possible	Maintenance (heavy deterioration)	Difficulty in obtaining a constant mechanical torque. Acoustic vibration and noise
Speed variation	Possible	Common	Always	Always
Speed variation Cost	Economic	Economic	Very economical	Very economical

Table 17.3 Specifications of the gear motor.

Type of electric motor	TRANSTECNO EMC-250 /A
Gearbox type and ratio	Worm and wheel gearbox with a reduction ratio of 1:000
Engine Torque	5.2 N m
Supply voltage	24 V
Motor speed	1200 rpm
Power	250 W
Maximum current	7.5 A
Protection mode	IP65
Fixing and coupling with the manifolds	Rigid fastening

Table 17.4 List of materials required for the realization of the prototype.

Material	description	Series
Sensors	RTC real-time clock: allows calculating the date and time used in the astronomical algorithm (SPA) Supply voltage: 3.3 V Pinout: I2C bus (two analog pins: SDA and SCL)	DS3231
	Position sensor: (Incremental Rotary Encoder): Acquires the angular position of the CSP manifold. Resolution: 2048 PPR; output frequency: 500 kHZ Supply voltage: 5–28 V; pinout: three digital pins: A, B, and Z	Eltra EL40A1024 Z5/28P6X
	Proximity sensor (inductive): Allows to detect the passage of the collector to perform the angle calibration before starting the tracking Supply voltage: 6–36 V; pinout: one digital pin	LJ12A3–4-Z/ BX
Display	OLED LCD screen: Allows to display the tracker data and the control mode (auto/manual) Power supply voltage: 5 V Pinout: SPI bus (three pins: MOSI, MISO, and CS)	2.4-inch OLED SSD1309
Actuators	DC motor with gear; supply voltage: 5 V; rated speed: 58 rpm	5 V DC geared motor 58 rpm
Motor control circuit	H-bridge DC–DC converter; supply voltage: 6–12 V Pinout: three digital inputs (IN1, IN2, and ENA) and two motor outputs	L298 Dual H-Bridge
Microcontroller	ESP32 DevkitC Microprocessor: Tensilica Xtensa LX6 Maximum; operating frequency: 240 MHz; operating voltage: 3.3 V Analog input pins:12-bit, 18 channel; DAC pins:8-bit, 2 channel Digital I/O pins: 39 (of which 34 is a normal GPIO pin); SRAM: 520 KB Communication: SPI (4), I2C (2), I2S (2), CAN, and UART (3) Wi-Fi: 802.11 b/g/n Bluetooth: V4.2—Supports BLE and Classic	ESP32 DevkitC V4

17.2.2.15   System implementation

The following flow chart illustrates (Fig. 17.5) the implementation of the proposed sun tracker. The main steps required to perform the precise tracking angle based on astronomical calculation as well as from the sensor data were described.

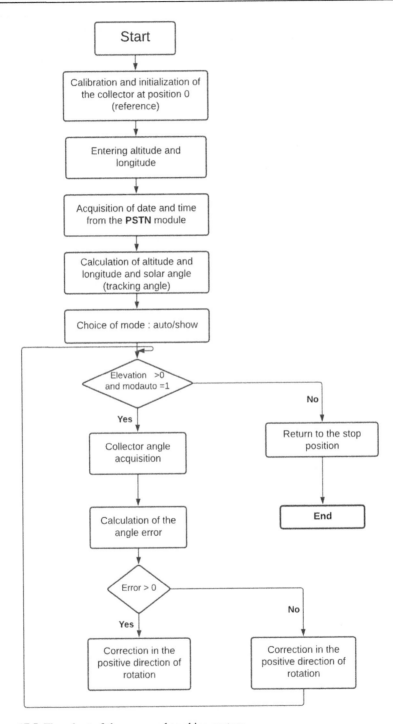

Figure 17.5 Flowchart of the proposed tracking system.

17.3   Results and discussions

The program used is based on an algorithm implemented in the control board of the solar tracker, consisting of a series of instructions that are interpreted and executed by the microcontroller in order to manage the proper functioning of the system. The Arduino Integrated Development Environment or Arduino software (IDE) is used in this work.

17.3.1   *Validation of the SPA algorithm*

Any reliable sun tracking system must be able to track the sun at the right angle, even during periods of cloud cover. Therefore, the implementation of a sun position algorithm (SPA) determining the position of the sun at a given time for a specific location is mandatory, which is based on astronomical calculations presented before. In order to ensure high accuracy of the elevation angle and azimuth calculated by the SPA algorithm, the SunEarthTools.com website was used as a reference in the first stage. This site can automatically calculate the position of the sun in the year according to the date and GPS coordinates. The test is performed for the four specific days of the year that determine in many countries the seasons of the year, the equinoxes days (March 20, 2021, and September 23, 2021), and solstice days (June 21, 2021, and December 23, 2021). It is observed from the validation curves of the solar elevation and azimuth that the results are convergent with high accuracy. Based on the simulation and validation results, a prototype is developed to ensure and test the interaction between all sensors used and implemented algorithms (Fig. 17.6).

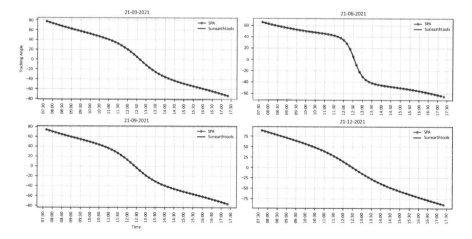

Figure 17.6 Validation of SPA outputs against SunEarthTools.

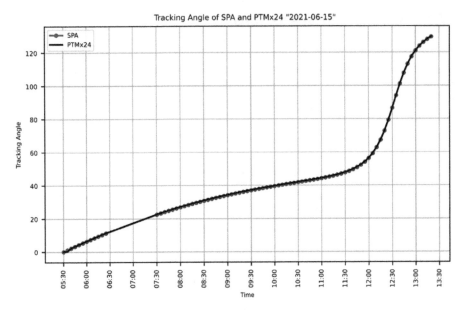

Figure 17.7 Validation tracking angle between SPA and PTMx24.

17.3.2 Tracking angle validation

In addition to elevation and azimuth validation, it is also crucial to calculate and validate the real tracking angle. For this purpose, the calculated tracking angle from SPA output was compared with a real tracking system already installed at the Green Energy Park research platform. The tracker used for validation is commercialized internationally by Soltigua company. The results obtained are presented in Fig. 17.7 where a clear convergence and accuracy are observed, which make the developed algorithm reliable.

17.3.3 Test and prototype

After validation of the SPA algorithm against SunEarthTools.com as well as the tracking angle with a real tracking system already installed at the GEP platform, a prototype is then designed and developed (shown Fig. 17.8). All the algorithms and sensors were implemented and tested. In addition, the important data is displayed in real time to well track the system outputs.

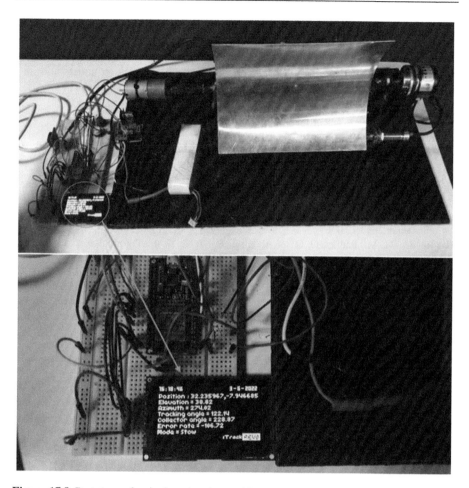

Figure 17.8 Prototype of a single-axis solar tracking system.

17.4 Conclusion

In this work, a low-cost solar tracker was presented in which the main steps to develop a tracking system were identified. The astronomical calculations taking into account all angles and parameters were presented. Following, the SPA algorithm was compared and validated with the SunEarthTools.com tracking method. Besides, the tracking angle was calculated and validated with real data from an already solar tracker installed at the Green Energy Park research platform. As the validation results showed high accuracies and lower errors, a first prototype was designed and developed before moving to a real solar sensor and tracker. The electronic sensors selected were implemented and tested as well as the main tracking data was displayed in real time. This report serves researchers to understand and develop their solar trackers using low-cost electronic hardware.

References

Abdallah, S. (2004). The effect of using sun tracking systems on the voltage-current characteristics and power generation of flat plate photovoltaics. *Energy Conversion and Management*, *45*(11–12), 1671–1679. Available from https://doi.org/10.1016/j.enconman.2003.10.006.

Abu-Khader, M. M., Badran, O. O., & Abdallah, S. (2008). Evaluating multi-axes sun-tracking system at different modes of operation in Jordan. *Renewable and Sustainable Energy Reviews*, *12*(3), 864–873. Available from https://doi.org/10.1016/j.rser.2006.10.005.

Arbab, H., Jazi, B., & Rezagholizadeh, M. (2009). A computer tracking system of solar dish with two-axis degree freedoms based on picture processing of bar shadow. *Renewable Energy*, *34*(4), 1114–1118. Available from https://doi.org/10.1016/j.renene.2008.06.017.

Bentaher, H., Kaich, H., Ayadi, N., Ben Hmouda, M., Maalej, A., & Lemmer, U. (2014). A simple tracking system to monitor solar PV panels. *Energy Conversion and Management*, *78*, 872–875. Available from https://doi.org/10.1016/j.enconman.2013.09.042.

Chang, T. P. (2009). The gain of single-axis tracked panel according to extraterrestrial radiation. *Applied Energy*, *86*(7–8), 1074–1079. Available from https://doi.org/10.1016/j.apenergy.2008.08.002, http://www.elsevier.com/inca/publications/store/4/0/5/8/9/1/index.htt.

De Castro, C., Mediavilla, M., Miguel, L. J., & Frechoso, F. (2013). Global solar electric potential: A review of their technical and sustainable limits. *Renewable and Sustainable Energy Reviews*, *28*, 824–835. Available from https://doi.org/10.1016/j.rser.2013.08.040.

Huang, B. J., Ding, W. L., & Huang, Y. C. (2011). Long-term field test of solar PV power generation using one-axis 3-position sun tracker. *Solar Energy*, *85*(9), 1935–1944. Available from https://doi.org/10.1016/j.solener.2011.05.001.

Ismail, M. S., Moghavvemi, M., & Mahlia, T. M. I. (2013). Analysis and evaluation of various aspects of solar radiation in the Palestinian territories. *Energy Conversion and Management*, *73*, 57–68. Available from https://doi.org/10.1016/j.enconman.2013.04.026.

Koussa, M., Cheknane, A., Hadji, S., Haddadi, M., & Noureddine, S. (2011). Measured and modelled improvement in solar energy yield from flat plate photovoltaic systems utilizing different tracking systems and under a range of environmental conditions. *Applied Energy*, *88*(5), 1756–1771. Available from https://doi.org/10.1016/j.apenergy.2010.12.002, http://www.elsevier.com/inca/publications/store/4/0/5/8/9/1/index.htt.

Lamm, L. O. (1981). A new analytic expression for the equation of time. *Solar Energy*, *26*(5), 465. Available from https://doi.org/10.1016/0038-092X(81)90229-2.

Lazaroiu, G. C., Longo, M., Roscia, M., & Pagano, M. (2015). Comparative analysis of fixed and sun tracking low power PV systems considering energy consumption. *Energy*

Conversion and Management, *92*, 143−148. Available from https://doi.org/10.1016/j. enconman.2014.12.046.

Li, Z., Liu, X., & Tang, R. (2011). Optical performance of vertical single-axis tracked solar panels. *Renewable Energy*, *36*(1), 64−68. Available from https://doi.org/10.1016/j. renene.2010.05.020.

Masters, G. M. (2013). *Renewable and efficient electric power systems*. John Wiley & Sons.

McFee, R. H. (1975). Power collection reduction by mirror surface nonflatness and tracking error for a central receiver solar power system. *Applied Optics*, *14*(7), 1493. Available from https://doi.org/10.1364/ao.14.001493.

Nsengiyumva, W., Chen, S. G., Hu, L., & Chen, X. (2018). Recent advancements and challenges in solar tracking systems (STS): A review. *Renewable and Sustainable Energy Reviews*, *81*, 250−279. Available from https://doi.org/10.1016/j.rser.2017.06.085.

Prinsloo, G., Dobson, R. (2015). Solar Tracking: High precision solar position algorithms, programs, software and source-code for computing the solar vector, solar coordinates & sun angles in Microprocessor, PLC, Arduino, PIC and PC-based sun tracking devices or dynamic sun following hardware. Available from https://doi.org/10.13140/RG. 2.1.4265.6329/1.

Reda, I., & Andreas, A. (2004). Solar position algorithm for solar radiation applications. *Solar Energy*, *76*(5), 577−589. Available from https://doi.org/10.1016/j.solener. 2003.12.003, http://www.elsevier.com/inca/publications/store/3/2/9/index.htt.

Sallaberry, F., Pujol-Nadal, R., Larcher, M., & Rittmann-Frank, M. H. (2015). Direct tracking error characterization on a single-axis solar tracker. *Energy Conversion and Management*, *105*, 1281−1290. Available from https://doi.org/10.1016/j.enconman. 2015.08.081.

Shi, L., & Chew, M. Y. L. (2012). A review on sustainable design of renewable energy systems. *Renewable and Sustainable Energy Reviews*, *16*(1), 192−207. Available from https://doi.org/10.1016/j.rser.2011.07.147.

Skouri, S., Ben Haj Ali, A., Bouadila, S., Ben Salah, M., & Ben Nasrallah, S. (2016). Design and construction of sun tracking systems for solar parabolic concentrator displacement. *Renewable and Sustainable Energy Reviews*, *60*, 1419−1429. Available from https:// doi.org/10.1016/j.rser.2016.03.006.

Sungur, C. (2009). Multi-axes sun-tracking system with PLC control for photovoltaic panels in Turkey. *Renewable Energy*, *34*(4), 1119−1125. Available from https://doi.org/ 10.1016/j.renene.2008.06.020.

Sun, J., Wang, R., Hong, H., & Liu, Q. (2017). An optimized tracking strategy for small-scale double-axis parabolic trough collector. *Applied Thermal Engineering*, *112*, 1408−1420. Available from https://doi.org/10.1016/j.applthermaleng.2016.10.187, http://www.journals.elsevier.com/applied-thermal-engineering/.

An internet of things—based intelligent smart energy monitoring system for solar photovoltaic applications

18

Challa Krishna Rao[1,2], Sarat Kumar Sahoo[2] and Franco Fernando Yanine[3]
[1]Department of Electrical and Electronics Engineering, Aditya Institute of Technology and Management, Tekkali, Andhra Pradesh, India, [2]Department of Electrical Engineering, Parala Maharaja Engineering College, Berhampur, Affiliated to Biju Patnaik University of Technology, Rourkela, Odisha, India, [3]School of Engineering of Universidad Finis Terrae, Providencia, Santiago, Chile

18.1 Introduction

The population of the world has risen steadily from 3200 million in 1962 to 7700 million in 2021, with a projected increase to 10000 million by 2050 (World population, n.d.). Consequently, rising standards of living, as well as increased energy, water, and food demands, place a strain on the ecosystem. The use of oil, gas, and coal is likely to diminish shortly. Among the aforementioned concerns, electricity production is one of the most important factors in many emerging countries (Mellit & Pavan, 2010). The energy demand is currently at an all-time high due to the growth of the commercial sector. Simultaneously, climate change is putting pressure on the electrical sector to shift away based on carbon energy sources and toward more environmentally sustainable options (Energy policy of India, n.d.). India currently produces over 1,037,185.0 GWh of energy through extensive use of carbon-based fuels, which is extremely harmful to the environment. The annual amount of energy produced in India from fossil fuels is shown in Table 18.1 (Energy policy of India, n.d.). Since fossil fuels are rapidly depleting, sustainable energy sources have steadily grown in importance as a significant energy source generation (Energy policy of India, n.d.). Renewable energy resources were the most efficient approach to meet the demand for energy while also creating environmentally friendly energy. It aids the environment in reducing greenhouse gas emissions and ozone depletion to succeed in long-term generating electricity sustainability (Izgi et al., 2012). Solar energy output is advancing at a quicker rate than any other renewable energy source because of lower costs, availability, maintenance, and installation effort. Table 18.2 (Energy Policy of India, n.d.) shows the annual generation of electrical energy in India using renewable energy sources. As a result, for today's clients, the availability, consistency, and quality of renewable energy systems'

Performance Enhancement and Control of Photovoltaic Systems. DOI: https://doi.org/10.1016/B978-0-443-13392-3.00019-0

Table 18.1 India's annual fossil fuel power generation (GWh) (Energy policy of India, n.d.).

Sl. No	Year	Gas (GWh)	Oil (GWh)	Coal (GWh)
1	2011−12	612,497	2,649	93,281
2	2012−13	691,341	2,449	66,664
3	2013−14	746,087	1,868	44,522
4	2014−15	835,838	1,407	41,075
5	2015−16	896,260	406	47,122
6	2016−17	944,861	275	49,094
7	2017−18	986,591	386	50,208
8	2018−19	1,021,997	129	49,886
9	2019−20	994,197	199	48,443
10	2020−21	981,239	129	51,027

Table 18.2 India's annual electrical energy production (GWh) from renewable energy sources (Energy policy of India, n.d.).

Sl. No	Year	Mini hydro	Solar	Wind	Biomass	Other
1	2014−15	8,060	4,600	28,214	14,944	414
2	2015−16	8,355	7,450	28,604	16,681	269
3	2016−17	7,673	12,086	46,011	14,159	213
4	2017−18	5,056	25,871	52,666	15,252	358
5	2018−19	8,703	39,268	62,036	16,325	425
6	2019−20	9,366	50,103	64,639	13,843	366
7	2020−21	10,258	60,402	60,150	14,816	1621

power supply are essential factors in overcoming the aforementioned difficulties (Mellit, Sağlam, et al., 2013).

Solar cells are used in a PV system to consume sunlight and convert it to energy. Silicon or other semiconductor materials that make up the PV cell use the PV effect to convert solar energy into DC power. When solar radiation interacts with the solar cell, it absorbs solar energy and generates electricity (Mellit & Pavan, 2010). When sunlight is mixed with a semiconductor, electricity will be created. Because of changes in parameter deviation, the solar system's electrical potential may differ from one location to the next location and from one technology to another. As a result, proper precautions must be taken during the installation of any PV system to maximize the system's potential energy. To minimize shortfalls and maintenance problems with the PV system's performance, extra caution is essential before or during the installation of PV panels (Chen, Gooi, et al., 2013) (Table 18.1).

Photovoltaic systems installed in far-flung or isolated sites are particularly vulnerable to these challenges. To address the aforementioned concerns, a remote monitoring strategy in the PV system (Izgi et al., 2012) should be chosen.

For a few years, smart sensors in the Internet of Things (IoT) have been recognized. In general, the IoT is a data-sharing ecosystem in which ordinary life is linked to wired and wireless networks. A terminal server will function as the hub for all sensors and devices. In recent years, there has been a lot of focus on updated utilization in every area (Yona et al., 2013). Due to the widespread employment of many devices, such as sensors, actuators, radio frequency identification, and Internet-connected mobiles and tablets that are examples of these technologies, the IoT is necessary (Ramu et al., 2021). The IoT allows important items to be considered, heard, seen, and "spoken," as well as share data. By utilizing communication architecture and cloud settings, the IoT makes things easier to operate since they are simpler, faster, and simpler (Almonacid et al., 2014). Today's world of electronics is changing. When the IoT was first introduced, RFID was favored in all systems. IoT is currently developing some technologies, including M2M, V2V, and NFC, which are similar to RFID (Almonacid et al., 2014). IoT integration with solar systems was a critical area for both commercial and residential consumers. Customers and IoT service providers will lose a lot of money as a result of this. The efficiency of a solar panel is impacted by changes in voltage, temperature, current, and irradiance. To avoid major failures, smart sensors and IoT must be used to monitor the solar PV system (Dahmani et al., 2014). So, in this chapter, we will look at how to use IoT to monitor and control a solar PV system utilizing a variety of smart sensors (Zhang et al., 2015).

Solar energy is a renewable resource that we can harness. This solar energy is readily available, has not been depleted, and may be able to be reused. Solar energy may be able to be used in a variety of ways and for a variety of uses (Lima et al., 2016). To burn a piece of paper with energy, for example, all we need is a focus on the paper with a magnifying glass. The paper begins to burn when sunlight is a source of power beams concentrated on it (Sharma et al., 2016).

A solar cell, also known as a PV cell, is a device that uses solar energy to generate electricity. Because of the differences in production methods and quality, the panels are made up of a variety of absorbent materials, and each one differs from the next. Silicon makes up the great majority of panels on the market, with thin-film solar cells being one of them. Depending on the design, every solar panel has a distinct cost and efficiency (Kaushika et al., 2014), and depending on the application, the appropriate solar energy panel must be utilized. Solar energy could be used straight or turned into other forms of energy using various methods. Fig. 18.1 depicts the commonly accessible technologies (Hossain et al., 2017).

18.2 Photovoltaic solar systems

Solar panels were commonly used to direct sun energy to electricity conversion. It is also excellent for commercial and industrial applications because it matches the required level of power. There are two types of solar panels: stand-alone and grid-connected systems (Deng et al., 2015) (Fig. 18.1).

Figure 18.1 Solar power technologies (Dahmani et al., 2014).

18.2.1 Stand-alone photovoltaic modules

Off-grid systems, often known as stand-alone systems, can be utilized for smaller purposes (Teo et al., 2015). Stand-alone systems include solar water pumps, home automation, energy-efficient lighting systems, and rural microgrids that are often smaller in size (Pawar et al., 2020) (Fig. 18.2).

Such a solar system may be present in a remote place without a grid or even when the cost of utility electricity is relatively expensive (Chen, Li, et al., 2013). Here, solar panels take on the role of the utility provider and provide the energy required by a person's home or any other energy-dependent system (Ekici, 2014). There could be no choice except to use an off-grid solar system. Off-grid systems take more upkeep and maintenance, but they may create a powerful sense of independence by removing the threat of utility grid outages (Pawar et al., 2018). Off-grid solar systems do not interface with the main grid at all because the solar energy is produced and consumed in the same location. A straightforward design of the off-grid solar PV system is shown in Fig. 18.3 (Izgi et al., 2012).

18.2.2 Photovoltaic systems connected to the grid

The panels are hooked into 230-V or 440-V power grids and then dispersed by joining a small portion of them above the ground. Many large-scale solar panel projects will necessitate a voltage-rated grid substation of 11 kV or greater. Fig. 18.3 depicts grid-connected PV systems.

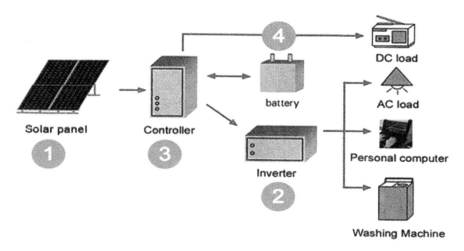

Figure 18.2 Stand-alone PV systems (Ramu et al., 2021).

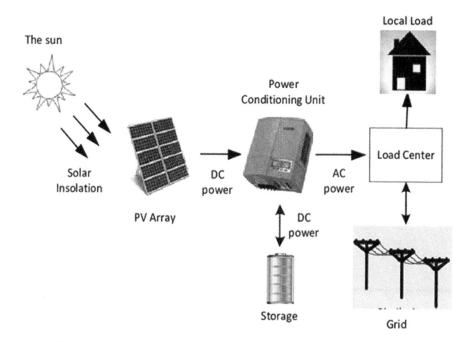

Figure 18.3 Grid-connected PV systems (Hossain et al., 2017).

It is also known as a utility-interactive system or on-grid system. Solar PV systems that are on-grid only produce electricity, while the utility grid is operational (Ramli et al., 2015). Additionally, for it to work, the grid must be linked. When the system is producing too much electricity, it may send the extra back to the utility grid so that it can be used later (Huang et al., 2013). It is also thought to be the easiest to design and install, although there is a worry that it won't work if the power goes out. A straightforward design for the grid-connected PV system is shown in Fig. 18.2.

18.2.3 Concentrated solar power

Concentrated solar power (CSP) is a method of utilizing PV energy in the form of PV sunshine to produce steam, by which the turbine is driven within the electricity generation procedure. In terms of steam output, CSP was equivalent to a thermal station (Ssekulima et al., 2016). This technology utilizes enormous mirrors to focus the energy to boil water in a centralized area (Chen, Li, et al., 2013). This system can be used to cook, heat, or generate electricity for the turbine. The sun-tracking approach is used with the concentrators to improve the system performance faster and more effectively throughout the day. Fig. 18.4 shows several examples of concentrated solar electricity.

Concentrated solar power technology employs concentrated sunlight. CSP facilities create electric power by employing mirrors to concentrate (focus) the sun's energy and convert it into high-temperature heat (Phinikarides et al., 2013). After that, a typical generator receives the heat. The plants are divided into two sections: one that gathers solar energy and transforms it into heat; the other transforms the heat energy into electricity. On the website of the Department of Energy's Solar Energy Technologies, there is a small video that demonstrates the operation of concentrating solar power using a parabolic trough system as an example (Yang et al., 2012). CSP plants have been successfully

Figure 18.4 Concentrated solar power (Lima et al., 2016).

running for over 15 years inside the United States. When utilized to generate energy at a commercial scale, all CSP technical techniques need enormous regions for solar radiation gathering. The three alternative technical methods used by CSP technology are trough systems, power tower systems, and dish/engine systems (Yang et al., 2015).

18.2.4   Solar water heating system

The use of a heater was advantageous and is becoming increasingly common in today's society. The goal is to use less nonrenewable energy and more easily available energy (Wang et al., 2016). Solar collectors, either flat plate or evacuated tubes, were utilized to heat the water in this system (Ekici, 2014). These collectors are not the same as solar panels that generate power. Fig. 18.5 depicts the usual solar water heater construction.

18.2.5   Passive solar design

It is yet another example, which is a building that is constructed in such a way that it makes use of solar energy. The fundamental idea behind this design is to allow solar energy into the building, capturing it before it enters during the summer. Fig. 18.6 depicts the layout of the passive solar architecture (Li et al., 2016).

By exposing living areas to light, the passive solar architecture uses the sun's energy to heat and cool them. Building components can transmit, reflect, or absorb solar energy as it hits the structure (Massidda & Marrocu, 2017). Additionally, the heat from the sun creates airflow that is predictable in places that have been built. These fundamental reactions to solar heat influence design decisions on materials, placements, and architectural features that might give heating and cooling effects in

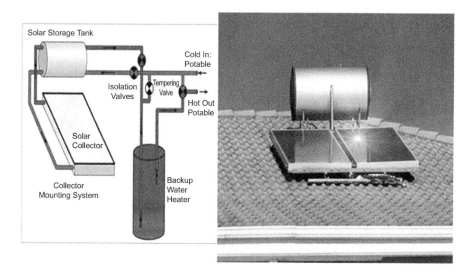

Figure 18.5 Solar water heater (Ha & Phung, 2019).

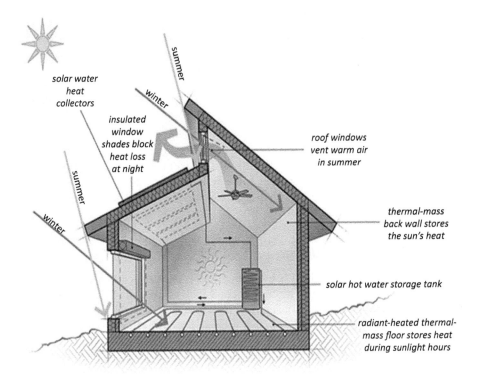

Figure 18.6 Passive solar designs (Pawar et al., 2020).

a home (Persson et al., 2017). Contrary to active solar heating systems, passive solar heating systems are straightforward and don't depend heavily on mechanical and electrical components like fans, pumps, or electrical controllers to transport solar energy (Inman et al., 2013; Rao et al., 2021) (Table 18.2).

Aperture/collector: The expansive glass area that lets light into the structure. During the heating season, the aperture(s) should be within 30 degrees of true south and not be blocked by any other structures or trees from 9 a.m. to 3 p.m. (Monteiro et al., 2013).

Absorber: The storage element's rough, dingy surface. The area is directly in the direction of the sun and maybe a stone wall, a floor, or a water container. Heat is created when sunlight hits a surface (Khan et al., 2015).

Thermal mass: Materials that hold on to or store solar heat. The thermal mass is the substance beneath and behind the absorber, which is an exposed surface (Lorenz & Heinemann, 2012).

Distribution: The process by which solar heat is distributed throughout the home from the places of collection and storage. Conduction, convection, and radiation are the only three natural heat transport modes that are used in a completely passive design (Mathiesen & Kleissl, 2011). In some situations, the heat may be dispersed around the home using fans, ducts, and blowers.

Control: During the summer, roof overhangs can be employed to provide shade over the aperture. Electronic sensing systems, such as a differential thermostat that instructs a fan to switch on, movable vents and dampers that let or limit heat flow, low-emissivity shutters, and awnings, are further components that regulate under- and/or overheating (Fernandez-Jimenez et al., 2012).

18.2.6 Solar microgrid system

It is a small solar power grid that can independently power a small total number of loads in a small area (Sittón-Candanedo et al., 2019). It also requires power distribution connections and additional security devices in addition to the basic components. Fig. 18.7 depicts a solar microgrid system schematic. The system's size is determined by the needs and desires of the consumer. Power of fewer than 100 kW is necessary for basic lights and home loans (Tuohy et al., 2015).

The IEC technical specifications of voltage levels are the following:

- AC systems' low voltage level is 208—415 V 3-phase/120—240 V 1-phase at either 50 Hz or 60 Hz domestic frequency.
- A DC system has a voltage level smaller than 120 V; it was referred to as extra lower voltage (Table 18.3).

18.2.6.1 Photovoltaic module

Today, there are many different types of modules; each with its own set of characteristics, such as the type of silicon used in manufacturing procedures and the component's characteristics (Tuohy et al., 2015). The bulk of commercially available

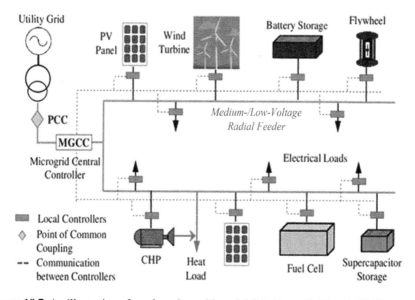

Figure 18.7 An illustration of a solar microgrid model (Mathiesen & Kleissl, 2011).

Table 18.3 IEC 60038 Specifications (Olatomiwa et al., 2015).

IEC range of voltage	A RMS voltage (V)	DC volts (V)
High voltage	>1000	>1500
Low voltage	50–1000	120–1500
Extra-low voltage	<50	<120

modules for solar power was made of silicon and fell into one of the three solar cell categories (Ekici, 2014). Thin-film solar cells such as monocrystalline and polycrystalline thin-film solar cells are examples. PV panels come in a variety of sizes and ratings, as shown in Table 18.3. PV modules are available in a range of forms and sizes, with each having its efficiency, cost, and size (Urquhart et al., 2014).

The positive terminal of one panel is connected to the negative terminal of another panel to create a series connection (Rao et al., 2020). When two or more solar panels are linked, this develops a PV source circuit. While the voltage of series-connected solar panels rises, the amperage remains constant. If two solar panels with rated voltage and amperage of 40 V and 5 A each were linked, the voltage of the series would thus be 80 V while the amperage would remain at 5 A. When panels are joined in series, the voltage of the array increases (Yang et al., 2014). This is crucial because the inverter needs a specific voltage to function effectively in a solar power system.

When solar panels are wired in parallel, the positive terminal from one solar panel is connected to the positive terminal of another panel, and the negative terminals of the two panels are linked (Rao, Sahoo, & Yanine, 2021). A positive connection joins the positive wires in a combiner box, and a negative connector joins the negative wires. Multiple solar panels are linked together in parallel using PV output circuits (Inman et al., 2013). When solar panels are connected in parallel, the amperage increases but the voltage remains constant. The voltage of the system would remain at 40 V, but the amperage would increase to 10 A if you wired the same panels in parallel as before. To generate additional power without exceeding the inverter's maximum operating voltage, add extra solar panels in tandem (Hossain et al., 2017). Additionally, you may connect your solar panels in parallel to satisfy any amperage requirements that inverters may have.

A PV system's connectors are a modest but crucial component. They are used to link solar panels, as their name indicates, to add solar panels, the inverter, or module-level devices like power optimizers (Mohammadi et al., 2015). There are several different types of solar panel connectors, including MC4, T4, and MC3. Some manufacturers employ generics, which are frequently interchangeable with common MC4 connectors and are simple to spot. Just scan the datasheet for terms like "MC4 compatible." Unfortunately, most of the time, such a combination will not result in a connection that is UL certified (Olatomiwa et al., 2015). The most common variety is unquestionably the MC4 connection. It is now almost standard on all solar panels and module-level electronics, including power optimizers and microinverters (Shamshirband et al., 2014). These connections meet the National Electric Code and have received UL certification. Multi-Contact, the manufacturer's brand name that came to be associated with the product, is represented by the letters MC. The connections are completely waterproof because they are constructed of UV-resistant material and have an IP67 rating (Mellit, Pavan, et al., 2013). For increased dependability, MC4 connectors have a locking mechanism that can only be opened with a certain tool. Male and female connections are included on each solar panel. They are situated at the ends of the wires in the junction box. The first is good, whereas the second is bad. Typically, the positive lead is connected to the female connection (Salam et al., 2010). The best course of action is to search for the marks or run a voltmeter test because there are certain exceptions.

Because of a flexible seal that shields them from precipitation, MC3 connections are waterproof (Sahin, 2020). To avoid incorrect connections, they are likewise split into male and female sorts like MC4. MC3 connections were once widely used; however, they are now largely obsolete. The only powers that keep them from separating are friction and suction. MC4 connections dominated the market once the National Electric Code mandated a positive locking mechanism. The MC3 is still simple to purchase, nonetheless (Akpolat et al., 2019). People utilize them for smaller installations or older solar systems that don't need to comply (Table 18.4).

18.2.7 Battery

In a stand-alone device, battery storage is essential when electrical loads must operate at night or during extended periods of cloudy or humid weather. Furthermore, the PV array is unable to generate enough power on its own (Pawar et al., 2018). The following are the major properties of a battery in a PV system:

- Capability and dependability of energy storage
- Surge currents in the power supply
- The stability of voltage and current

"Autonomy" in stand-alone PV systems refers to the number of days when the direct usage of storing power is possible without the use of electricity relying on PV energy (Adineh et al., 2021). The two most common types of electric batteries are both primary and secondary batteries. Because primary batteries are not refillable, they were not preferred in PV systems. Secondary batteries have the potential

Table 18.4 PV panel rating and sizing (Ramu et al., 2021).

Capacity in Wp	I_{sc} in Amp	I_{mp} in Amp	V_{oc} in Volt	V_{mp} in Volt	Length in cm	Width in cm	Weight in kg
50	3.04	2.80	21.77	17.89	60.8	66.6	4.6
100	6.11	5.57	21.84	17.89	115.2	66.6	8.0
200	8.10	7.48	32.65	26.74	148.6	98.2	15.5
250	8.71	8.18	37.55	30.58	163.9	98.2	17.45
300	8.74	8.05	45.10	37.28	195.6	98.2	27

to store as well as generate energy (Ramli et al., 2015). By reversing the discharge and charge currents, it might be charged. Batteries that are commercially viable for use in solar systems are now available.

The rising importance of solar and wind energy will define the kind of energy systems used in the future. Long-term price increases for crude oil and gas are anticipated, and CO_2 emission fines will start to matter economically (Şahin & Blaabjerg, 2020). Electricity generated by solar and wind power will become substantially less expensive, making electrolyzed hydrogen comparably priced to natural gas-produced hydrogen. Energy storage methods that cover the complete range of power (in MW) and energy storage quantities (in MW/h) are needed to accommodate the unpredictability of the system input from variable energy sources (Cuce et al., 2016). Particularly for storing, transmitting, and dispersing massive and extremely massive amounts of energy at the gigawatt-hour and terawatt-hour scales, hydrogen is a viable secondary energy vector. But we also talk about energy storage at the 120−200 kWh scale, for instance, for onboard hydrogen storage in fuel cell cars employing compressed hydrogen storage (Shi et al., 2012). This article focuses on the traits and future possibilities of hydrogen storage systems in light of the difficulties associated with an evolving energy system. Therefore, emphasis is placed particularly on technological issues that affect the dynamics, flexibility, and operational costs of unstable operations. Additionally, the possibility of employing renewable hydrogen in the transportation, manufacturing, and heating markets is examined (Zeng & Qiao, 2013). This potential might have a big impact on the future economic worth of hydrogen storage technology as it relates to various sectors of the economy. This analysis clarifies tried-and-true hydrogen storage methods and might serve as a roadmap for the advancement of more recent, less mature technologies (Chen, Li, et al., 2013).

18.2.7.1   Flooded lead acid battery

The common lead-acid batteries have vents that allow hydrogen gas to escape electrically. Over and around the flooded batteries, adequate ventilation is required (Ekici, 2014).

18.2.7.2   VRLA battery

The electrolyte in these batteries is known as a captive electrolyte, and under normal operating conditions, the electrolyte was immobilized and the battery screwed. When subjected to extreme overload, normally preserved devices become available underneath the pressure of gas via the control of pressure method (Ramli et al., 2015).

18.2.7.3   Lithium-ion battery

It's a one-of-a-kind technology with numerous advantages over existing batteries. These are typically smaller and lighter for the equivalent volume, charge faster, and are small and prone to deterioration during storage and discharge (Wolff et al., 2016). Conversely, rising upstream costs are high, and they're vulnerable to high voltages and temperatures.

18.2.8   MPPT controller

The MPPT controller is the component that controls the PV to maximize power production. If the controller consistently functions at MPP, no matter the weather, the PV system's efficiency is boosted (Wolff et al., 2016). It should be possible to legally match the PV source with the load in every climatic condition to generate the most power. Either electrical tracking or mechanical tracking can be used to fully power the PV array. During the year, the PV panels' orientation changes with mechanical tracking as the months and seasons change, but with electrical tracking, the MPP is identified using the IV curve (Dahmani et al., 2014; Khan et al., 2015; Lorenz & Heinemann, 2012). To ensure that the most power is sent to the load, batteries, motors, and the power grid for off-grid and on-grid applications, respectively, modern power systems must feature MPPT. Since solar radiation is not always uniform and PV arrays still have a low conversion rate of solar energy to electrical energy, the MPPT controller is commonly utilized in PV facilities (Hocaoglu & Serttas, 2016).

Maximum power point tracking is a collection of algorithms used in PV modules to harvest energy in specific situations. Peak power voltage, also known as maximum power point, is the voltage at which a PV system can produce the most electricity (MPP). MPPT looks into the output of PV voltage and how it connects to battery voltage (Bhardwaj et al., 2013). To charge the battery, the system maximizes the power generated by the PV module and converts it to an optimum voltage, facilitating the supply of maximum current to the battery (Hocaoglu & Serttas, 2016). The maximum power varies depending on the ambient temperature, solar irradiance, and solar cell temperature. The basic goal of MPPT is to draw as much power as possible in all scenarios to study the PV generation voltage and provide it to different loads such as resistive and inductive loads. The MPPT is used to incorporate electric power converters that perform voltage or current transformations, as well as the regulation of various loads such as batteries, motors, and the electric grid (Bhardwaj et al., 2013).

18.2.8.1   MPPT controller requirement

Any changes to the environment impose demands on the production of energy from a sustainable power source. In solar and wind energy systems, the impact is particularly severe. Additionally, there are problems with (1) grid integration for wind and solar systems and (2) changing meteorological conditions (Izgi et al., 2012; Sanjari & Gooi, 2016). To provide a sustainable power output, solar PV and wind energy conversion systems now use MPPT techniques (Box et al., 2015; Huang et al., 2013). The existence of an MPP in the $I-V$ and $P-V$ curves for varied irradiation and temperature must thus be ensured. This MPP constantly alters its location in response to environmental changes. Therefore, MPPT controllers are a crucial component of the PV system since they are made to maintain the tracking of MPP. The panel is prompted to work closer to MPP by the presence of a controller because it effectively adjusts the resistance it perceives (Sanjari & Gooi, 2016). To change the operating point of the load associated with modifying the converter's duty cycle, efficient MPPT controllers are crucial.

18.2.8.2 Choosing factors for the MPPT controller

Numerous MPPT approaches have been given in various study literature for tracking the PV system's actual MPP. The MPPT controller's selection parameters are crucial for determining which option is best among the alternatives (Sanjari & Gooi, 2016). The selection factors offer crucial knowledge on which approach is superior for a certain application. These criteria are not used to categorize the techniques; rather, they are used to compare the ways each categorizes the MPPT method. Here, several techniques from each category are compared using eleven selection criteria. Table 18.1 (Phinikarides et al., 2013; Ssekulima et al., 2016; Yang et al., 2012) provides brief descriptions of each selection parameter, and Table 18.5 compares commonly used analog/digital MPPT ICs or microcontrollers.

Table 18.5 Description of several selection parameters of MPPT controllers.

Selection parameters	Descriptions
Design complexity	The complexity of an algorithm's design and efficiency has a significant impact on how accurate it is. A user is always more drawn to a straightforward, user-friendly procedure than one that is difficult. As a result, choosing an optimal MPPT method is greatly influenced by this selection parameter.
Tracking true MPP	Panels that are partially shaded by other panels may interfere with the MPPT's regular performance. The PV system's tracking efficiency is directly impacted by the many peaks on its P–V characteristic curve (Mellit and Pavan, 2010). As a result, monitoring the actual MPP is an important factor to consider when choosing the best MPPT controller.
Cost	The cost of a tracking system is influenced by the system's features, programming difficulty, needed number of sensors, decision between analog and digital structure, and hardware cost (Ssekulima et al., 2016). Therefore, when choosing the appropriate MPPT controller, these aspects should be taken into consideration.
PV array dependency	An MPPT algorithm should track the MPP without the knowledge of the array size and settings. However, other techniques, like FLC and ANN, require an array. Some numerical techniques can solve these issues since they are dependable, independent of PV arrays, and can numerically track the location of MPP.
Previous training	Before being utilized in the scheme, the ANN, look-up table approach, etc., require prior training. Even before creating the MPPT, the FLC needs certain background data (Ssekulima et al., 2016).

(*Continued*)

Table 18.5 (Continued)

Selection parameters	Descriptions
Convergence speed	To track the genuine MPP of a PV system, a very sensitive algorithm should have a fast rate of convergence rather than variations in solar isolation or temperature. This parameter is essential in preventing energy losses. As a result, before choosing a precise MPPT approach, this parameter is always checked while constructing a PV system.
Analog/digital	To achieve MPPT, either analog or digital control can be used. Techniques like fractional SCC or OCV that use analog circuits are known as analog MPPT methods, whereas those that use digital circuits are known as digital MPPT methods. While digital procedures are expensive and accurate, analog methods may be less accurate but less expensive. As a result, this value is crucial for selecting the appropriate MPPT schemes. Table 18.2 (Phinikarides et al., 2013) lists and contrasts the most popular digital microcontrollers and analog MPPT-integrated circuits (ICs).
Sensed parameter periodic tuning	The MPPT controller's key parameters are the voltage and current sensors. Current sensors are expensive and cumbersome when compared to voltage sensors. Therefore, these aspects have to be taken into account when developing MPPT. The performance of MPPT intentionally declines due to the uneven temperature of the PV panel, the impacts of dust, and these factors. A solar panel's projected lifespan is currently 25 years or so. As a result, some approaches could need regular adjustment (Ssekulima et al., 2016).
Stability	The MPP of traditional techniques typically deviates from their stable state, which adversely alters the result. As a result, stability becomes a crucial factor to consider when selecting the approaches.
Efficiency	By taking into account the system's persistent steady-state response, the effectiveness of the MPPT approaches was evaluated subjectively in light of simulations Yang et al. (2012). One of the crucial criteria for choosing the right MPPT controller is this characteristic.

18.2.9 Inverter and other electronic equipment

DC voltage could be generated by the battery and PV array. The inverter's purpose is to convert DC to alternative current electricity and export excess power to the alternative current grid (Huang et al., 2013). In a household or light commercial facility, the standard low-voltage (LV) supply source was either 1 phase 230 V or 3 phase 415 V. Higher voltage may be used in larger commercial buildings, which is then converted to 230 V or 415 V (Sanjari & Gooi, 2016). Stand-alone inverters were typically voltage sensitive, meaning they were planned to run at a specific average voltage such as 12 V, 24 V, or 48 V DC. Because the PV array in stand-alone systems is often not connected to the inverter, but rather to batteries via a system controller, stand-alone inverters are not the same as grid-based inverters. The inverter is a self-contained power device that draws power from the batteries to power the AC circuits (Huang et al., 2013). The voltage controller may be MPPT itself. The MPPT controllers are in charge of battery charging. This function had no bearing on the power supply to an alternating current circuit.

A grid-tied PV network and its arrays were directly connected to a grid-connected inverter. Solar electricity was transferred to the AC grid by the grid-tied converter. The PV array is designed to work with a limited range of DC voltages for grid-connected inverter requirements (Box et al., 2015). This inverter turns solar energy into an alternating current sine wave with the same frequency and voltage as the rest of the system. The inverter will not work if there isn't an AC grid there.

18.2.10   Charge controller

A primary PV system's purpose was the supervision and regulation of the energy produced. Throughout a PV array, the following are the characteristics of battery charge controllers:

- Make load-controlling capabilities available
- Prevent overcharging and discharging of batteries
- Shift PV energy to a different load
- Users should be informed about their status and operators of the system
- Sources of energy
- Backup control and user interface

18.2.11   Equipment for additional systems

In addition to PV modules, inverters, and charge controllers, a sun-fueled PV microgrid structure requires several distinct parts (Ssekulima et al., 2016). These are individuals:

- Deployment system for solar arrays: This framework was utilized to safely associate the PV module system with the sensing layer or floor.
- Cabling: Components should be connected via direct current and alternative current cabling.
- Array junction box: This box was used to combine multiple array strings.
- Switches for protection and disconnection: These are components that ensure the security of the system.
- System monitoring: This is used to detect system failures as well as the current state of the system.
- Metering: This is a technique for determining how much electricity is produced by solar energy and how much electricity is consumed by domestic consumers.

18.3 Internets of things

The IoT is a network of physical objects like instruments, buildings, cars, and other devices that are covered with electronics, sensors, software, and actuators to enable data collection and exchange (Graditi et al., 2016; Sharma et al., 2016). As a result, an integrated system that continually transmits and retains larger volumes of data about electrical appliances is made available. Additionally, this technology starts both simple and complex analyses while giving users access to real-time intuition. Recently, three important IoT components were taken into consideration. Examples include big data and cloud computing, artificial intelligence and machine learning, and smart sensors (Rao et al., 2020).

Today's world has made electricity a need for daily living. Every home now uses three times as much electricity as there are available resources, and this trend has been growing with each passing year (Yang et al., 2014). People are flocking to solar energy as a sustainable way to meet these soaring requirements. Due to its wide availability, low cost, and rapid installation, PV solar technology has grown significantly in popularity. Energy production, however, still poses a significant obstacle to the widespread use of solar energy (Rao, Sahoo, & Yanine, 2021). This is why EPC businesses are searching for methods that may reduce the increased maintenance costs and manage the power consumption of solar panels. They may then assist consumers in realizing a solar system's full potential and maximizing benefits from the installation (Hossain et al., 2017).

Current, voltage, irradiance, and temperature are the fundamental variables that have an impact on a solar solution's output. As a result, the real-time solar output monitoring system is crucial to enhancing the PV system's efficacy by comparing it to the results of experiments to initiate preventative actions (Mohammadi et al., 2015). The IoT can be useful in this situation. Smart sensors connected to production, transmission, and distribution equipment are among the applications of the IoT that generate renewable energy. These tools enable commercial clients and solar investors to remotely monitor and control the functioning of the complete solar system in real time. It lowers the cost of running and lessens our dependency on fossil fuels (Olatomiwa et al., 2015).

18.3.1 Applications of IoT in solar industry

An intriguing idea is the IoT. Since it touches practically every area of our modern life, it is fairly comparable to the Internet. The IoT is now a part of everyone's lives, whether they realize it or not, in everything from entertainment to transportation to shopping (Shamshirband et al., 2014). As a result, several sectors have emerged that are currently utilizing the potential of the Internet. Countless businesses, from service providers and merchants to manufacturers and distributors of hardware, are beginning to look into the potential of utilizing the IoT (Mellit, Pavan, et al., 2013).

The solar industry is one of those industries. Things have changed significantly since the early 1990s when solar panels were large and unsightly to most customers.

As of right now, the IoT has made it feasible to produce portable solar cells that can be mounted on cars and roofs (Salam et al., 2010). Homeowners and other solar energy investors would be able to do this, reducing or eliminating their need for regional utility corporations. As a result, the Internet of Objects (IoT) is currently developing as a cutting-edge technology that uses a cloud platform and protocol stack to make things highly intelligent and user-friendly. The PV system can gather, monitor, and share data thanks to these cutting-edge IoT systems that are equipped with software, network connectivity, and sensors (Sahin, 2020).

IoT-based solar panel tracking is intended for online performance improvement and simulation. It makes it possible to start preventative maintenance, identify the accident's root cause, and pinpoint the breakdown's location. Additionally, by using the capabilities of IoT, EPC providers, power suppliers, solar investors, and utility companies can manage their solar energy resources (Akpolat et al., 2019). In essence, this gives them useful information for decision-making based on facts. By using the IoT-generated data, you may also determine and look into user power use patterns (Adineh et al., 2021).

18.3.2 Machine learning and artificial intelligence

It supports the identification and analysis of data developed within the face of energy forecasting, prognostic maintenance, and intelligent control, giving analytical solutions. This is how you may increase earnings while also increasing efficiency (Phinikarides et al., 2013). The network of gadgets and technology is known as the IoT. Hardware, middleware, cloud, and application are the four essential components of a typical IoT setup.

18.3.2.1 Hardware

This is a term used to describe the actuators and sensors that make up a system connecting a device to the IoT platform. It continuously senses data from entity equipment, such as temperature, current, wetness, voltage, light, and so on. After that, they moved to middleware (Şahin & Blaabjerg, 2020).

18.3.2.2 Middleware

The cloud and sensor connections are made possible via middleware, which is part of the IoT platform. This covers both hardware and software components that interact with sensors conceptually or electrically. They convert the raw sensor data into a usable format that can be used in a subsequent process (Yang et al., 2012).

18.3.2.3 Cloud

The cloud is a critical component of any IoT system. Middleware collects data, which could then be merged and stored on the cloud. Because it can generate and respond to signals from a large number of interconnected devices, the cloud is critical (Yang et al., 2015). This is how current and historical data are created.

18.3.3 Big data and cloud technology

To implement advanced analytics on sensor information, the big information system provides a quick, real-time basic, and complicated event processing capability. It uses cloud computing technology, which enables an unlimited amount of information to be kept in a remote region, making data more accessible, reducing storage requirements, and ensuring data information security (Cuce et al., 2016).

18.3.4 Smart sensors

The hardware connected to gadgets detects ambient conditions and converts them to signals. Fig. 18.8 depicts a few sensors. They employ a fixed microcontroller with wireless communication capabilities for automated data collection (Zaidan & Zaidan, 2018).

18.3.4.1 Temperature sensor

They are usually thermocouples or resistance temperature detectors. Its job is to take temperature data from a specific device and convert it into a format that a device or observer can understand. It is made to measure surface temperatures on the backside of PV modules (Shi et al., 2012) (Fig. 18.9).

18.3.4.2 Humidity sensor

It detects the presence of moisture, the amount of moisture, and the temperature in the air. The percentage of air moisture content to maximal moisture content at the present temperature is known as relative humidity (Fig. 18.10). The humidity levels vary as the temperature increases; whenever the air warms, it retains greater

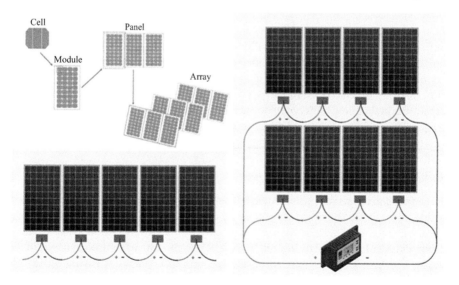

Figure 18.8 PV modules and their connections (Hossain et al., 2017).

Figure 18.9 Connectors and cables (Ssekulima et al., 2016).

moisture. The relative humidity sensor is made up of a humidity sensor and a temperature thermistor (Yang et al., 2015). This sensor is used to determine the amount of moisture in PV modules (Fig. 18.10).

18.3.4.3   Tilt sensor

This sensor is a device that measures the orientation and inclination of a reference plane in several axes. They utilize a stationary microcontroller with wireless communication capabilities for the automated collection of data (Zeng & Qiao, 2013). It is made out of a ball mechanism that detects the object's motion along any axis (Fig. 18.10). The electrical signal is proportional to, using one or more axes, the degree of inclination. This device generates electrical signals that change in response to angular motion (Agoua et al., 2017). Single-axis tilt sensors are the most common tilt sensors used for single-axis tracking (Figs. 18.11–18.19).

18.3.4.4   Carbon dioxide sensor

The carbon dioxide sensor measures the amount of infrared radiation absorbed by carbon dioxide molecules to determine the amount of carbon dioxide present in the environment (Chen, Li, et al., 2013). These sensors are employed in the exact measurement of gases. Carbon dioxide detects signals that are proportional to the amount of absorbing gas in the sample. The light tubes are injected or diffused with CO_2 (Fumo & Biswas, 2015). IR radiation is absorbed by atmospheric components, influencing electronic properties and contributing to changes in CO_2 concentration (Fig. 18.20).

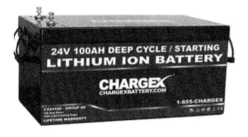

Figure 18.10 Different batteries (Khan et al., 2015).

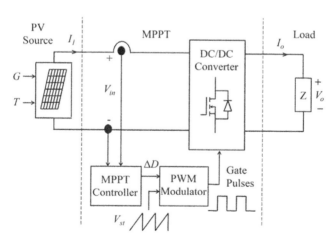

Figure 18.11 MPPT controller (Rao et al., 2021).

18.3.4.5 Voltage and current sensor

Voltage and current sensors are used in voltage and current monitoring, logging, and verification applications. Both AC and DC voltage levels can be

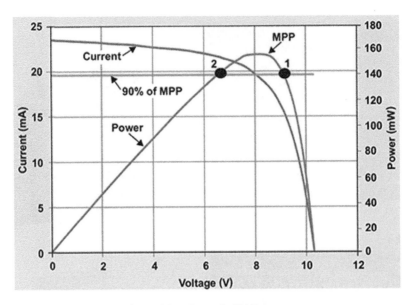

Figure 18.12 MPPT characteristics (Hossain et al., 2017).

Figure 18.13 Inverter (Hossain et al., 2017).

determined by the voltage sensor (Hossain et al., 2017). The current sensor produces an analog or digital output that is comparable to the current in the wire Fig. 18.21.

Figure 18.14 Charge controller (Persson et al., 2017).

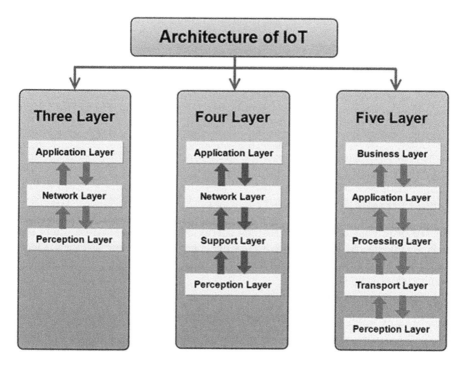

Figure 18.15 IoT architecture (Fumo & Biswas, 2015).

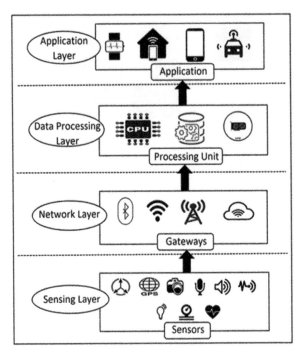

Figure 18.16 IoT architecture layers and components (Hossain et al., 2017).

Figure 18.17 Temperature sensor (Hossain et al., 2017).

18.3.4.6 Light sensor

This sensor is a device that converts light energy into electrical energy. A light sensor is a photoelectric device that converts the energy of light into electrical energy (Trapero et al., 2015). Available light sensors include photoresistors, photodiodes, and phototransistors (Fig. 18.22).

18.3.4.7 Microelectromechanical systems sensor

Microelectromechanical systems (MEMS) sensor is the technology that combines electrical and mechanical components to generate small integrated devices. These have been utilized to detect, regulate, and actuate microscale devices. It can also be

Figure 18.18 Humidity sensor (Ramu et al., 2021).

Figure 18.19 Tilt sensor (Ramu et al., 2021).

utilized to create macroscale effects and is compatible with integrated circuit technology (Jain et al., 2020). They were made in compact structures that allowed an electromechanical system with a large number of moving parts to be implemented at the microlevel (Ekici, 2014) (Fig. 18.23).

18.3.4.8 Ultrasonic sensor

An ultrasonic sensor consists of a piezoelectric crystal that operates at an extremely high frequency. These sensors were also known as 11 PM traducers since they

Figure 18.20 Carbon dioxide sensor (Ramu et al., 2021).

Figure 18.21 Voltage sensor and current sensor (Hossain et al., 2017).

Figure 18.22 Light sensor (Hossain et al., 2017).

Figure 18.23 MEMS sensor (Khan et al., 2015).

Figure 18.24 Ultrasonic sensor (Ramu et al., 2021).

send, receive, and convert ultrasonic pulses into electrical signals or the other way around. It emits a high-frequency sound that is used to locate an object underwater. It consists of a piezoelectric crystal that operates at an extremely high frequency (Wang et al., 2016) (Fig. 18.24).

18.3.4.9 IR sensor

Certain elements in its range are sensed by an infrared sensor. Infrared radiation may detect an item through thermally dependent processes, making photonic detectors extremely sensitive. The heat released by the object can be used to detect motion in this infrared band. It's an electrical sensor that detects distance using light emitted by objects (Li et al., 2016). It's the spectral sensitivity of distinct infrared wavelength optoelectronic components (Fig. 18.25).

Figure 18.25 IR sensor (obstacle type) and IR sensor (reflective type) (Ramu et al., 2021).

Figure 18.26 Proximity sensor (Ramu et al., 2021).

18.3.4.10   Proximity sensor

This sensor detects an object without physical contact by emitting an electromagnetic field, otherwise an infrared light ray (Ramli et al., 2015). It senses the existence of surrounding items. A specific distance was calculated, and this distance was utilized to control the opening and closing of switches for electrical circuits. In addition, the distance between the shaft and the support bearings must be measured (Massidda & Marrocu, 2017). It is capable of detecting metallic elements, coils, polymers, and other materials (Fig. 18.26).

18.3.5   Additional control and communication equipment

Controllers translate the sensor information into multiple logic representations utilizing controllers. These were commonly accustomed to changing or managing the claimed outputs of different processes over acceptable ranges using controllers. Arduino and Raspberry Pi were the most widely utilized controllers in IoT-based applications. They are the most popular, user-friendly, and budget-friendly alternatives (Persson et al., 2017). The figure depicts a few of the controllers utilized in an IoT-based solar PV system.

18.3.5.1 Arduino Uno

Arduino Uno is an open-source sophisticated microcontroller that may be used in a variety of systems. It can be used in low-voltage applications ranging from 3.3 to 5.5 V (Mathiesen & Kleissl, 2011). It is utilized to keep the solar panels running. The universal serial bus (USB) is used to connect Arduino to a laptop or computer. C and C++ language principles were commonly employed in Arduino (Inman et al., 2013). The user could utilize Arduino to make numerous alterations in the IoT utilizing different programming languages (Fig. 18.27).

18.3.5.2 Raspberry Pi

Raspberry Pi is one board that can be plugged into any device and programmed or implemented to meet the needs of those who want to learn about computing. Users could use an SD card to turn the device into a hard drive. It contains the best characteristics, such as the quickest clock speed, RAM, and features that will help you learn how to write in Scratch and Python (Monteiro et al., 2013). It has the potential to allow people to build their home automation systems, which is well known in the open-source world as it gives people control over a closed system (Wolff et al., 2016) (Fig. 18.28).

18.3.5.3 Transducer

An electrical device known as a transducer transforms energy from one form to another. Thermometers, loudspeakers, position and pressure sensors, and antennae are a few typical examples. Even though they aren't typically thought of as transducers, photocells, LEDs, and regular light bulbs are all types of transducers (Hocaoglu & Serttas, 2016) (Fig. 18.29).

Figure 18.27 Arduino uno (Hossain et al., 2017).

Figure 18.28 Raspberry Pi (Monteiro et al., 2013).

Figure 18.29 Transducer (Monteiro et al., 2013).

Following is a discussion of transducer-type applications.

- In other electromagnetic applications, the transducer types are used in antennas, among magnetic cubes, hall-effect sensors, and disk read/write heads.
- Electromechanical devices, including accelerometers, LVDTs, galvanometers, load cells, MEMS, potentiometers, airflow sensors, linear motors, and rotary motors, use a variety of transducer types.
- Oxygen sensors, hydrogen sensors, pH meters, and other electrochemical applications employ the transducer types.
- The transducer kinds are utilized in electroacoustic devices such as sonar, piezoelectric crystals, microphones, speakers, and others.
- LEDs, photodiodes, laser diodes, photoelectric cells, LDRs, fluorescent lighting, incandescent lighting, and phototransistors are a few examples of photoelectric applications that use the transducer types.

- The transducer types are used in thermoelectric applications like thermostats, thermocouples, and resistance temperature detectors as well as radio acoustic applications like Geiger-Muller tubes, radio transmitters, and receivers (RTD).

18.3.5.4 Global system for mobile communications

This module is a type of device for communicating that is found in mobile phones and computers. It transmits data from beginning to end through channels that digitize and decrease the data transmitted in two ways: by such a multiple access system using time division and a client data stream. It uses extremely high frequencies ranging from 850 to 1900 MHz to deliver mobile voice and data. A GSM module (also called a GPRS) is a chip or circuit that organizes the communication channel between a cellular phone and GSM (Bhardwaj et al., 2013). It was also preferred in Arduino and Raspberry Pi. GSM refers to a global data transfer system and is a mobile connection modem. The GPS in a solar PV system delivers information to the user indicating the panel's current state (Hossain et al., 2017; Mohammadi et al., 2015) (Fig. 18.30).

18.3.5.5 GSM module specifications

The following are some of the GSM module's features:

- Increased spectrum effectiveness
- Mobile roaming abroad
- Compatibility with the digital network for integrated services
- Assistance with new services
- SIM phonebook administration
- Fixed dialing number (FDN)
- Real-time clock with alarm management

Figure 18.30 GSM modules (Khan et al., 2015).

- High-quality speech
- Uses encryption to make phone calls more secure
- Short message service (SMS)

18.3.6   IoT and renewable energy in the energy sector

The sources of energy had improved over time while additional types of demand technologies declined. Using renewable energy to generate electrical energy might attract people and save money; however, those technologies were frequently more expensive than other, more widely available technologies. Renewable energy could be produced reliably if the power grid is run at a significantly higher energy level. Although there was considerable initial capital investment, the building industry found this emerging sector of renewable energy to be tremendously profitable. Renewable energy creates fewer emissions; it helps to improve the atmosphere. The use of renewable energy is increasing (Rao, Sahoo, Balamurugan et al., 2021). The renewable energy source met peak demand while also lowering electricity expenses (Hocaoglu & Serttas, 2016). The capital investments in generating transmission after the resources have been installed are far too large. Wind and solar farms on a large scale increase the generation of renewable electrical energy, making the notion of higher quality load across traditional grids more appealing.

The distribution of energy produced by the sun and wind is not continuous and is not widely spread. The primary goal is to generate renewable resources in remote areas. Solar and wind power, for example, can be installed in buildings to provide excess capacity to nearby structures (Olatomiwa et al., 2015; Shamshirband et al., 2014). When there is not enough surplus energy to go around and there isn't enough bidirectional flow, the grid plays a big role. When bidirectional electrical flows fulfill their expectations, the grid may experience voltage difficulties. There are numerous renewable energy sources and generations, but they are not operationally controlled due to voltage spikes higher than the grid voltage (Sanjari & Gooi, 2016).

18.3.7   IoT application

The application layer is the point where the web, a stand-alone operating system, or a mobile phone app meets. The application's decisive goal is to assemble data obtained from the cloud and provide it to users in certain formats, such as information or graphs. It is a model platform meant for continuous remote monitoring (Huang et al., 2013). It would improve resource presentation and increase profitability. IoT applications are unusually complex, necessitating a complete infrastructure to reliably transmit, process, and integrate data into systems. However, once in place, it will be extremely useful. Even if there are issues, they do not have to be crippling (Pawar et al., 2020).

18.3.7.1 Application to renewable energy systems

In residential regions, IoT was also used in electricity generation and consumption. The usage of IoT in renewable resources would be an aid in the efficient search for energy demands (Box et al., 2015). Its contemporary monitoring and control systems might fully replace conventional energy sources. Sensors connected to generating, transmission, and distribution equipment are used in these IoT applications in renewable energy resources. This reduces our reliance on previously scarce fossil fuels. Primarily, the use of renewable sources has several advantages, and when combined with IoT, it will aid in the development of large-scale clean energy sources to a greater level.

Smart automation, cost-effectiveness, improved grid management, and high-home-demand solutions are some of the advantages of merging IoT with renewable energy resources (Ha & Phung, 2019). An IoT gadget aids in the discovery of favorable energy production conditions. The arrangement can be tweaked to get the most out of it. The main advantage of IoT components is that it produces data in real time, which reduces any waste. With automated controls, the power plant may operate more effectively, and integrating with IoT would be a useful, helpful resource for solar energy companies to meet consumer requests and increase overall efficiency (Ssekulima et al., 2016). The setup was used to compare output current and voltage, as well as the temperature of the environment, to earlier data.

When a device in a solar PV system was linked to the cloud network system, the source of the problem could be identified and fixed before the entire system was disrupted (Phinikarides et al., 2013). Customers may notice that the system is running smoothly, but there might be a difficulty with the devices; without IoT, determining where the fault originated, whether in hardware or software, would be difficult (Yang et al., 2012). The IoT enables the detection of errors in real time, and problems can be resolved rapidly because the basis of the error is identified. When implementing IoT, a gadget would be less influenced by outages as well as difficulties. Furthermore, IoT in solar energy can improve electricity quality and performance. As a result, energy production is both cost and systemization efficient (Mellit, Pavan, et al., 2013).

Real-time monitoring, rectification, and predictive analytics are among the many advantages. On a single panel, an IoT sensor may monitor many measures of success, such as electricity production, temperature, cardinal direction, and tilt angle. It ensures that any problems with specific panels are resolved as soon as possible. Customers could check the status of their PV system in a variety of ways (Yang et al., 2015).

- Tracking and analyzing data from the sensor would be useful in monitoring the solar plant's physical health. This aids in the understanding of PV modules, inverters, and other similar devices.
- Through IoT, past data may be accessed to learn about the device's performance. When performance issues are addressed regularly, apparatus efficiency and reliability improve.
- Data analysis assists with the detection of device issues such as PV arrays and inverters. It may aid in reducing the amount of energy lost due to device failures.
- With the use of data analytics, preventive maintenance would reduce device failures and expenses. This will contribute significantly to increased effectiveness and productivity.

18.3.7.2 Application to grid management

Grid management advances allow additional distributed resources to be integrated into the grid while also enhancing grid management. Real-time data on power usage are provided by sensors located at distribution systems along with distribution systems, which can be used to make choices about voltage management, load switching, and other concerns (Agoua et al., 2017). For load forecasting, power usage data are used as a source. This information can be collected via IoT devices, which can be used for a variety of applications. Sensors allow for precise energy creation based on precise projections, which improves production and control. This also contributes to the compilation of a comprehensive energy usage report, which allows clients to save money on their power bills (Inman et al., 2013). Using the IoT to leverage the power of digital transformation can help solve common challenges with complex energy networks and make monitoring panels and energy production much easier.

18.3.8 Remote solar photovoltaic system monitoring methods

Solar energy output must be conditioned, which enhances the power quality, efficiency, and productivity of PV panels. As a consequence, in a PV system, remote monitoring and accurate load forecasting are required (Yang et al., 2015). There are a variety of remote solutions that have been used in the past, but a prediction and an efficient load schedule must be developed (Shi et al., 2012). The remote monitoring scheme is shown in (Fig. 18.31) (Mathiesen & Kleissl, 2011; Mellit & Pavan, 2010).

18.3.8.1 Wireless monitoring

Using automated transmission and measurement techniques, data might be processed and gathered from the remote transmitter to the receiving point without wires.

18.3.8.2 Wired monitoring

After being received from the remote transmitter to the receiving site via automated transmission and measurement operations employing wires, data may be examined and recorded.

18.3.8.3 SCADA monitoring

SCADA monitoring was essential to both the hardware and the software. It is used in industrial organizations to remotely or locally monitor industrial operations.

18.3.8.4 Monitoring using cloud computing

In general, "cloud computing" refers to an information center that is accessible to multiple users over the Internet. It as well provides data about computer systems that require resources such as data and computational storage but do not require active user administration. Functions are spread across several places from central servers in today's massive clouds (Khan et al., 2015). Cloud computing allows for real-time monitoring and management of the Internet.

18.3.8.5 Monitoring using IoT

The IoT refers to the continuation of a physical device's Internet connection. These gadgets, which are usually embedded with electrical devices, Internet connectivity, and other physical apparatus, can interact and communicate with one another over the Internet and are routinely monitored and operated remotely (Yang et al., 2012) (Fig. 18.31).

18.3.9 IoT-based remote monitoring

When AC power is required, the PV system contains a solar panel that would be linked to a battery for energy storage, and a charge controller as well as an inverter.

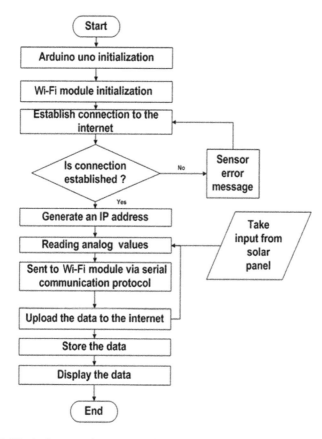

Figure 18.31 Block diagram of remote monitoring system (Mathiesen & Kleissl, 2011; Mellit & Pavan, 2010).

PV arrays are made up of several PV modules connected. PV modules are made up of PV arrays grouped in parallel and serial configurations. DC electricity can be created using PV modules (Phinikarides et al., 2013). The adjustable DC voltage can be created utilizing DC to DC converters such as boost, buck-boost, SEPIC, and Luo converters (Ssekulima et al., 2016). Then, using three-phase voltage source inverters transforms AC to DC. Depending on the requirements, energy is used by specific loads or transmitted to the grid via a transmission network. Even though electricity was generated and provided, the viability of battery storage was crucial within particular circumstances. The IoT-based PV monitoring system is depicted in Fig. 18.28. Three layers make up an IoT-based PV system for remote monitoring (Zeng & Qiao, 2013). The levels are installation of the PV system, networking entry point for connections, and then remote monitoring and control (Lorenz & Heinemann, 2012). The PV system design environment is the first layer, where all of the components are connected according to the required configurations to meet the user's needs. Communication connections are the second layer, and they offer a link between using an Internet firewall and router, different hardware components of the PV system, and a remote server. The Arduino server is a necessary component for connecting your web server to your PC using an Ethernet or wireless router module. The PV system hardware components are monitored, controlled, and managed by the microcontroller on the Arduino server. The second layer's data is transferred to the final layer (Mathiesen & Kleissl, 2011). This manages and monitors the PV system and provides users with frequent reports. Users can also utilize the Android app with cloud service to view this data in reports or visual charts over a Wi-Fi network (Fig. 18.32).

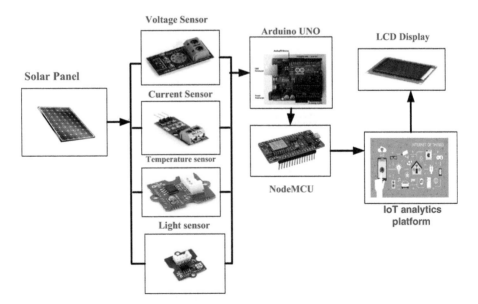

Figure 18.32 Monitoring of PV system using IoT (Mathiesen & Kleissl, 2011).

18.4   Challenges and issues of implementation of IoT on renewable energy resources

In the field of renewable energy, the use of IoT technology is extremely popular. Implementing IoT provides advantages in risky data assortments, wise choices, and offering users or consumers extended facilities. Integrating IoT devices into a renewable energy system, on the other hand, is a difficult undertaking for any energy provider, and they confront several obstacles along the road (Mathiesen & Kleissl, 2011).

18.4.1   Challenges
18.4.1.1   Complexity

Integrating IoT-based system solutions into a traditional energy system is quite tough. Because the traditional energy sector is so complex, integrating IoT devices and energy system components will be extremely tough (Box et al., 2015).

18.4.1.2   Security

The most difficult aspect of implementing a complete IoT-based system will be data security. There will be a great deal of sensitive and involved customer information, and any errors or omissions would be terrible for the consumers and the company's reputation. Data security policies will be an issue as a result (Ssekulima et al., 2016).

18.4.1.3   Data collection

Sunrays and wind statistics, for example, are inaccurate and unexpected, and they vary often, making it impossible to produce accurate projections.

18.4.1.4   Data handling

The bulk of renewable energy collection systems is located in rural areas, whereas city load centers account for the majority of power demand. As a result, data handling between these two places will be difficult.

18.4.2   Solutions
18.4.2.1   Enhanced energy storage system

Improved energy storage devices that can collect and store renewable energy when it is abundant are required. The integration of these enhancements into the electricity system will be crucial; however, challenges must be overcome to effectively address the unpredictability of renewable energy. Solutions include improvements in energy generation and storage, as well as interlocation energy transmission (Mathiesen & Kleissl, 2011).

18.4.2.2   Rapid energy transmission system

The development of a high-capacity electrical transmission system for ultrahigh-ways, swiftly transmitting power resources, and ensuring uniform distribution within a defined time frame is required (Ssekulima et al., 2016).

18.4.2.3   Supply-oriented power consumption

The implementation of an IoT-enabled smart grid design in the electricity sector will facilitate power usage that aligns with supply dynamics. This would allow for more accurate energy usage forecasting.

18.5   Conclusions

This chapter examines IoT-enabled smart sensors for solar PV systems, as well as remote monitoring solutions. Accessibility, free energy, and low environmental impact are just a few of the advantages of solar energy. One of the most recent trends in the PV industry is the IoT. IoT monitoring of a PV system enables automated solar power monitoring from any location with an Internet connection. It is critical for acquiring control of PV systems that are placed in remote locations or far away from the control center. One suggestion for minimizing the influence on the environment is to use renewable energy technology. The use of renewable energy and monitoring it are crucial due to frequent power outages. The user is guided by monitoring while analyzing the utilization of renewable energy. This method is economical. The system's effectiveness is around 95%. This makes it possible to use renewable energy effectively. The temperature sensor aids in the study of solar energy storage. Thus, the problems with electricity are decreasing. The outcomes, i.e., the monitored values obtained, which are valuable for forecasting the future values of the factors under consideration, can be leveraged to enhance this paper further. For example, they can contribute to the estimation of the battery's capacity to store solar energy. Customers may also easily distinguish between various types of problems and track the system's overall success. This chapter discusses the importance of solar energy as a requirement for IoT implementation in PV systems through remote monitoring via sensors and controllers. By completing an extensive study on remote monitoring, it is possible to improve the performance of solar PV systems. This will help to improve the level of remote monitoring for load scheduling and estimating in real time.

References

Adineh, B., Habibi, M. R., Akpolat, A. N., & Blaabjerg, F. (2021). Sensorless voltage estimation for total harmonic distortion calculation using artificial neural networks in microgrids. *IEEE Transactions on Circuits and Systems II: Express Briefs, 68*(7), 2583−2587, July.

Agoua, X. G., Girard, R., & Kariniotakis, G. (2017). Short-term spatio-temporal forecasting of photovoltaic power production. *IEEE Transactions on Sustainable Energy.*

Akpolat, A. N., Dursun, E., Kuzucuoğlu, A. E., Yang, Y., Blaabjerg, F., & Baba, A. F. (2019). Performance analysis of a grid-connected rooftop solar photovoltaic system. *Electronics, 8*(8), 905.

Almonacid, F., Pérez-Higueras, P., Fernández, E. F., & Hontoria, L. (2014). A methodology based on dynamic artificial neural network for short-term forecasting of the power output of a PV generator. *Energy Conversion and Management, 85*, 389−398.

Bhardwaj, S., Sharma, V., Srivastava, S., Sastry, O., Bandyopadhyay, B., Chandel, S., et al. (2013). Estimation of solar radiation using a combination of Hidden Markov Model and generalized Fuzzy model. *Solar Energy, 93*, 43−54.

Box, G. E., Jenkins, G. M., Reinsel, G. C., & Ljung, G. M. (2015). *Time series analysis: forecasting and control.* John Wiley & Sons.

Chen, S., Gooi, H., & Wang, M. (2013). Solar radiation forecast based on fuzzy logic and neural networks. *Renewable Energy, 60*, 195−201.

Chen, J.-L., Li, G.-S., & Wu, S.-J. (2013). Assessing the potential of support vector machine for estimating daily solar radiation using sunshine duration. *Energy Conversion and Management, 75*, 311−318.

Cuce, E., Harjunowibowo, D., & Cuce, P. M. (2016). *Renewable and sustainable energy saving strategies for greenhouse systems: A comprehensive review. Renewable and Sustainable Energy Reviews* (64, pp. 34−59). Elsevier BV.

Dahmani, K., Dizene, R., Notton, G., Paoli, C., Voyant, C., & Nivet, M. L. (2014). Estimation of 5-min time-step data of tilted solar global irradiation using ANN (artificial neural network) model. *Energy, 70*, 374−381.

Deng, R., Yang, Z., Chow, M.-Y., & Chen, J. (2015). A Survey on demand response in smart grids: mathematical models and approaches. *IEEE Transactions on Industrial Informatics, 11*(3), 570−582. Available from https://doi.org/10.1109/tii.2015.2414719.

Ekici, B. B. (2014). A least squares support vector machine model for prediction of the next day solar insolation for effective use of PV systems. *Measurement, 50*, 255−262.

Energy policy of India. (n.d.) https://en.wikipedia.org/wiki/Energy_policy_of_India.

Fernandez-Jimenez, L. A., Muñoz-Jimenez, A., Falces, A., Mendoza-Villena, M., Garcia-Garrido, E., Lara-Santillan, P. M., et al. (2012). Short-term power forecasting system for photovoltaic plants. *Renewable Energy, 44*, 311−317.

Fumo, N., & Biswas, M. R. (2015). Regression analysis for prediction of residential energy consumption. *Renewable and Sustainable Energy Reviews, 47*, 332−343.

Graditi, G., Ferlito, S., & Adinolfi, G. (2016). Comparison of photovoltaic plant power production prediction methods using a large measured dataset. *Renewable Energy, 90*, 513−519.

Ha, Q., & Phung, M. D. (2019). *IoT-enabled dependable control for solar energy harvesting in smart buildings. IET Smart Cities* (1, pp. 61−70). Institution of Engineering and Technology (IET), 2.

Hocaoglu, F. O., & Serttas, F. (2016). A novel hybrid (Mycielski-Markov) model for hourly solar radiation forecasting. *Renewable Energy.*

Hossain, M., Mekhilef, S., Danesh, M., Olatomiwa, L., & Shamshirband, S. (2017). Application of extreme learning machine for short term output power forecasting of three gridconnected PV systems. *Journal of Cleaner Production, 167*, 395−405.

Huang, J., Korolkiewicz, M., Agrawal, M., & Boland, J. (2013). Forecasting solar radiation on an hourly time scale using a coupled autoregressive and dynamical system (CARDS) model. *Solar Energy, 87*, 136−149.

Inman, R. H., Pedro, H. T., & Coimbra, C. F. (2013). Solar forecasting methods for renewable energy integration. *Progress in Energy and Combustion Science, 39*, 535−576.

Izgi, E., Öztopal, A., Yerli, B., Kaymak, M. K., & Şahin, A. D. (2012). Short−mid-term solar power prediction by using artificial neural networks. *Solar Energy, 86*, 725−733.

Jain, M., Kain, P., Gupta, D., & Rodrigues, J. J. P. C. (2020). *The role of intelligent grid technology in cloud computing. Applications of cloud computing* (pp. 83−99). Chapman and Hall/CRC.

Kaushika, N., Tomar, R., & Kaushik, S. (2014). Artificial neural network model based on interrelationship of direct, diffuse and global solar radiations. *Solar Energy, 103*, 327−342.

Khan, M. A., Javaid, N., Mahmood, A., Khan, Z. A., & Alrajeh, N. (2015). A generic demand-side management model for smart grid. *International Journal of Energy Research, 39*(7), 954−964.

Li, Y., He, Y., Su, Y., & Shu, L. (2016). Forecasting the daily power output of a grid-connected photovoltaic system based on multivariate adaptive regression splines. *Applied Energy, 180*, 392−401.

Lima, F. J., Martins, F. R., Pereira, E. B., Lorenz, E., & Heinemann, D. (2016). Forecast for surface solar irradiance at the Brazilian Northeastern region using NWP model and artificial neural networks. *Renewable Energy, 87*, 807−818.

Lorenz, E., & Heinemann, D. (2012). Prediction of solar irradiance and photovoltaic power. *Comprehensive Renewable Energy, 1*, 239−292.

Massidda, L., & Marrocu, M. (2017). Use of multilinear adaptive regression splines and numerical weather prediction to forecast the power output of a PV plant in Borkum, Germany. *Solar Energy, 146*, 141−149.

Mathiesen, P., & Kleissl, J. (2011). Evaluation of numerical weather prediction for intra-day solar forecasting in the continental United States. *Solar Energy, 85*, 967−977.

Mellit, A., & Pavan, A. M. (2010). A 24-h forecast of solar irradiance using artificial neural network: application for performance prediction of a grid-connected PV plant at Trieste, Italy. *Solar Energy, 84*, 807−821.

Mellit, A., Sağlam, S., & Kalogirou, S. (2013). Artificial neural network-based model for estimating the produced power of a photovoltaic module. *Renewable Energy, 60*, 71−78.

Mellit, A., Pavan, A. M., & Benghanem, M. (2013). Least squares support vector machine for short-term prediction of meteorological time series. *Theoretical and Applied Climatology, 111*, 297−307.

Mohammadi, K., Shamshirband, S., Tong, C. W., Arif, M., Petković, D., & Ch, S. (2015). A new hybrid support vector machine−wavelet transform approach for estimation of horizontal global solar radiation. *Energy Conversion and Management, 92*, 162−171.

Monteiro, C., Santos, T., Fernandez-Jimenez, L. A., Ramirez-Rosado, I. J., & Terreros-Olarte, M. S. (2013). Short-term power forecasting model for photovoltaic plants based on historical similarity. *Energies, 6*, 2624−2643.

Olatomiwa, L., Mekhilef, S., Shamshirband, S., Mohammadi, K., Petković, D., & Sudheer, C. (2015). A support vector machine−firefly algorithm-based model for global solar radiation prediction. *Solar Energy, 115*, 632−644.

Persson, C., Bacher, P., Shiga, T., & Madsen, H. (2017). Multi-site solar power forecasting using gradient boosted regression trees. *Solar Energy, 150,* 423−436.

Pawar, P., Sampath, S., Ghosh, T., & Vittal, K.P. (2018). Load Scheduling Algorithm Design for Smart Home Energy Management System. 2018 IEEE 7th International Conference on Power and Energy (PECon). 2018 IEEE 7th International Conference on Power and Energy (PECon).

Pawar, P., TarunKumar, M., & Vittal, K. P. (2020). An IoT based intelligent smart energy management system with accurate forecasting and load strategy for renewable generation. *Measurement, 152,* 107187.

Phinikarides, A., Makrides, G., Kindyni, N., Kyprianou, A., & Georghiou, G. E. (2013). ARIMA modeling of the performance of different photovoltaic technologies. In: 2013 IEEE 39th photovoltaic specialists conference (PVSC). IEEE; pp. 0797−801.

Ramli, M. A., Twaha, S., & Al-Turki, Y. A. (2015). Investigating the performance of support vector machine and artificial neural networks in predicting solar radiation on a tilted surface: Saudi Arabia case study. *Energy Conversion and Management, 105,* 442−452.

Ramu, S. K., Irudayaraj, G. C. R., & Elango, R. (2021). *An IoT-based smart monitoring scheme for solar PV applications. Electrical and electronic devices, circuits, and materials* (pp. 211−233). Wiley. Available from https://doi.org/10.1002/9781119755104.ch12.

Rao, C. K., Sahoo, S. K., Balamurugan, M., Satapathy, S. R., Patnaik, A., & Yanine, F. F. (2020). Applications of sensors in solar energy systems. *ICREISG-2020, Bhubaneswar, 14−15 Feb,* 2020.

Rao, C. K., Sahoo, S. K., Balamurugan, M., & Yanine, F. F. (2021). *Design of smart socket for monitoring of IoT-based intelligent smart energy management system. Intelligent Computing in Control and Communication. Lecture Notes in Electrical Engineering* (p. 702) Singapore: Springer.

Rao, C. K., Sahoo, S. K., & Yanine, F. F. (2021). *Demand response for renewable generation in an IOT based intelligent smart energy management system.* Chennai: Innovations in Power and Advanced Computing Technologies (i-PACT).

Şahin, M. E., & Blaabjerg, F. (2020). A hybrid PV-battery/supercapacitor system and a basic active power control proposal in MATLAB/Simulink. *Electronics, 9*(1), 129.

Sahin, M. (2020). *Energy management and measurement of computer controlled solar house model for Rize city. Gümüşhane Üniversitesi Fen Bilimleri Enstitüsü Dergisi.* Gumushane University Journal of Science and Technology Institute.

Salam, Z., Ishaque, K., & Taheri, H. (2010). *An improved two-diode photovoltaic (PV) model for PV system. 2010 Joint International Conference on Power Electronics, Drives and Energy Systems & 2010 Power India. 2010 Power India.* IEEE.

Sanjari, M. J., & Gooi, H. B. (2016). Probabilistic forecast of PV power generation based on higher-order Markov chain. *IEEE Transactions on Power Systems.*

Shamshirband, S., Petković, D., Saboohi, H., Anuar, N. B., Inayat, I., Akib, S., et al. (2014). Wind turbine power coefficient estimation by soft computing methodologies: Comparative study. *Energy Conversion and Management, 81,* 520−526.

Sharma, V., Yang, D., Walsh, W., & Reindl, T. (2016). Short term solar irradiance forecasting using a mixed wavelet neural network. *Renewable Energy, 90,* 481−492.

Shi, J., Lee, W.-J., Liu, Y., Yang, Y., & Wang, P. (2012). Forecasting power output of photovoltaic systems based on weather classification and support vector machines. *IEEE Transactions on Industry Applications, 48,* 1064−1069.

Sittón-Candanedo, I., Alonso, R. S., García, Ó., Muñoz, L., & Rodríguez-González, S. (2019). Edge computing, IoT and social computing in smart energy scenarios. *Sensors, 19*(15), 3353. Available from https://doi.org/10.3390/s19153353.

Ssekulima, E. B., Anwar, M. B., Al Hinai, A., & El Moursi, M. S. (2016). Wind speed and solar irradiance forecasting techniques for enhanced renewable energy integration with the grid: a review. *IET Renewable Power Generation*.

Teo, T., Logenthiran, T., & Woo, W. (2015). *Forecasting of photovoltaic power using extreme learning machine. Smart grid technologies-Asia (ISGT ASIA), 2015 IEEE innovative* (pp. 1–6). IEEE.

Trapero, J. R., Kourentzes, N., & Martin, A. (2015). Short-term solar irradiation forecasting based on dynamic harmonic regression. *Energy, 84*, 289–295.

Tuohy, A., Zack, J., Haupt, S. E., Sharp, J., Ahlstrom, M., Dise, S., et al. (2015). Solar forecasting: Methods, challenges, and performance. *IEEE Power & Energy Magazine, 13*, 50–59.

Urquhart, B., Kurtz, B., Dahlin, E., Ghonima, M., Shields, J., & Kleissl, J. (2014). Development of a sky imaging system for short-term solar power forecasting. *Atmospheric Measurement Techniques, 7*, 4859–4907.

Wang, G., Su, Y., & Shu, L. (2016). One-day-ahead daily power forecasting of photovoltaic systems based on partial functional linear regression models. *Renewable Energy, 96*, 469–478.

Wolff, B., Kühnert, J., Lorenz, E., Kramer, O., & Heinemann, D. (2016). Comparing support vector regression for PV power forecasting to a physical modelling approach using measurement, numerical weather prediction, and cloud motion data. *Solar Energy, 135*, 197–208.

World population. (n.d.) https://en.wikipedia.org/wiki/World_population.

Yang, D., Jirutitijaroen, P., & Walsh, W. M. (2012). Hourly solar irradiance time series forecasting using cloud cover index. *Solar Energy, 86*, 3531–3543.

Yang, H., Kurtz, B., Nguyen, D., Urquhart, B., Chow, C. W., Ghonima, M., et al. (2014). Solar irradiance forecasting using a ground-based sky imager developed at UC San Diego. *Solar Energy, 103*, 502–524.

Yang, C., Thatte, A. A., & Xie, L. (2015). Multitime-scale data-driven spatio-temporal forecast of photovoltaic generation. *IEEE Transactions on Sustainable Energy, 6*, 104–112.

Yona, A., Senjyu, T., Funabashi, T., & Kim, C.-H. (2013). Determination method of insolation prediction with fuzzy and applying neural network for long-term ahead PV power output correction. *IEEE Transactions on Sustainable Energy, 4*, 527–533.

Zaidan, A. A., & Zaidan, B. B. (2018). A review on intelligent process for smart home applications based on IoT: coherent taxonomy, motivation, open challenges, and recommendations. *Artificial Intelligence Review, 53*, 1.

Zeng, J., & Qiao, W. (2013). Short-term solar power prediction using a support vector machine. *Renewable Energy, 52*, 118–127.

Zhang, Y., Beaudin, M., Taheri, R., Zareipour, H., & Wood, D. (2015). Day-ahead power output forecasting for small-scale solar photovoltaic electricity generators. *IEEE Transactions on Smart Grid, 6*, 2253–2262.

Index

Printed and bound by CPI Group (UK) Ltd, Croydon, CR0 4YY

03/10/2024

01040847-0010